Sustainable Design for Global Equilibrium

Md. Faruque Hossain

Sustainable Design for Global Equilibrium

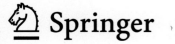

Springer

Md. Faruque Hossain
College of Architecture and Construction Management
Kennesaw State University
1100 South Marietta Parkway, Marietta, GA, USA

ISBN 978-3-030-94817-7 ISBN 978-3-030-94818-4 (eBook)
https://doi.org/10.1007/978-3-030-94818-4

This Springer imprint is published by the registered company Springer Nature Switzerland AG
The registered company address is: Gewerbestrasse 11, 6330 Cham, Switzerland

To
Nowshin
Shafin
Faria
I love them more than their imagination. Hope at least one of them will be the captain planet to fulfill my dream—Build a Better World

Contents

Part I Introduction

1 **Green Science and Technology for Designing Sustainable World** 3
Introduction 3
Methods and Simulation 4
CO_2 Emissions 5
CO_2 Sink 6
Growth Rate of the Atmospheric CO_2 Concentration (G_{ATM}) 7
Results and Discussion 7
CO_2 Emission 7
CO_2 Sink 8
Growth Rate of the Atmospheric CO_2 Concentration (G_{ATM}) 8
Conclusion 10
References 10

Part II Sustainable Energy

2 **Harvesting Global Solar Energy** 15
Introduction 16
Material, Methods, and Simulation 17
Calculation of Net Solar Energy on Earth 17
Modeling of Net Electricity Energy Generation from Total Solar Irradiance on Earth 23
Results and Discussion 26
Calculation of Net Solar Energy on Earth 26
Modeling of Net Electricity Energy Generation from Total Solar Irradiance on Earth 34
Conclusion 38
References 38

3 Reconfiguration of Bose–Einstein Photonic Structure to Produce Clean Energy 41
Introduction 41
Methods and Simulations 42
Photon Dynamics Transformation 42
Photovoltaic (PV) Modeling 45
Results and Discussion 47
Photon Production Proliferation 47
Electricity Conversion by a Photovoltaic Panel 53
Conclusions 54
References 54

4 Transformation of Building's Biowaste into Electricity Energy to Mitigate the Global Energy Vulnerability 57
Introduction 57
Materials and Methods 58
Results and Discussion 65
Conclusion 70
References 71

Part III Sustainable Building, Water, and Transportation Technology

5 Photonic Thermal Control to Cool and Heat the Housing Naturally 77
Introduction 77
Methods and Simulation 78
Cooling Mechanism 78
Heating Mechanism 82
Results and Discussion 84
Cooling Mechanism 84
Heating Mechanism 89
Conclusions 96
References 98

6 Implementation of Bose–Einstein (B–E) Photon Energy Reformation for Cooling and Heating the Building Naturally 101
Introduction 101
Methods and Simulation 102
Cooling Mechanism 102
Heating Mechanism 105
Results and Discussion 107
Cooling Mechanism 107
Heating Mechanism 112
Conclusions 119
References 121

7 **Photon Application in the Design of Sustainable Buildings to Console Global Energy and Environment** . . . 125
Introduction . . . 125
Methods and Materials . . . 127
Results and Discussion . . . 134
Conversion of Electricity . . . 136
Savings on Energy Cost . . . 137
Conclusions . . . 139
References . . . 139

8 **Rerouting the Transpiration Vapor of Trees to Mitigate Global Water Supply in Rural Area Naturally** . . . 143
Introduction . . . 143
Materials, Methods, and Simulations . . . 144
Static Electric Force Generation . . . 144
In Site Water Treatment . . . 146
Results and Discussion . . . 147
Electrostatic Force Analysis . . . 147
Conclusions . . . 149
References . . . 149

9 **Application of Hybrid Wind and Solar Energy in the Transportation Section** . . . 151
Introduction . . . 151
Material and Methods . . . 152
Modeling of Wind Turbine . . . 152
Conversion of Wind Energy . . . 155
Modeling of Solar Energy . . . 156
Solar Energy Conversion . . . 158
Electrical Subsystem . . . 159
Results and Discussions . . . 160
The Wind Turbine Model . . . 160
Conversion of Wind Energy . . . 164
Solar Energy Modeling . . . 165
Solar Energy Conversion . . . 168
Electrical Subsystem . . . 169
Conclusions . . . 171
References . . . 172

10 **Flying Transportation Technology to Console Global Communication Crisis** . . . 175
Introduction . . . 175
Materials, Methods, and Simulation . . . 176
Solving the Concept Numerically . . . 176
The Flying Car's Wind Energy Modeling Sequence . . . 178

Conversion of Wind Energy . . . 180
Modeling the Generator . . . 181
Battery Modeling . . . 182
Results, Optimization, and Discussion . . . 183
Wind Energy Modeling for the Flying Vehicles . . . 186
Wind Energy Conversion . . . 190
Electrical Subsystem . . . 191
Generator Modeling . . . 194
Battery Modeling . . . 197
Conclusion . . . 197
References . . . 199

Part IV Sustainable Society, and Environment

11 Photophysical Reaction Technology to Eliminate Pathogens Naturally from Earth . . . 203
Introduction . . . 203
Materials and Methods . . . 204
UVGI Production by the Exterior Glazing Wall Skin . . . 204
Electromagnetic Radiation . . . 207
Killing Pathogens . . . 208
Results and Discussion . . . 210
UVGI Production by the Exterior Glazing Wall Skin . . . 210
Electromagnetic Radiation . . . 215
Killing Pathogens . . . 215
Conclusions . . . 217
References . . . 217

12 Air Pollution and the Survival Period of Earth . . . 219
Introduction . . . 219
Methods and Simulation . . . 220
CO_2 Emissions . . . 220
CO_2 Sink . . . 221
Atmospheric CO_2 Concentration (G_{ATM}) Increasing Rate . . . 222
Results and Discussion . . . 222
CO_2 Emission . . . 222
CO_2 Sink . . . 223
Atmospheric CO_2 Concentration (G_{ATM}) Increasing Rate . . . 223
Conclusion . . . 225
References . . . 225

13 Modeling of Climate Control to Secure Global Environmental Equilibrium . . . 227
Introduction . . . 228
Methods and Simulation . . . 228
Cooling Mechanism of Earth Surface . . . 228
Heating Mechanism of Earth Surface . . . 233

Results and Discussion 236
Cooling Mechanism of Earth Surface 236
Heating Mechanism of Earth Surface 242
Conclusions 249
References 251

Index 255

About the Author

Md. Faruque Hossain has more than 20 years of industry experience in the field of sustainability research, development, and project management under global top public agencies and fortune-listed companies. He worked and/or consulted in diverse global top-tier companies to conduct research and development for million dollars to over billion dollars projects for ensuring global sustainability. Faruque also worked for the NYC Department of Citywide Administrative Services as Senior Management team and interacted with the heads of all public agencies, highest level government officials of local, state, federal, and international organization leaders for developing global sustainability policy. During his tenure in NYC Department of Environmental Protection as Acting Director, Faruque managed a world-class team of scientists, consultants, architects, engineers, and contractors from AECOM, Fluor, Skanska, and maintained highest level professional relationship to conduct global sustainability research and practice. Hossain received his Ph.D. from Hokkaido University. He did post-graduate research in Chemical Engineering at the University of Sydney and Executive Education in Architecture at Harvard University. He is an LEED-certified professional and editor of several International Journal of Global Sustainability-related field. Dr. Hossain is renowned as the industry leader and notable scientist to conduct innovative research and project development for energy, environment, building, infrastructure sustainability field for building a better Earth. He has hundreds of world-class publications in very high impact journals, and he wrote four books (published by Springer, Elsevier) and seven book chapters (published by Springer, Elsevier, and Francis & Taylor) in the field of global sustainability. Currently, he is working at the school of architecture and construction management at Kennesaw State University as an Assistant Professor and simultaneously running his own company "Green Globe Technology" with his motto to practice sustainability for building a better planet.

Part I
Introduction

Chapter 1
Green Science and Technology for Designing Sustainable World

Abstract The conventional application of science and technology in our daily life will cause deadly environmental vulnerabilities to the comfortable living of mankind on Mother Earth in the near future. Simply, the application of traditional science and technology is becoming obsolete since it is causing the accumulation the CO_2 into the atmosphere at alarming rate; thus, it needs to be replaced by green science and technology to secure a balanced Earth. The notion of green science and technology which is the application of sustainable science and technology might be wide, but if we think about a calculative solution to protect our mother Earth by reducing CO_2 emission, we only need to focus on few sectors that needs to be holistic research and development approaches for green science and technology in the field of energy, building, water, transportation, environment, and society (EBWTES) to achieve a best-balanced Earth. Simply, this notional term of "Green Science and Technology" must be implemented in every sector of EBWTES for confirming sustainable world which would be the "autonomous adaptation," responds to conditions change by the adaptive capacity in all sectors of EBWTES system to level the CO_2 emission into the atmosphere. Indeed, it will without a doubt make the global environment green and clean as a result of its versatility, adaptability, and manageability, which won't result in maladjustment simultaneously, but will present a sustainable world for our future generation.

Keywords Green science and technology · Energy · Building · Water · Transportation technology · Environment · Sustainable world

Introduction

Since the 1960s, there has been an expansion in the huge development of conventional energy, building, water, and transportation (EBWT) systems throughout the world, quickening the accumulation of atmospheric CO_2 concentration heavily, which is indeed a clear and present danger for the survival of our planet earth in the near future [7, 21, 30]. For hundreds of years, architects and engineers have been designing conventional energy, building, water, and transportation (EBWT) systems for the betterment of our daily lives, which are in fact releasing nearly 91% CO_2 of

M. F. Hossain, *Sustainable Design for Global Equilibrium*,
https://doi.org/10.1007/978-3-030-94818-4_1

total CO_2 emission worldwide [3, 24]. Now, however, those are instantly becoming obsolete since they are causing heavy accumulation of CO_2 into the atmosphere which is 400 ppm into the atmosphere [6, 12, 25]. Since the atmospheric CO_2 is increasing rapidly due to the conventional design of EBWT system and thus emitted CO_2 will reach soon at the toxic level of 1200 ppm into the atmosphere, results in, all human being on Earth will face severe respiratory problem [4, 13, 24]. Therefore, the conventional EBWT system is presently required by nature to be environmentally friendly by *Designing Sustainable World* which only can be achieved by the application of green science and technology. Henceforth, design a sustainable world can be illustrated as "*which meets the needs of the present without compromising the ability of future generations to meet their own needs to secure a resilient global environmental system.*" Naturally, application of green science and technology strategies must need to work with the climate change mitigation to keep the global environment physically, chemically, biologically, and socially balanced to reduce the global CO_2 emission comfort level [1, 13, 22]. Simply, we are the first generation to see how the global environmental system is running toward danger due to the practice of traditional science and technology, at the same time, we are the last generation with the tools to have the last opportunity to prevent these dangers [8, 14]. Thus, we must do great work individually and/or collectively by applying green science and technology in every sector of EBWT sectors to create a total view for implementing these technologies to confirm a sustainable global design to reduce CO_2 emission throughout the world. Thus, in the research, a detailed calculation of global CO_2 emission and sequestration has been conducted in order evaluate the current danger in order to practices of green science and technology in every sector of EBWT. Simply, it is an urgent demand to define the context of instigation to scientific theories and practical application of this green science and technology to secure a sustainable world to mitigate global CO_2 by the application of green science and technology in all sectors of energy, building, water, and transportation technology.

Methods and Simulation

Global CO_2 emissions released by energy, building, water, and transportation (EBWT) sectors were analyzed from 1960 to 2029 by conducting 10 years period of each experimental data by interpreting reports from several organizations (DEP, USDOE, IPCC, CFC, CDIAC, IEA, UNEP, NEAA, NEDO, NOAA, and NASA). Then these data were incorporated into MATLAB software to accurately calculate annual emissions considering CO_2 releases from the fossil energy burning, and CO_2 releases from the conventional design of building, water, transportation sector throughout the world. Then these data were integrated together in order to calculate the net CO_2 emission per the unit time period (yr^{-1}) to determine the final average annual rate CO_2 emission into the atmosphere.

CO_2 Emissions

The yearly growth rate in CO_2 emissions due to the burning fossil fuel was estimated from the difference between two consecutive years from the period of 10 years set of 1960s, 1970s, 1980s, 1990s, 2000s, 2010s, and 2020s, which was divided by the first-year emissions per the following equation:

$$\mathrm{FF} = \left[\frac{E_{\mathrm{FF}\ (t_{0+1})} - E_{\mathrm{FF}\ (t_0)}}{E_{\mathrm{FF}(t_0)}}\right] * 100\%\mathrm{yr}^{-1} \tag{1.1}$$

Here, a simple calculation is being analyzed in yearly CO_2 emissions growth rate. To accurately determine the CO_2 growth rate over multiple decades, a leap-year factor is also being applied to confirm the net annual growth rate of carbon (E_{FF}) by using its logarithm equivalent in the following equation:

$$\mathrm{FF} = \frac{1}{E_{\mathrm{FF}}} \frac{d(\ln E_{\mathrm{FF}})}{dt} \tag{1.2}$$

Here, the pertinent CO_2 emission growth rates have been calculated considering multi-decadal periods by implementing a nonlinear drift into ln(E_{FF}) in Eq. (1.2) and by calculating the yearly growth percentage [4, 5, 11]. Thus, the logarithm of E_{FF} into the equation has been fitted into MATLAB algorithm rather than directly using E_{FF} to ensure an accurate growth rate estimate and satisfy Eq. (1.1).

Similarly, the CO_2 emission calculated here (E_{LUC}) due to the conventional design of building, water transportation sectors (BWT) throughout the world was calculated by implementing dynamic global environmental modeling (DGVM) simulations in MATLAB considering the difference between two consecutive years from the period of 10 years set of 1960s, 1970s, 1980s, 1990s, 2000s, 2010s, and 2020s, which was divided by the first-year emissions per the following equation [19, 24, 25]. Then, the simulations are being clarified to determine the yearly changes of atmospheric CO_2 concentrations for the past several decades of 1960s–2020s [9, 26, 27]. Therefore, a time series is being implemented in this simulation by allocating the dynamic emission of CO_2 due to the conventional design of building, water transportation technology between two consecutive years, which was divided by the first-year emissions per the following equation:

$$\mathrm{LUC} = \left[\frac{E_{\mathrm{LUC}\ (t_{0+1})} - E_{\mathrm{LUC}\ (t_0)}}{E_{\mathrm{LUC}\ (t_0)}}\right] * 100\%\mathrm{yr}^{-1} \tag{1.3}$$

Here, the equation is being calculated in yearly CO_2 emissions growth rate [12, 15, 29]. However, to accurately determine the growth rate over multiple decades, a leap-year factor is also being applied to confirm the net annual growth rate of carbon (E_{LUC}) by using its logarithm equivalent in the following equation:

$$\mathrm{LUC} = \frac{1}{E_{\mathrm{LUC}}} \frac{d(\ln E_{\mathrm{LUC}})}{dt} \tag{1.4}$$

Here, the CO_2 emission increasing rates have been estimated corresponding to multi-decadal periods by applying a nonlinear drift into $\ln(E_{\mathrm{LUC}})$ in Eq. (1.4) and by calculating the yearly growth percentage [16, 18, 29]. Thus, the logarithm of E_{FF} into the equation has been integrated into MATLAB algorithm rather than directly using E_{LUC} to ensure an accurate emission rate of CO_2.

Finally, the global total CO_2 emission per year has been calculated by combing these four equations as follows:

$$\begin{aligned}\mathrm{FL} = & \left[\frac{E_{\mathrm{FF}\ (t_{0+1})} - E_{\mathrm{FF}\ (t_0)}}{E_{\mathrm{FF}\ (t_0)}}\right] * 100\%\mathrm{yr}^{-1} + \frac{1}{E_{\mathrm{FF}}} \frac{d\,(\ln E_{\mathrm{FF}})}{dt} \\ & + \left[\frac{E_{\mathrm{LUC}\ (t_{0+1})} - E_{\mathrm{LUC}\ (t_0)}}{E_{\mathrm{LUC}\ (t_0)}}\right] * 100\%\mathrm{yr}^{-1} + \frac{1}{E_{\mathrm{LUC}}} \frac{d\,(\ln E_{\mathrm{LUC}})}{dt}\end{aligned} \tag{1.5}$$

Then, the global CO_2 sequestration considering all (1) ocean sink and (2) terrestrial sink has been calculated throughout the world from 1960 to 2029 by conducting 10 years period each experimental set and then converted the time period into average annual rate.

CO_2 Sink

Consequently, the CO_2 sequestered by the ocean is being calculated for the past years and the next years from the decadal set of 1960s, 1970s, 1980s, 1990s, 2000s, 2010s, and 2020s by implementing seven global oceans biogeochemical cycle models [3, 20, 31]. This approach is being used to comprehensively analyze the physical, chemical, and biological processes of global ocean that directly involve sequestering CO_2 by the ocean surface as well as the air-sea CO_2 fluxes [2, 6, 10]. Thus, the ocean CO_2 sequestration is being normalized by the accurate observational values by dividing the individual yearly values by the modeled average for years; therefore, the oceanic CO_2 sequestration per year (t) in GtC yr^{-1} is being calculated as follows:

$$S_{\mathrm{OCEAN}}\ (t) = \frac{1}{n} \sum_{m=1}^{m=n} \frac{S^m_{\mathrm{OCEAN}}\ (t)}{S^m_{\mathrm{OCEAN}(t10-t1)}} * 2.2 \tag{1.6}$$

where n represents the number of variables, m represents the factors, and t represents the period of time. The normalization is being considered when the ratio (2.2) assumed to be relied on the CO_2 gradient as a natural time dependence of CO_2 sequestration [2, 4, 13].

Then the absorption of CO_2 per year by terrestrial vegetation and the Earth are also being determined the net sequestration of CO_2 by the terrestrial vegetation (S_{LAND}) similarly from the decadal set of 1960s, 1970s, 1980s, 1990s, 2000s, 2010s, and 2020s. Here, the net sequestration by the terrestrial vegetation and the Earth is being computed as the CO_2 remaining from the mass balanced budget, which is expressed as follows:

$$S_{LAND} = E_{FF} + E_{LUC} - (G_{ATM} + S_{OCEAN}) \tag{1.7}$$

Here, S_{LAND} is computed from the remainder of the estimates where G_{ATM} is the presence of CO_2 in the atmosphere, (E_{FF}) is the carbon from fossil fuels burning, and the E_{LUC} is the CO_2 from conventionally designed building, water, and transportation sectors [2, 29].

Then, the computation of S_{LAND} in Eq. (1.7) with the budget from the DGVM$_S$ is being used to calculate E_{LUC} by subtracting the (G_{ATM} + S_{OCEAN}) CO_2, which suggested an independent calculation of a consistent S_{LAND} [17, 23, 28]. Thus, it can represent an appropriate understanding of the role of the terrestrial vegetation in determining the response to CO_2 sequestration by S_{LAND}.

Finally, the total CO_2 sequestration in a year period has been calculated by combing these two equations (Eqs. (1.6) and (1.7)) as follows:

$$S_{OCEAN}(t) + S_{LAND} = \frac{1}{n}\sum_{m=1}^{m=n} \frac{S^m_{OCEAN}(t)}{S^m_{OCEAN^{(t10-t1)}}} * 2.2 + (E_{FF} + E_{LUC}) - (G_{ATM} + S_{OCEAN}) \tag{1.8}$$

Growth Rate of the Atmospheric CO_2 Concentration (G_{ATM})

The net yearly growth rate of the concentration of atmospheric CO_2 was then determined, that is, the annual increase in the concentration of CO_2 into the atmosphere. Simply, the growth rate unit, ppm yr^{-1}, is then converted from GtC yr^{-1} by the total global carbon emission and sequestration rates to determine accumulation of CO_2 concentration increasing rate each year in the atmosphere.

Results and Discussion

CO_2 Emission

The average global CO_2 emissions from 1960 to 2029 during this time scale showed that total CO_2 emissions from combined burning fossil fuels, and building, water, transportation sectors are at an average of 1.7 GtC yr^{-1} of 1.7 ± 0.7 GtC yr^{-1} per decade in the 1960s (1960–1969); 2.2 GtC yr^{-1} of 1.7 ± 0.8 GtC yr^{-1} per decade in the 1970s (1970–1979); 1.5 GtC yr^{-1} of 1.6 ± 0.8 GtC yr^{-1} per decade in the 1980s (1980–1989); 2.45 GtC yr^{-1} of 2.6 ± 0.8 GtC yr^{-1} per decade in the 1990s (1990–1999); 2.45 GtC yr^{-1} of 2.6 ± 0.8 GtC yr^{-1} per decade in the 2000s (2000–2009); 3.26 GtC yr^{-1} of 3.26 ± 0.5 GtC yr^{-1} per decade in the 2010s (2010–2019), and expected to be increased to 3.26 GtC yr^{-1} of 3.26 ± 0.5 GtC yr^{-1} per decade in the 2020s (2020–2029) (Table 1.1).

CO_2 Sink

Subsequently, the results of CO_2 sequestration by ocean and the terrestrial vegetation and land suggested that the average global CO_2 sink from 1960 to 2029 during this time scale showed that total CO_2 emissions from combined burning fossil fuels, building, water, transportation sectors are at an average of 1.5 GtC yr^{-1} of 1.5 ± 0.2 GtC yr^{-1} per decade in the 1960s (1960–1969); 1.3 GtC yr^{-1} of 1.3 ± 0.5 GtC yr^{-1} per decade in the 1970s (1970–1979); 1.4 GtC yr^{-1} of 1.4 ± 0.6 GtC yr^{-1} per decade in the 1980s (1980–1989); 1.4 GtC yr^{-1} of 1.6 ± 0.4 GtC yr^{-1} per decade in the 1990s (1990–1999); 1.15 GtC yr^{-1} of 1.15 ± 0.5 GtC yr^{-1} per decade in the 2000s (2000–2009); 1.15 GtC yr^{-1} of 1.15 ± 0.5 GtC yr^{-1} per decade in the 2010s (2010–2019); and expected to be increased to 1.15 GtC yr^{-1} of 1.15 ± 0.5 GtC yr^{-1} per decade in the 2020s (2020–2029) (Table 1.1).

Growth Rate of the Atmospheric CO_2 Concentration (G_{ATM})

Then, the rate of growth of the atmospheric CO_2 concentration is being calculated by comparing the decadal and individual annual values for 10 years periodical set which suggested that the average global CO_2 annual growth from 1960 to 2029 are 0.2% at the decade 1960s; 0.9% at the decade 1970s; 0.1% at the decade 1980s; 1.15% at the decade 1990s; 1.3% at the decade 2000s; 2.11% at the decade 2010s; and expected to be 2.11% at the decade 2020s. The projected growth rate of atmospheric CO_2 concentration is presumably suggested that the increased rate of CO_2 will remain the same 2.11% per year for next several decades if we do not curb this acceleration of CO_2 emissions.

Since the current CO_2 concentration into the atmosphere is 400 ppm and is being growing at a rate of 2.11% per year, the following equations confirmed that it will attain a toxic level of 1200 ppm in 53 years.

Table 1.1 The results from DGVM simulation in MATLAB, implemented from the data of DEP, USDOE, IPCC, CFC, CDIAC, IEA, UNEP, NEAA, NEDO, NOAA, and NASA to determine the Annual Growth Rate of Atmospheric CO_2 (%). The results described the variation of the total CO_2 emissions from fossil fuel and BWT sectors and the Total CO_2 sink (Ocean and Terrestrial Vegetation and Land) from the years of 1960–1969, 1970–1979, 1980–1989, 1990–1999, and 2000–2009, 2010–2019, and 2020–2029 shown in GtC yr^{-1}

Mean (GtC yr^{-1})	1960–1969	1970–1979	1980–1989	1990–1999	2000–2009	2010–2019	2020-2029
Total CO_2 emission (ocean and terrestrial vegetation and land)							
DGVM simulations and the mean average data of DEP, USDOE, IPCC, CFC, CDIAC, IEA, UNEP, NEAA, NEDO, NOAA, and NASA	1.7 ± 0.7	1.7 ± 0.8	1.6 ± 0.8	2.6 ± 0.8	2.6 ± 0.8	3.26 ± 0.5	3.26 ± 0.5
	1.7	2.2	1.5	2.45	2.45	3.26	3.26
Total CO_2 sink (ocean and terrestrial vegetation and land)							
DGVM simulations and the mean average data of DEP, USDOE, IPCC, CFC, CDIAC, IEA, UNEP, NEAA, NEDO, NOAA, and NASA	1.5 ± 0.5	1.3 ± 0.5	1.4 ± 0.6	1.6 ± 0.4	1.15 ± 0.5	1.15 ± 0.5	1.0 ± 0.5
	1.5	1.3	1.4	1.4	1.15	1.15	1.15
Annual growth rate of atmospheric CO_2 (%)							
G_{ATM}	0.2	0.9	0.1	1.05	1.3	2.11	2.11

$$1200 = 400(1 + 0.0211)^{\text{Year}} \tag{1.9}$$

$$3 = (1 + 0.0211)^{\text{Year}} \tag{1.10}$$

$$\text{Log } 3 = \text{Year Log}(1.0211) \tag{1.11}$$

$$\text{Year} = 52.61 = 53 \text{ (round figure)} \tag{1.12}$$

Consequently, all human being on earth will be perished in few days due to the toxic level of CO_2 into the atmosphere. Simply, the importance of the application of green science and technology must be enforced by rules, regulations, and laws in all sectors of environment, building infrastructure to reduce the CO_2 emission in order to protect human lives as well as build a better planet for our next generation.

Conclusion

The total global CO_2 emissions due to the burning fossil fuel and EBWT system estimated for the past several decades as well as total CO_2 sequestration by the atmosphere, ocean, and terrestrial vegetation and Earth were also calculated to determine the increasing rate in CO_2 into the atmospheric each year. The annual growth rate of the global atmospheric CO_2 concentration over the last several years confirmed that currently atmospheric CO_2 is increasing at a rate of 2.11% annually. If the current annual CO_2 growth rate is not copped now, the atmospheric CO_2 concentration will eventually reach a toxic level of 1200 ppm in 53 years. Consequently, entire human race will be extinct in few days, and this will be the end of the story of the human civilization on Earth. So, it is time to enforce Green Science and Technology for Designing Sustainable World to protect our Earth and the humane civilization on it.

Acknowledgments This research was supported by Green Globe Technology under the grant RD-02021-03. Any findings, conclusions, and recommendations expressed in this paper are solely those of the author and do not necessarily reflect those of Green Globe Technology.

References

1. F. Achard et al., Determination of tropical deforestation rates and related carbon losses from 1990 to 2010. Glob. Change Biol. **20**, 2540–2554 (2014)
2. N. Arnell et al., Climate change 1995, in *Impacts, Adaptations, and Mitigation of Climate Change*, ed. by R. T. Watson, M. C. Zinyowera, R. H. Moss, (Cambridge University Press, Cambridge, 1996), pp. 325–363
3. A.P. Ballantyne, C.B. Alden, J.B. Miller, P.P. Tans, J.W.C. White, Increase in observed net carbon dioxide uptake by land and oceans during the past 50 years. Nature **488**, 70–72 (2012)
4. J.E. Bauer et al., The changing carbon cycle of the coastal ocean. Nature **504**, 61–70 (2013)

5. R.A. Betts, C.D. Jones, J.R. Knight, R.F. Keeling, J.J. Kennedy, El Nino and a record CO_2 rise. Nat. Clim. Change **6**, 806–810 (2016)
6. J.G. Canadell et al., Contributions to accelerating atmospheric CO_2 growth from economic activity, carbon intensity, and efficiency of natural sinks. Proc. Natl. Acad. Sci. USA **104**, 18866–18870 (2007)
7. F. Chevallier, On the statistical optimality of CO_2 atmospheric inversions assimilating CO_2 column retrievals. Atmos. Chem. Phys. **15**, 11133–11145 (2015)
8. P. Ciais, C. Sabine, Chapter 6: Carbon and other biogeochemical cycles, in *Climate Change 2013. The Physical Science Basis*, ed. by T. Stocker, D. Qin, G. K. Platner, (Cambridge University Press, Cambridge, 2013)
9. J.C. van Dam, *Impacts of Climate Change and Climate Variability on Hydrological Regimes* (Cambridge University Press, Cambridge, 1999)
10. K.L. Denman et al., *Couplings Between Changes in the Climate System and Biogeochemistry* (Cambridge University Press, Cambridge, 2007)
11. E. Dietzenbacher, J. Pei, C. Yang, Trade, production fragmentation, and China's carbon dioxide emissions. J. Environ. Econ. Manage. **64**, 88–101 (2012)
12. E. Dlugokencky, P. Tans, *Trends in Atmospheric Carbon Dioxide* (National Oceanic & Atmospheric Administration, Earth System Research Laboratory (NOAA/ESRL), Boulder, 2021). http://www.esrl.noaa.gov/gmd/ccgg/trends/global
13. R.A. Duce et al., Impacts of atmospheric anthropogenic nitrogen on the open ocean. Science **320**, 893–897 (2008)
14. K.-H. Erb et al., Bias in the attribution of forest carbon sinks. Nat. Clim. Change **3**, 854–856 (2013)
15. B. Gonzalez-Gaya et al., High atmosphere-ocean exchange of semivolatile aromatic hydrocarbons. Nat. Geosci. **9**, 438–442 (2016)
16. M.F. Hossain, Solar energy integration into advanced building design for meeting energy demand and environment problem. Int. J. Energy Res. **40**, 1293–1300 (2016)
17. M.F. Hossain, Theory of global cooling. Energy Sustain. Soc. **6**, 24 (2016)
18. M.F. Hossain, Green science: Independent building technology to mitigate energy, environment, and climate change. Renew. Sustain. Energy Rev. **73**, 695–705 (2017)
19. M.F. Hossain, Green science: Advanced building design technology to mitigate energy and environment. Renew. Sustain. Energy Rev. **81**(2), 3051–3060 (2018)
20. R. Houghton, Balancing the global carbon budget. Annu. Rev. Earth Planet. Sci. **35**, 313–347 (2007)
21. C. Le Quéré et al., Global carbon budget 2016. Earth Syst. Sci. Data **8**, 605–649 (2016)
22. W. Li et al., Reducing uncertainties in decadal variability of the global carbon budget with multiple datasets. Proc. Natl. Acad. Sci. USA **113**, 13104–13108 (2016)
23. Z. Liu et al., Reduced carbon emission estimates from fossil fuel combustion and cement production in China. Nature **524**, 335–338 (2015)
24. J. Mason Earles, S. Yeh, K.E. Skog, Timing of carbon emissions from global forest clearance. Nat. Clim. Change **2**, 682–685 (2012)
25. J. Milliman, R. Mei-e, in *Climate Change: Impact on Coastal Habitation*, ed. by D. Eisma, (CRC Press, Boca Raton, 1995), pp. 57–83
26. S.L. Postel, G.C. Daily, P.R. Ehrlich, Human appropriation of renewable fresh water. Science **271**, 785 (1996)
27. J. Prietzel, L. Zimmermann, A. Schubert, D. Christophel, Organic matter losses in German Alps forest soils since the 1970s most likely caused by warming. Nat. Geosci. **9**, 543–548 (2016)

28. S. Schwietzke et al., Upward revision of global fossil fuel methane emissions based on isotope database. Nature **538**, 88–91 (2016)
29. B.B. Stephens et al., Weak northern and strong tropical land carbon uptake from vertical profiles of atmospheric CO_2 science. Science **316**, 1732 (2007)
30. C. Vörösmarty, B. Fekete, M. Meybeck, R. Lammers, Global system of rivers: Its role in organizing continental land mass and defining land to ocean linkages. Global Biogeochem. Cycles **14**, 599–621 (2000)
31. G.R. van der Werf et al., Climate regulation of fire emissions and deforestation in equatorial Asia. Proc. Natl. Acad. Sci. USA **105**, 20350–20355 (2008)

Part II
Sustainable Energy

Chapter 2
Harvesting Global Solar Energy

Abstract The present level of consumption of the reversed fuel on Earth for powering the world is quickening fossil fuel to a limited level and all the while fossil fuel burning is causing deadly environmental vulnerability as well. Inevitably, the alternative energy source must be needed in order to drive the modern world which is abundant and benign to the environment. The use of solar energy in every sectors of our modern life will be an interesting option to power the modern world which is renewable and abundant everywhere in the world. In this research is, therefore, a calculative mechanism has been conducted to estimate global solar energy in order to present a clean and renewable energy system for the world. Interestingly, the calculative result depicted that the average energy density of sunlight irradiance energy on the surfac0e of Earth is 1366 W/m^2 which has been determined by the calculated diameter of Earth of 10,000,000 of meridian at the North Pole to the equator and the radius of Earth is $2/\pi \times 10^7$ m. Therefore, the net energy of solar energy coming to Earth is calculated as $1366 \times (4/\pi) \times 10^{14} \cong 1.73 \times 10^{17}$ W by computing a day consisting of 86,400 s, and a year consisting of 365.2422 days in average. Therefore, the net solar irradiance reaching on Earth annually is $1.73 \times 10^{17} \times 86{,}400 \times 365.2422 \cong 5.46 \times 10^{24}$ J that is equivalent to 5,460,000 EJ energy which is 10,000 times higher than the net current energy demand on Earth annually. Simply, utilization of solar energy could be the primary source of energy to satisfy the net power demand of the whole world, which is clean and environmentally friendly.

Keywords Fossil fuel reserve limitation · Solar radiation · Alternative source of energy · Net solar energy on earth · Clean energy technology · Meeting global energy demand

M. F. Hossain, *Sustainable Design for Global Equilibrium*,
https://doi.org/10.1007/978-3-030-94818-4_2

Highlights
- Estimation of net fossil fuel reserve on Earth
- Calculation of global energy demand annually
- Calculation of global solar energy reaching on Earth per year
- Estimation of net electricity power generation from the net solar energy reaching on Earth annually
- Solar energy implementation to satisfy the world energy requirement

Introduction

The conventional energy consumption for powering the modern civilization throughout the world is indeed accelerating the finite level of current fossil fuel reserve of 36,630 EJ [12, 19]. The global fossil fuel energy consumption was 283 EJ/Yr in the year 1980, 348 EJ/Yr in the year 1995, 405 EJ/Yr in 2005, and 515 EJ/Yr in 2015 and will reach 610 EJ/Yr in 2025, 705 EJ/Yr in 2035, 860 EJ/Yr in the year 2045, and 990 EJ/Yr in the year 2050 [13, 19]. The utilization of fossil fuel globally in the year 2018 was 2.236×10^{20} EJ which is responsible for creating 8.01×10^{11} ton CO_2 into the atmosphere and accounted for acceleration of deadly climate change rapidly [7, 9, 25]. Consequently, adverse environmental impact such as acidic rain, flood, and climate change is occurring unpredictably throughout the world [13, 32, 36]. Recent study shows that the concentration of CO_2 into the atmosphere is currently 400 ppm which needs to be lowered to a standard level grade of 300 ppm CO_2 for the wellness of clean breathing and respiratory system for all mammals [39, 50, 51]. Another research revealed that greenhouse gas emission accelerating the global diurnal mean temperature fluctuating rapidly which posing a serious threat to natural ecosystem and mankind well-being due to the utilization of burning fossil fuel since it create radioactive CO_2 into the atmosphere in a certain period of time [10, 13].

Unfortunately, the consumption of conventional energy currently is still accelerating rapidly throughout the world; the situation shall remain unchanged until a renewable source of energy is developed to utilize sustainable energy [5, 8, 26]. Simply it is an urgent demand to develop sustainable energy technology to mitigate fossil fuel consumption where the "new source that fulfills the requirement of the present without disrupting the ability of future demand of energy to fulfill the complete needs for the future generations." Hence, the solar energy utilization globally can be the interesting source to fulfill the net energy requirement throughout the world. It is a natural renewable energy source that is generated by the sun which is created by nuclear fusion that takes place in the sun which is constantly flowing away from the sun to the solar system and part of it reaches on Earth which is tremendous source of clean and renewable energy [21, 22, 25, 37]. If only a 0.001% of the annual solar energy coming on Earth is used, it will meet the net energy demand for the whole planet which is clean and abundant everywhere. In this study, therefore, a research has been performed to harvest the total global solar energy reaching on Earth in order to mitigate global net energy need which is clean and environmentally friendly.

Material, Methods, and Simulation

Calculation of Net Solar Energy on Earth

The total Earth surface is being clarified by characterizing various directional angles considering the Cartesian coordinate system, where x denotes to skyline convention, y denotes for east-west, z denotes for zenith in order to measure the total solar irradiance during the day entire year (Fig. 2.1). The position of the celestial body in this framework is thus chosen by h which is denoted for height and A denoted for the azimuth angle while the central framework utilized it as the convention factor which is z hub. It focuses toward the North Pole, the y hub indistinguishably focuses on the horizon of the skylight, and x pivot opposite to both the North pole and horizon. Therefore, the angles and coordinate frequencies are being encountered mathematically by calculating the latitude and longitude in order to implement correct angles to trap the solar irradiance most efficiently. Here, the zero point of latitude is considered the primary meridian which controls the function of meridian of Eastern Hemisphere and Western Hemisphere angle of the Earth surface and so does the north of the equator is the Northern Hemisphere and south of the equator is the Southern Hemisphere are also being controlled by this Earth surface modeling to trap solar energy more efficiently. Finally, the δ and ω point hours are being clarified accurately considering this analytical Cartesian coordinates in order to decide the position of solar irradiance vector to clarify the solar energy reaching on the Earth surface to determine net solar energy calculation into the Earth surface [17, 22, 27].

Once the angle of the Earth surface is being modeled, then the Earth surface is being considered as the net areal dimension of solar energy emission by considering the peak hours solar radiation generation from the sun [21, 28, 53]. Thus, the radiation of the solar irradiance released by the sun and accosted by the Earth is computed by the *solar constant which is* defined by the measurement of the solar energy flux density perpendicular to the ray direction per unit area per unit time [3, 4, 6]. Thus, the calculation of this amount of net solar energy includes all types of radiations of scattered and reflected ones and both are being modeled by using MATLAB software to calculate *total global solar radiation emission on Earth.*

Then, the sunlight is being clarified as the motion of the photon flux by considering the first function of the fundamental solar thermal energy and anti-reflective coatings of solar cells and then it is modified into the second order of function of the solar energy [11, 14, 23]. The integration of these two functions is computed by implementing the solar quantum dynamics which is clarified as the most acceptable quantum technology to calculate the net solar energy emission on Earth [35, 39, 49]. This is because the Earth surface can emit solar energy accurately at a given

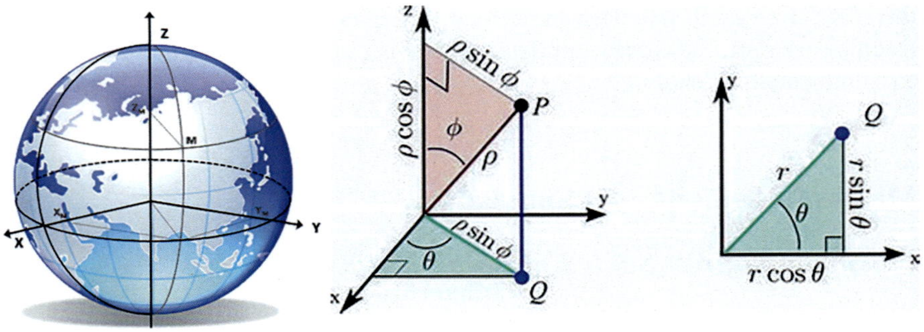

a

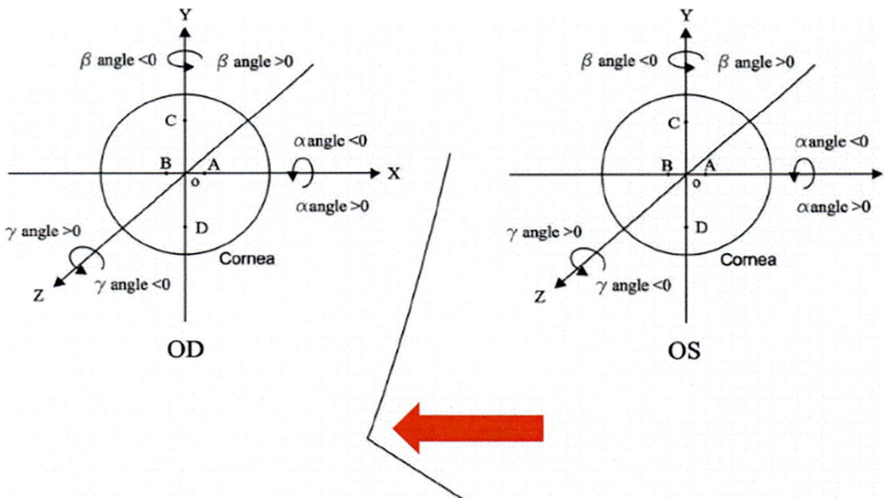

b

Fig. 2.1 (**a**) Cartesian coordinates clarification of southern axis of x, western axis of y, and the zenith axis of z clarified to calculate the total solar energy reaching on Earth considering the average energy density of sunlight on the surface of Earth is 1366 W/m^2 by implementing the diameter of the Earth as 10,000,000 of meridian at the North Pole to the equator and the radius of Earth as $2/\pi \times 10^7$ m. The location of this celestial body is analyzed by determining two angles of sin θ and cos θ. (**b**) The longitudinal and latitudinal equatorial angles have been clarified where the convention z-axis point denotes the North Pole and the east-west axis y-axis denotes the identical angle of the horizon

temperature of approximately 700 °C where the energy density of the solar radiation is being derived from the peak solar irradiance generation from a single solar photon flux [10, 13, 43].

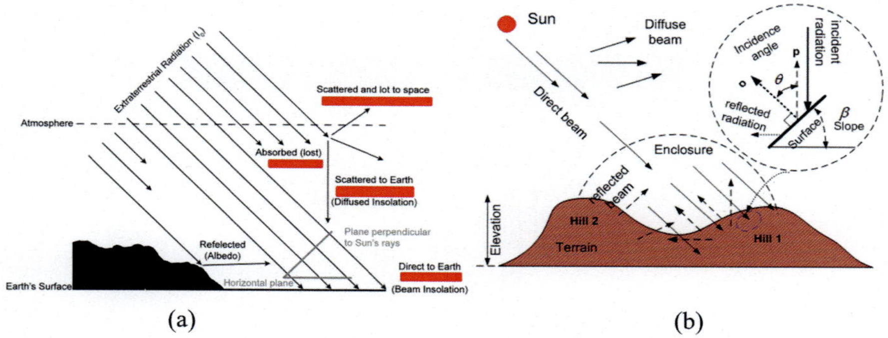

Fig. 2.2 (**a**) Shows the emission of solar energy on the Earth surface, (**b**) various kinds of irradiance on Earth surface; direct beam, reflected beam, diffuse beam at various angles

The estimation of global solar irradiance calculation on the Earth surface is being further clarified considering the three background solar data calculation by using *Pyrheliometer* to measure direct beam irradiance approaching from the sun and radius of the earth surface [4, 40]. Then, the *Pyranometer* is also utilized to determine net hemispherical irradiance beam along with the diffused beam on horizon and, thus, the net global solar radiation (W/m^2) is being determined considering the horizon by a pyranometer which is denoted as:

$$I_{\text{tot}} = I_{\text{beam}} \cos\theta + I_{\text{diffuse}}$$

where θ represents the zenith angle which has been implemented to calculate the net solar energy reaching on Earth by the clarification of Electron Energy Level of Hydrogen (Fig. 2.3).

This measurement is being then confirmed against modest pyrheliometers using the thermo-coupler detector and the PV detectors recorder considering the determination of wavelength of the solar spectrum [3, 7]. Eventually *Photoelectric sunshine recorder has been used* which is intermittent and varied by the solar irradiance intensity [12, 18, 46]. Since the solar radiation is related to the photon charge, the attributes of photon energy on Earth surface is being computed considering the quantum flow of photon radiation in global scale by using MATLAB 9.0 Classical Multidimensional Scaling [1, 15, 20]. Consequently, a computational model of photon radiation is being quantified to demonstrate the solar energy generation from sunlight considering radiation emission. Thereafter, the mode of the solar quanta absorbance by Earth surface is being determined by the peak solar radiation output tracking into the Earth surface [4, 31, 38]. Naturally, the induced solar irradiance is, thereafter, computed by the Earth surface area by implementing the parameters of solar energy proliferation on it, and transformation rate of solar energy into electricity energy generation. Thus, the accurate calculation of the current–voltage (*I*–*V*) characteristic is being subsequently conducted by the conceptual

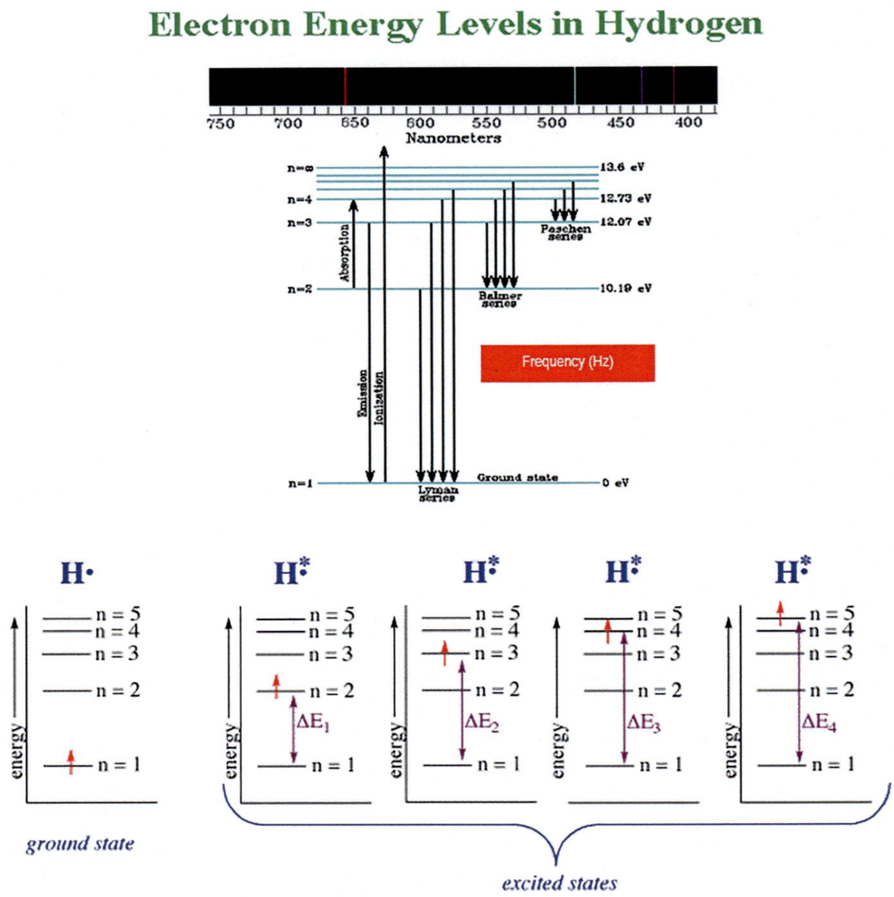

Fig. 2.3 The solar energy state hydrogen depicting the absorption and emission modes, and energy deliberation rate revealing the energy density by clarifying the optimum solar irradiance deliberation from a photon particle at various wavelength and frequencies of 10^{17}–10^{14} Hz. Then the electron energy level H2 is being determine by conducting a down path transition involves in accost of a solar energy with respect ground state and excited state of solar energy

model of net solar radiation intake into the Earth surface by computing the net active solar volt (I_{v+}) generation into the Earth surface [16, 24, 26].

Then, the mathematical determination of the net current formation via I_{pv} on earth surface has been modeled out, by calculating *I*–*V*–*R* correlationship within the Earth surface in order to use this energy commercially throughout the world (Fig. 2.4).

Hence, the following equation is being computed as the energy deliberation from the Earth surface, whose origin is the photon irradiance and ambient temperature of the solar energy:

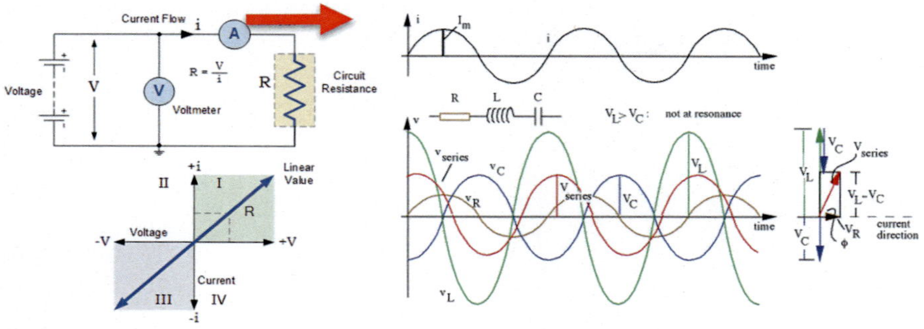

Fig. 2.4 The conceptual circuit diagram of the whole Earth surface depicted the (**a**) net photophysical current generation into the Earth surface by detailing model of *I–V–R* relationship in order to use electricity throughout the world, respectively

$$P_{pv} = \eta_{pvg} A_{pvg} G_t \tag{2.1}$$

Here, η_{pvg} denotes the Earth surface performance rate, A_{pvg} denotes the Earth surface array (m^2), and G_t denotes the photon irradiance intaking rate on the plane (W/m^2) of Earth surface, and thus the η_{pvg} could be rewritten as follows:

$$\eta_{pvg} = \eta_r \eta_{pc}[1 - \beta(T_c - T_{cref})] \tag{2.2}$$

η_{pc} denotes the energy formation efficiency, when maximum power point tracking (MPPT) is being implemented which is close to 1; Here β denotes the temperature cofactor (0.004–0.006 per °C); η_r denotes the mode of energy efficiency; and T_{cref} denotes the condition of temperature at °C. The reference Earth temperature (T_{cref}) can be rewritten by calculating from the equation below:

$$T_c = T_a + \left(\frac{\text{NOCT} - 20}{800}\right) G_t \tag{2.3}$$

T_a denotes the encircling temperature in °C, G_t denotes the solar radiation in Earth surface (W/m^2), and thus it denotes the modest optimum Earth temperature in °C. Considering this temperature condition, the net solar radiation Earth surface can be calculated by the equation below:

$$I_t = I_b R_b + I_d R_d + (I_b + I_d) R_r \tag{2.4}$$

The solar energy here is necessarily working as a conceptual P–N junction superconductor in order to form electricity through the Earth surface, which is interlinked in a parallel series connection [6, 29, 47]. Thus, a unique conceptual circuit model, as shown in Fig. 2.4, with respect to the N_s series of Earth surface and

N_p parallel arrays has been computed by the following Earth surface solar energy equation based on current and volt relationship

$$I = N_p\left[I_{ph} - I_{rs}\left[\exp\left(\frac{q(V + IR_s)}{AKTN_s} - 1\right)\right]\right] \tag{2.5}$$

where

$$I_{rs} = I_{rr}\left(\frac{T}{T_r}\right)^3 \exp\left[\frac{E_G}{AK}\left(\frac{1}{T_r} - \frac{1}{T}\right)\right] \tag{2.6}$$

Hence, in Eqs. (2.25) and (2.26), q denotes the electron-charge (1.6×10^{-19} C), K denotes the Boltzmann's constant, A denotes the diode standardized efficiency, and T denotes the Earth temperature (K). Accordingly, I_{rs} denotes the Earth surface reversed current motion at T, where T_r denotes the Earth condition temperature, I_{rr} denotes the reverses current at T_r, and E_G denotes the photonic bandgap energy of the superconductor utilized for the Earth surface. Thus, the photonic current I_{ph} will be generated in accordance with the Earth surface temperature and radiation condition which can be expressed by:

$$I_{ph} = \left[I_{SCR} + k_i(T - T_r)\frac{S}{100}\right] \tag{2.7}$$

Here, I_{SCR} denotes the current-motion considering the optimum temperature of the Earth and solar radiation dynamic on the Earth surface, k_i denotes the short-circuited current-motion, and S denotes the solar radiation calculation in a unit area (mW/cm^2). Subsequently, the I–V features of the Earth surface shall be deformed from the conceptual model of the circuit which can be expressed as:

$$I = I_{ph} - I_D \tag{2.8}$$

$$I = I_{ph} - I_0\left[\exp\left(\frac{q(V + R_sI)}{AKT} - 1\right)\right] \tag{2.9}$$

I_{ph} denotes the photonic current-dynamic (A), I_D denotes the diode originated current-dynamic (A), I_0 denotes the inversed current-dynamic (A), A denotes the diode induced constant, q denotes the charge of the electron (1.6×10^{-19} C), K denotes the Boltzmann's constant, T denotes the Earth temperature (°C), R_s denotes the series-resistance (ohm), R_{sh} denotes the shunt-resistance (Ohm), I denotes the cell current-motion (A), and V denotes the Earth voltage-motion (V). Therefore, the net current flow into the Earth surface can be determined by conducting the following equation:

$$I = I_{PV} - I_{D1} - \left(\frac{V + IR_S}{R_{SH}}\right) \tag{2.10}$$

where

$$I_{D1} = I_{01}\left[\exp\left(\frac{V + IR_s}{a_1 V_{T1}}\right) - 1\right] \tag{2.11}$$

Here, I and I_{01} are being denoted as the reversed current flow into the conceptual circuit, and V_{T1} and V_{T2} are being denoted as the optimum thermal volts into the circuit. Thus, the circuit standard factor is being presented by a_1 and a_2 and then it has been normalized by the mode of Earth surface by expressing the following equation:

$$v_{oc} = \frac{V_{oc}}{cK\,T/q} \tag{2.12}$$

$$P_{\max} = \frac{\frac{V_{oc}}{cKT/q} - \ln\left(\frac{V_{oc}}{cKT/q} + 0.72\right)}{\left(1 + \frac{V_{oc}}{KT/q}\right)}\left(1 - \frac{V_{oc}}{\frac{V_{oc}}{I_{SC}}}\right)\left(\frac{V_{oc0}}{1 + \beta \ln \frac{G_0}{G}}\right)\left(\frac{T_0}{T}\right)^{y} I_{sc0}\left(\frac{G}{G_0}\right)^{a} \tag{2.13}$$

where ν_{oc} denotes the standard point of the open-circuit voltage, V_{oc} denotes the thermal voltage $V_t = nkT/q$, c denotes the constant current motion, K denotes the Boltzmann's constant, T denotes the temperature into the Earth surface PV cell in Kelvin, α denotes the function which represents the nonlinear motion of photocurrents, q denotes the electron charge, γ denotes the function acting for all nonlinear temperature-voltage currents, while β denotes the Earth surface mode for specific dimensionless function for enhancing current flowing rate. Subsequently, Eq. (2.13) represents the peak energy generation from the Earth surface module which is interlined in both series and parallel connection. Thus, the equation for the net energy formation into the array of N_s has been interlinked in series and N_p has been interlinked in parallel considering the power P_M of each mode of connection and which is finally expressed by using the following equation:

$$P_{\text{array}} = N_s N_p P_M \tag{2.14}$$

Modeling of Net Electricity Energy Generation from Total Solar Irradiance on Earth

To convert global solar energy into electricity energy, a model is also being prepared by integrating global Albanian symmetries of gauge field scalar [28, 38, 43]. Naturally, the net solar energy particle will functionally be acted as the dynamic photons of particle T^{α} at the global symmetrical array of Earth surface by initiating the gauge field of $A_{\mu}^{\alpha}(x)$, and then the local Albanian will subsequently be started to activate at the global $U(1)$ phase symmetry to deliver net electricity energy [28, 44]. Thus, the model is being considered as a complex vector field of $\Phi(x)$ of Earth surface where electric charge q will couple with the EM field of $A^{\mu}(x)$, and thus the equation can be expressed by $\mathfrak{h}$:

$$\mathfrak{h} = -\frac{1}{4}F_{\mu\nu}F^{\mu\nu} + D_{\mu}\Phi^{*}\,D^{\mu}\Phi - V(\Phi^{*}\Phi) \tag{2.15}$$

where

$$\begin{aligned} D_{\mu}\Phi(x) &= \partial_{\mu}\Phi(x) + iqA_{\mu}(x)\Phi\,(x) \\ D_{\mu}\Phi^{*}(x) &= \partial_{\mu}\Phi^{*}(x) - iqA_{\mu}(x)\Phi^{*}(x) \end{aligned} \tag{2.16}$$

and

$$V(\Phi^{*}\Phi) = \frac{\lambda}{2}\,(\Phi^{*}\Phi)^{2} + m^{2}(\Phi^{*}\Phi) \tag{2.17}$$

Here $\lambda > 0$ $m^2 < 0$, therefore $\Phi = 0$ is a local optimum vector quantity, while the minimum form of degenerated scalar circle is clarified as $\Phi = \frac{v}{\sqrt{2}} * e^{i\theta}$,

$$v = \sqrt{\frac{-2m^{2}}{\lambda}}, \quad \text{any real } \theta \tag{2.18}$$

Subsequently, the vector field Φ of the global Earth surface will form a non-zero functional value $\langle\Phi\rangle \neq 0$, which will simultaneously determine the $U(1)$ symmetrical net solar energy generation. Therefore, the global $U(1)$ net symmetrical electrical energy of $\Phi(x)$ will be delivered as expected value of $\langle\Phi\rangle$ by confirming the x-dependent state of the symmetrical $\Phi(x)$ array of Earth surface and can be expressed by the following equation:

$$\Phi(x) = \frac{1}{\sqrt{2}}\Phi_{\mathrm{r}}(x) * e^{i\Theta(x)}, \quad \text{real } \Phi_{\mathrm{r}}(x) > 0, \text{real } \Phi(x) \tag{2.19}$$

Thus, the net calculation of the electricity energy generation from the net Earth surface solar energy is being determined considering the vector $\Phi(x) = 0$, and it is first order function of $\langle\Phi\rangle \neq 0$, considering the peak level of solar energy emission on the Earth surface of $\Phi\langle x\rangle \neq 0$ [10, 19, 31]. Thus, the net electricity energy generation is from the global solar energy calculation $\phi_{\mathrm{r}}(x)$ and $\Theta(x)$; its vector on the Earth surface field ϕ_{r} has been confirmed by conducting the following equation:

$$V(\phi) = \frac{\lambda}{8}\,\left(\phi_{\mathrm{r}}^{2} - v^{2}\right)^{2} + \text{const}, \tag{2.20}$$

or the resultant electricity energy generation is shifted by its VEV, $\Phi_r(x) = v + \sigma(x)$,

$$\phi_r^2 - v^2 = (v+\sigma)^2 - v^2 = 2v\sigma + \sigma^2 \tag{2.21}$$

$$V = \frac{\lambda}{8}\left(2v\sigma - \sigma^2\right)^2 = \frac{\lambda v^2}{2} * \sigma^2 + \frac{\lambda v}{2} * \sigma^3 + \frac{\lambda}{8} * \sigma^4 \tag{2.22}$$

Simultaneously, the functional derivative $D_\mu\phi$ will become

$$\begin{aligned} D_\mu\phi &= \frac{1}{\sqrt{2}}\left(\partial_\mu\left(\phi_r e^{i\Theta}\right) + iqA_\mu * \phi_r e^{i\Theta}\right) \\ &= \frac{e^{i\Theta}}{\sqrt{2}}\left(\partial_\mu\phi_r + \phi_r * i\partial_\mu\Theta + \phi_r * iqA_\mu\right) \end{aligned} \tag{2.23}$$

$$\begin{aligned} \left|D_\mu\phi\right|^2 &= \frac{1}{2}\left|\partial_\mu\phi_r + \phi_r * i\partial_\mu\Theta + \phi_r * iqA_\mu\right|^2 \\ &= \frac{1}{2}\left(\partial_\mu\phi_r\right) + \frac{\phi_r^2}{2} * \left(\partial_\mu\Theta qA_\mu\right)^2 \\ &= \frac{1}{2}\left(\partial_\mu\sigma\right)^2 + \frac{(v+\sigma)^2}{2} * \left(\partial_\mu\Theta + qA_\mu\right)^2 \end{aligned} \tag{2.24}$$

Altogether,

$$\mathfrak{h} = \frac{1}{2}\left(\partial_\mu\sigma\right)^2 - v(\sigma) - \frac{1}{4}F_{\mu\nu}F^{\mu\nu} + \frac{(v+\sigma)^2}{2} * \left(\partial_\mu\Theta + qA_\mu\right)^2 \tag{2.25}$$

To determine the formation of this net electricity generation referred as ($\mathfrak{h}_{\text{sef}}$) into the Earth surface, the function of the electro-static fields has been quantified by conducting the quadratic calculation and described by the following equation:

$$\mathfrak{h}_{\text{sef}} = \frac{1}{2}\left(\partial_\mu\sigma\right)^2 - \frac{\lambda v^2}{2} * \sigma^2 - \frac{1}{4}F_{\mu\nu}F^{\mu\nu} + \frac{v^2}{2} * \left(qA_\mu + \partial_\mu\Theta\right)^2 \tag{2.26}$$

Here this net electricity generation ($\mathfrak{h}_{\text{free}}$) function certainly will admit a realistic vector particle of positive mass$^2 = \lambda v^2$ integrating the areal $A_\mu(x)$ function and the electricity energy generation fields $\Theta(x)$ to confirm to determine the net electricity energy from the global solar energy calculation into the Earth surface (Fig. 2.4).

Results and Discussion

Calculation of Net Solar Energy on Earth

To calculate the net solar energy on the Earth surface, the net irradiance of photon emission has been calculated by integrating Eqs. (2.22) and (2.23). Necessarily, the functional Earth surface area $J(\omega)$, the photonic quantum field and the unit area $J(\omega)$ are being calculated considering the constant irradiance coupling point, and the Weisskopf-Winger approximation mechanism to confirm the accurate solar energy reaching on the Earth surface [12, 34].

The computed results show that the distribution of solar irradiance on the Earth's sphericity and orbital parameters is the application of the unidirectional beam incident to a rotating sphere of Milankovitch cycles from spherical Earth law of cosines:

$$\cos(c) = \cos(a)\cos(b) + \sin(a)\sin(b)\cos(C) \tag{2.27}$$

where a, b, and c are being considered as the arc lengths, in radians, of the sides of a spherical triangle. C represents the angle of the vertex which has arc length of c. Determining this calculation of solar zenith angle Θ, the following equation is being clarified considering the application of spherical law of cosines:

$$C = h$$

$$c = \Theta$$

$$a = \frac{1}{2}\pi - \phi$$

$$b = \frac{1}{2}\pi - \delta$$

$$\cos(\Theta) = \sin(\phi)\sin(\delta) + \cos(\phi)\cos(\delta)\cos(h) \tag{2.28}$$

In order to simplify this equation, it has been further clarified as a general derived as follows:

$$\begin{aligned}\cos(\theta) = {} & \sin(\phi)\sin(\delta)\cos(\beta) + \sin(\delta)\cos(\phi)\sin(\beta)\cos(\gamma) \\ & + \cos(\phi)\cos(\delta)\cos(\beta)\cos(h) - \cos(\delta)\sin(\phi)\sin(\beta)\cos(\gamma)\cos(h) \\ & - \cos(\delta)\sin(\beta)\sin(\gamma)\sin(h)\end{aligned}$$

where β denotes the angle from the horizon and γ denotes the azimuth angle.

The sphere of Earth from the sun here is denoted by R_E where the average distance is represented as R_0, with approximation of one astronomical-unit (AU). Here, the solar constant is being represented as s_0 where the solar irradiance density onto an Earth plane tangent is calculated as :

$$Q = \begin{cases} s_0 \dfrac{R_0^2}{R_E^2} \cos(\theta) & \cos(\theta) > 0 \\ 0 & \cos(\theta) \leq 0 \end{cases}$$

The mean Q over a day is the average of Q over one rotation, or the hour angle progressing from $h = \pi$ to $h = -\pi$: Thus, the equation has been rewritten as

$$Q^{-\text{day}} = -\frac{1}{2\pi} \int_{\pi}^{-\pi} Q dh$$

Since h_0 is the hour angle when Q becomes positive; thus, it could occur at sunrise when $\Theta = 1/2\pi$, or for h_0 as a solution of

$$\sin(\phi) \sin(\delta) + \cos(\phi) \cos(\delta) \cos(h_0) = 0$$

or

$$\cos(h_0) = -\tan(\phi) \tan(\delta)$$

Once $\tan(\varphi)\tan(\delta) > 1$, then the sun does not set and the sun is already risen at $h = \pi$, so $h_0 = \pi$. Then the $\tan(\varphi)\tan(\delta) < -1$, the sun does not rise and

$$Q^{-\text{day}} = 0$$

$\frac{R_0^2}{R_E^2}$ is nearly constant over the course of a day, and can be taken outside the integral

$$\begin{aligned} \int_{\pi}^{-\pi} Q dh = \int_{h_0}^{-h_0} Q dh &= s_0 \frac{R_0^2}{R_E^2} \int_{h_0}^{-h_0} \cos(\theta) dh \\ &= s_0 \frac{R_0^2}{R_E^2} [h \sin(\phi) \sin(\delta) + \cos(\phi) \cos(\delta) \sin(h)]_{h = h_0}^{h = -h_0} \\ &= -2 s_0 \frac{R_0^2}{R_E^2} [h_0 \sin(\phi) \sin(\delta) + \cos(\phi) \cos(\delta) \sin(h_0)] \end{aligned}$$

Therefore:

$$Q^{-\text{day}} = \frac{s_0}{\pi} \frac{R_0^2}{R_E^2} [h_0 \sin(\phi) \sin(\delta) + \cos(\phi) \cos(\delta) \sin(h_0)]$$

Since the θ is being considered as the conventional polar angle describing a planetary orbit, $\theta = 0$ at the vernal equinox and the declination δ as a function of orbital position would be

$$\delta = \varepsilon \sin(\theta)$$

where ε is the obliquity and the conventional longitude of perihelion ϖ shall be related to the vernal equinox, so for the elliptical orbit it can be rewritten as:

$$R_{\mathrm{E}} = \frac{R_0}{1 + e\cos(\theta - \omega)}$$

or

$$\frac{R_0}{R_{\mathrm{E}}} = 1 + e\cos(\theta - \omega)$$

Here the ϖ, ε, and e are being calculated from astrodynamical laws, so that a consensus of observations of Q-day can be determined from any latitude φ and θ. However, $\theta = 0°$ is considered as duration of the vernal equinox, $\theta = 90°$ is exactly the time of the summer solstice, $\theta = 180°$ is exactly the time of the autumnal equinox, and $\theta = 270°$ is exactly the time of the winter solstice. Therefore, the equation can be simplified for irradiance on a given day as follows:

$$Q = s_0\left(1 + 0.034\cos\left(2\pi\frac{n}{365.25}\right)\right)$$

where n is the number of a day of the year and thus, the solar characteristics for both theoretical function of optimum and modular to generate electricity can be shown per unit area.

Eventually, a peak high-frequency cutoff Ω_{C} of solar irradiance is being calculated to keep away the bifurcation of DOS from the Earth surface. Necessarily, a tipped high-frequency cutoff of Earth surface Ω_{d} is being determined by controlling the positive DOS in 2D and 1D of the photon irradiance. Hence $\mathrm{pi}_2(x)$ acted as an algorithm function and $\mathrm{erfc}(x)$ acts as an additional function [11, 42]. Thus, the DOS of Earth surface, here represented as $\varrho_{\mathrm{PC}}(\omega)$, is determined by calculating photonic energy frequencies of Maxwell's rules into the Earth surface [10, 12, 40]. For a 1D on Earth surface, the represented DOS is thus being expressed as $\varrho_{\mathrm{PC}}(\omega) \propto \frac{1}{\sqrt{\omega - \omega_{\mathrm{e}}}}\Theta(\omega - \omega_{\mathrm{e}})$, where $\Theta(\omega - \omega_{\mathrm{e}})$ represents the Heaviside Functional Step and ω_{e} expresses the frequency of the net solar energy generation.

This DOS is thus determined to confirm a 3D isentropic function in the Earth surface to acquire an accurate net qualitative state of solar energy by inducing the non-Weisskopf-Winger mode of photons in the Earth surface [8, 21, 30]. Naturally, this 3D state will be the functional DOS into the PBE area of DOS: $\varrho_{\mathrm{PC}}(\omega) \propto \frac{1}{\sqrt{\omega - \omega_{\mathrm{e}}}}\Theta(\omega - \omega_{\mathrm{e}})$, and thus it has been integrated to the net electricity (EF) vector of Earth surface to determine the net electricity energy generation accurately on the Earth surface [26, 41]. Considering the 2D and 1D, the photonic energy DOS are being clarified by the pure algorithm of divergence which is close to the PBE, and thus expressed as $\varrho_{\mathrm{PC}}(\omega) \propto -[\ln|(\omega - \omega_0)/\omega_0| - 1]\Theta(\omega - \omega_{\mathrm{e}})$, where ω_{e} denotes the mid-point of tip algorithm. The functional area $J(\omega)$ is thus clarified as the photon energy generation on the Earth surface where the solar energy generation of $V(\omega)$ depends on the total solar irrdiance on the Earth surface [14, 16],

$$J(\omega) = \varrho(\omega)|V(\omega)|^2 \tag{2.29}$$

Hence, the PB frequency ω_c and proliferative solar energy are being considered as the function $u(t, t_0)$ for photon energy generation in the relation $\langle a(t)\rangle = u(t, t_0)\langle a(t_0)\rangle$. It is, therefore, determined using the functional integral equation and expressed as

$$u(t, t_0) = \frac{1}{1 - \Sigma'(\omega_b)} e^{-i\omega(t-t_0)} + \int_{\omega_e}^{\infty} d\omega \frac{J(\omega)e^{-i\omega(t-t_0)}}{[\omega - \omega_c - \Delta(\omega)]^2 + \pi^2 J^2(\omega)} \tag{2.30}$$

where $\Sigma'(\omega_b) = [\partial\Sigma(\omega)/\partial\omega]_{\omega=\omega_b}$ and $\Sigma(\omega)$ denote the storage induced PB photonic energy proliferations,

$$\Sigma(\omega) = \int_{\omega_e}^{\infty} d\omega' \frac{J(\omega')}{\omega - \omega'} \tag{2.31}$$

Here, the frequency ω_b in Eq. (2.17) denoted the photon energy frequency module in the PBG ($0 < \omega_b < \omega_e$) and thus it is calculated using the areal condition: $\omega_b - \omega_c - \Delta(\omega_b) = 0$, where $\lesssim\Delta(\omega) = \mathcal{P}\left[\int d\omega' \frac{J(\omega')}{\omega-\omega'}\right]$ is a primary-value integral.

Therefore, the net photon energy, considering the proliferation magnitude $|u(t, t_0)|$, has been calculated and is being shown in Fig. 2.6a for 1D, 2D, and 3D of Earth surface with respect to PBG function [16, 23, 44]. The solar energy dynamic rate $\kappa(t)$ is being depicted in Fig. 2.4b, neglecting the function $\delta = 0.1\omega_e$. The result revealed that emitted photons are being generated at a high rate once ω_c crosses from the PBG to PB area. Because the range in $u(t, t_0)$ is $1 \geq |u(t, t_0)| \geq 0$, the crossover area as related to the condition is being denoted as $0.9 \gtrsim |u(t \to \infty, t_0)| \geq 0$ where this corresponded to$-0.025\omega_e \lesssim \delta \lesssim 0.025\omega_e$, with a production rate $\kappa(t)$ within the PBG ($\delta < -0.025\omega_e$) and in the area of the PBE($-0.025\omega_e \lesssim \delta \lesssim 0.025\omega_e$) of the Earth surface.

The generation of solar energy emission is almost exponential for $\delta \gg 0.025\omega_e$, which is a Markov factor. It is shown in Fig. 2.5a as the dash-doted black curves with $\delta = 0.1\omega_e$. In the crossover area ($-0.025\omega_e \lesssim \delta \lesssim 0.025\omega_e$), the PB frequency of the PBE of Earth surface, which sharply increases the mode of the emission of photon energy generation [2, 48]. Thus, this proliferation of emitted solar photon confirms the net energy state photon in the Earth surface of the PBG where the photons are in a non-equilibrium photonic energy state [7, 50].

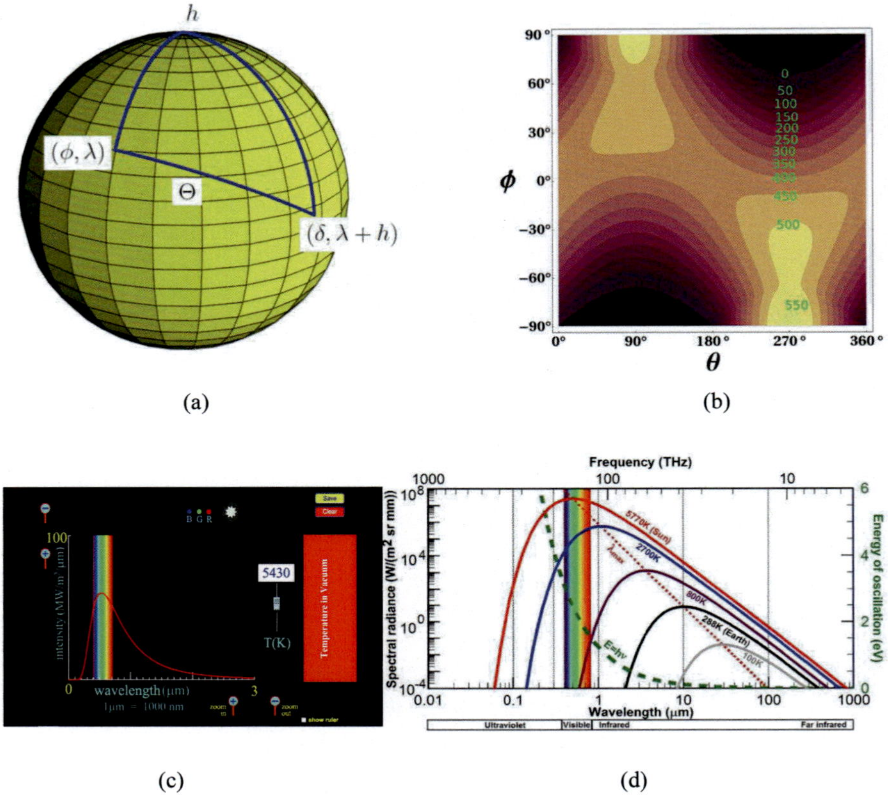

Fig. 2.5 (**a**) The sphere triangular for cosines is being clarified as the solar zenith angle Θ considering latitude φ and longitude λ. (**b**) The average daily irradiation at the top of the atmosphere, where θ is the polar angle of the Earth's orbit, and $\theta = 0$ at the vernal equinox, and $\theta = 90°$ at the summer solstice; φ is the latitude of the Earth. (**c**) Shows the solar irradiance at various frequencies, and (**d**) the peak temperature which suggest the calculative power to determine the net solar energy

Then, the solar irradiance on entire Earth surface is being clarified considering thermal variation with respect to the solar energy concentration function $v(t,t)$ by determining the non-equilibrium solar energy scattering and reflecting calculation globally [25, 52],

$$v(t,t) = \int_{t_0}^{t} dt_1 \int_{t_0}^{t} dt_2 u^*(t_1,t_0)\widetilde{g}(t_1,t_2)u(t_2,t_0) \tag{2.32}$$

Here, the two-time correlation function of Earth surface $\widetilde{g}(t_1,t_2) = \int d\omega J(\omega)\overline{n}(\omega,T)e^{-i\omega(t-t')}$ reveals the solar energy generation variations induced

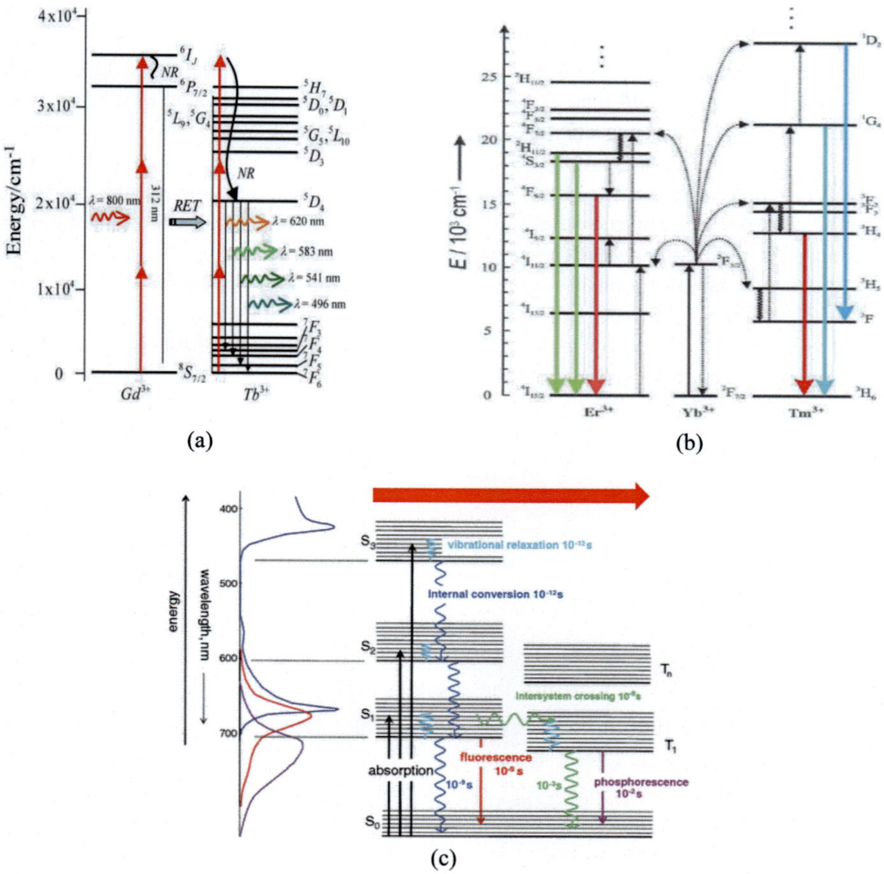

Fig. 2.6 (**a**) Shows the above stands for the solar energy function of optimum working point for energy generation per unit area, (**b**) depicted the theoretical function of modular point of energy generation per unit area, and (**c**) energy generation at various wavelength of photon spectrum

by the thermal relativistic condition of Earth surface, where $\overline{n}(\omega, T) = 1/\left[e^{\hbar\omega/k_B T} - 1\right]$ is the proliferation of the photon energy emission in the Earth surface at the optimum temperature T and expressed as

$$v(t, t \to \infty) = \int_{\omega_e}^{\infty} d\omega \mathcal{V}(\omega) \quad \text{with} \quad \mathcal{V}(\omega) = \overline{n}(\omega, T)[\mathcal{D}_1(\omega) + \mathcal{D}_d(\omega)] \tag{2.33}$$

Here, Eq. (2.18) is simplified to determine the non-equilibrium condition: $\mathcal{V}(\omega) = \overline{n}(\omega, T)\mathcal{D}_{\mathrm{d}}(\omega)$. Under low-temperature conditions on Earth surface, Einstein's photon energy fluctuation dissipation is not dynamically viable at the PB on Earth surface, but also connecting the photonic energy state which has been measured as the field intensity of solar energy induction $n(t) = \langle a^{\dagger}(t)a(t)\rangle = |u(t, t_0)|^2 n(t_0)v(t, t)$, where $n(t_0)$ represents the primary PB of Earth surface. Therefore, in Fig. 2.7, the plotted net amount of photon energy versus temperature on Earth surface has been clarified as the non-equilibrium proliferated photon energy generation, as shown by the solid-blue curve (Fig. 2.6). To be more specific, the first PB of Earth surface has been considered as the Fock state photon number n_0, i. e. $\rho(t_0) = |n_0\rangle\langle n_0|$, which is obtained mathematically through the quantum dynamics of the photon energy and then by solving the Equation (2.33), respecting the state of net photon energy production at time t:

$$\rho(t) = \sum_{n=0}^{\infty} \mathcal{P}_n^{(n_0)}(t)|n_0\rangle\langle n_0| \tag{2.34}$$

$$\mathcal{P}_n^{(n_0)}(t) = \frac{[v(t,t)]^n}{[1 + v(t,t)]^{n+1}}[1 - \Omega(\mathrm{t})]^{n_0} \ \mathrm{x} \sum_{k=0}^{\min\{n_0, n\}} \binom{n_0}{k}\binom{n}{k}\left[\frac{1}{v(t,t)}\frac{\Omega(\mathrm{t})}{1 - \Omega(\mathrm{t})}\right]^k \tag{2.35}$$

where $\Omega(\mathrm{t}) = \frac{|u(t,t_0)|^2}{1+v(t,t)}$. Therefore, the result reveales that an electron state photon energy will evolve into different Fock states of $|n_0\rangle is\ \mathcal{P}_n^{(n_0)}(t)$ on the Earth surface. The proliferation of net photon energy dissipation $\mathcal{P}_n^{(n_0)}(t)$ in the primary state $|n_0 = 5\rangle$ and steady-state limit, $\mathcal{P}_n^{(n_0)}(t \to \infty)$ is thus shown in Fig. 2.7. Therefore, the generation of net photon energy generation on Earth surface emit will ultimately reach the thermal non-equilibrium state, which is being expressed as

$$\mathcal{P}_n^{(n_0)}(t \to \infty) = \frac{[\overline{n}(\omega_{\mathrm{c}}, T)]^n}{[1 + \overline{n}(\omega_{\mathrm{c}}, T)]^{n+1}} \tag{2.36}$$

To probe this huge photon energy generation on Earth surface, a further calculation of the photon energy distribution within the quantum field of Earth surface has been conducted through the high temperature coherent states and solving Eq. (2.17) considering the energy state of photons, and expressed by

$$\rho(t) = \mathcal{D}[\alpha(t)]\rho_{\mathrm{T}}[v(t,t)]\mathcal{D}^{-1}[\alpha(t)] \tag{2.37}$$

where $\mathcal{D}[\alpha(t)] = \exp\left\{\alpha(t)a^{\dagger} - \alpha^*(t)a\right\}$ denotes the displacement functions with $\alpha(t) = u(t, t_0)\alpha_0$ and

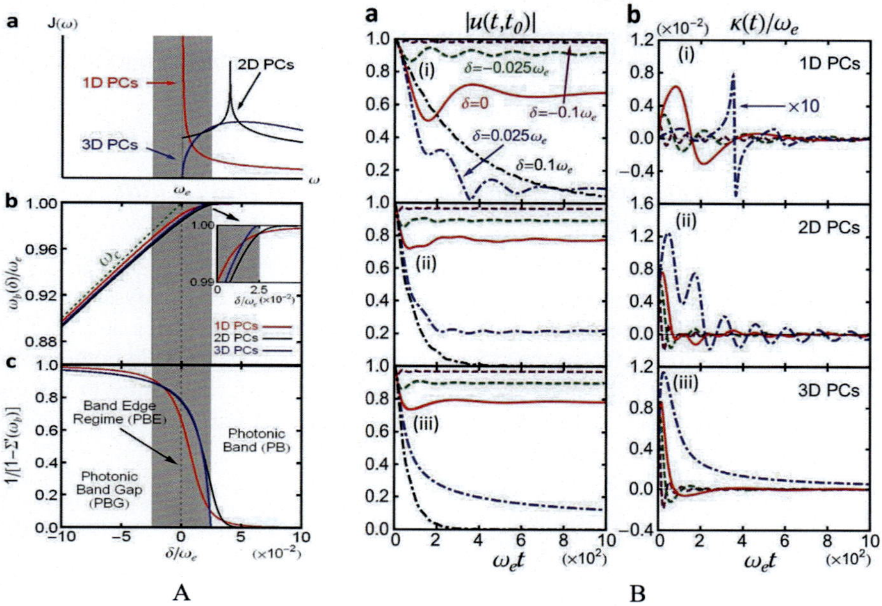

Fig. 2.7 (**A**) The structural composition of photon and rate of energy deliberation in the acting PV panel. (a) Functional area at different DOS magnitude of 1D, 2D, and 3D PV cells. (b) Photon frequency rate at the functional photonic band edge regime (PBE) and photonic bandgap (PBG). (c) Photon's magnitude to deliver high energy into the functional photonic band edge regime (PBE) and photonic bandgap (PBG). (**B**) Proliferation of photon dynamics into photovoltaic cells. (a) Considering the PB area, $\langle a(t)\rangle = 5u(t, t_0)\langle a(t_0)\rangle$, and (b) photon dynamic rate $k(t)$, functional variable for (i) 1D, (ii) 2D, and (iii) 3D quantum field into the PV cells. (Courtesy: Ping-Yuan Lo, Heng-Na Xiong & Wei-Min Zhang (2015); Scientific Reports, volume 5, Article number: 9423)

$$\rho_{\mathrm{T}}[v(t,t)] = \sum\nolimits_{n=0}^{\infty} \frac{[v(t,t)^{n}]}{[1+v(t,t)]^{n+1}} \mid n\rangle\langle n| \tag{2.38}$$

Here, ρ_{T} denotes a thermal state with an average particle quantum $v(t, t)$, where Eq. (2.11) suggests that the peak point photon energy generation state will be evolved into a thermal state [45, 50], which is considered as the functional state of the photon $\mathcal{D}[\alpha(t)] \mid n\rangle^{37}$ in the Earth surface. Thus, the net photon energy generation calculation is being represented by the following equation:

$$\begin{aligned}\langle m|\rho(t)|n\rangle = J(\omega) &= e^{-\Omega(t)|\alpha_0|^2} \frac{[\alpha(t)]^m[\alpha^*(t)]^n}{[1+v(t,t)]^{m+n+1}} \\ &= \sum_{k=0}^{\min\{m,n\}} \frac{\sqrt{m!n!}}{(m-k)!(n-k)!k!}\left[\frac{v(t,t)}{\Omega(t)|\alpha_0|^2}\right]^k\end{aligned} \tag{2.39}$$

where the emission of the net photons energy ($(\langle m| \rho(t)| n\rangle)$) into the Earth surface, and its conversion of photon energy into electricity $[1 + v(t, t)]^{m + n + 1}$ and non-equilibrium condition $[\alpha(t)]^m[\alpha^*(t)]^n$ of the Earth surface have been calculated.

Modeling of Net Electricity Energy Generation from Total Solar Irradiance on Earth

To transform this tremendous amount of photon energy into electricity energy, the net solar energy is being computed on a conceptual model of series and parallel circuit of Earth surface. The conceptual Earth surface is being then hypothetically implemented into the *I–V* single diode circuit of Earth surface in order to get the precise *I–V* relationship of the net solar energy reaches on Earth surface by calculating from the following equation:

$$I = I_{\mathrm{L}} - I_{\mathrm{o}}\left\{\exp^{\left[\frac{q(V+I_{\mathrm{RS}})}{AkT_{\mathrm{c}}}\right]} - 1\right\} - \frac{(V + I_{\mathrm{RS}})}{R_{\mathrm{Sh}}} \tag{2.40}$$

Here, I_{L} denotes the photon formation current, *Io* denotes the ideal current flow into the diode, R_{s} denotes the resistance in a series, *A* denotes the diode function, $k\,(= 1.38 \times 10^{-23}\ \mathrm{W/m^2\ K})$ denotes the Boltzmann's constant, $q\,(=1.6 \times 10^{-19}\ \mathrm{C})$ denotes the charge amplitude of the electron, and T_{c} denotes the Earth temperature. Consequently, the *I–q* linked in the Earth surface is being varied into the diode cell which is expressed as the dynamic current as follows [25] (Table 2.1).

$$I_{\mathrm{o}} = I_{\mathrm{RS}}\left(\frac{T_{\mathrm{c}}}{T_{\mathrm{ref}}}\right)^3 \exp\left[\frac{qEG\left(\frac{1}{T_{\mathrm{ref}}} - \frac{1}{T_{\mathrm{c}}}\right)}{KA}\right] \tag{2.41}$$

where I_{RS} denotes the dynamic current representing the functional transformation of solar radiation and *qEG* denotes the bandgap solar radiation into the conceptual Earth surface at different DOS dimensional modes of 1D, 2D, and 3D.

Here, considering this conceptual Earth surface, the *I–V* relationship with the exception of *I–V* curve, a calculative result of linked *I–V* curves among all of the conceptual solar cells has been determined [41, 42]. Thus, the equation is being rewritten as follows in order to determine the *V–R* relationship much more accurately:

$$V = -IR_{\mathrm{s}} + K\log\left[\frac{I_{\mathrm{L}} - I + I_{\mathrm{o}}}{I_{\mathrm{o}}}\right] \tag{2.42}$$

where K denotes the constant $\left(=\frac{AkT}{q}\right)$ and I_{mo} and V_{mo} are being denoted as the net current and voltage in the conceptual Earth surface. Subsequently, the relationship between I_{mo} and V_{mo} shall remain motional in the I–V Earth surface and can be written as:

$$V_{mo} = -I_{mo}R_{Smo} + K_{mo}\log\left(\frac{I_{Lmo} - I_{mo} + I_{Omo}}{I_{Omo}}\right) \tag{2.43}$$

where I_{Lmo} denotes the photon-induced current, I_{Omo} denotes the dynamic current into the diode, R_{Smo} denotes the resistance in series, and K_{mo} denotes the factorial constant.

Once all non-series (NS) cells are being interlinked in the series, then the series resistance is being calculated as the sum of each solar cell series resistance $R_{Smo} = N_s \times R_s$ current considering the functional coefficient of the constant factor $K_{mo} = N_s \times K$. The flow of current dynamics into the circuit is lined to the cells in a series connection [40, 41]. Thus, the current dynamics in Eq. (2.42) remains the same in each part of $I_{Omo} = I_o$ and $I_{Lmo} = I_L$. Thus, the mode of $I_{mo}{-}V_{mo}$ relationship for the N_s series of connected cells can be expressed by:

$$V_{mo} = -I_{mo}N_sR_s + N_sK\log\left(\frac{I_L - I_{mo} + I_o}{I_o}\right) \tag{2.44}$$

Naturally, the current–voltage relationship can be further modified considering all parallel links into N_p cells connection in all parallel mode and can be described as follows [31, 39]:

$$V_{mo} = -I_{mo}\frac{R_s}{N_p} + K\log\left(\frac{N_{sh}I_L - I_{mo} + N_pI_o}{N_pI_o}\right) \tag{2.45}$$

Since the photon-induced current primarily depends on the solar radiation and optimum temperature configuration, the net current dynamic is being calculated as:

$$I_L = G[I_{SC} + K_I(T_c - T_{ref})] * V_{mo} \tag{2.46}$$

where I_{sc} denotes the current at 25 °C and kW/m^2, K_I denotes Earth surface coefficient factor, T_{ref} denotes the optimum temperature, and G denotes the solar energy in mW/m^2 [48, 49].

Finally, the electricity energy generation around the Earth surface has been computed in order to confirm the net emitted photon utilization by integrating local Albanian electric fields; thus, the global $U(1)$ gauge field will allow to add a mass-term of the functional particle of $\varnothing' \rightarrow e^{i\alpha(x)}\varnothing$. It is then further clarified by explaining the variable derivative of transformation law of scalar field using the following equation [10, 41]:

Table 2.1 The solar energy in various DOS dimensional mode reaching on Earth Surface (ES) with corresponding various unit area $J(\omega)$ and self-energy induced reservoir of $\Sigma(\omega)$. The functional C, η, and χ act like coupled forces between the solar energy at Earth surface of 1D, 2D, and 3D areal surface (Courtesy: Ping-Yuan Lo, Heng-Na Xiong & Wei-Min Zhang (2015); Scientific Reports, volume 5, Article number: 9423)

Solar energy (ES)	Unit area $J(\omega)$ for different DOS	Solar energy correction in Earth surface $\Sigma(\omega)$
1D	$\frac{\text{ES}}{2\pi r}\frac{1}{\sqrt{\omega-\omega_e}}\Theta(\omega-\omega_e)$	$-\frac{\text{ES}}{\sqrt{2\omega_e-\omega}}$
2D	$-\text{ES}\left[\ln\left\lvert\frac{\omega-\omega_0}{2\omega_0}\right\rvert-1\right]\Theta(\omega-\omega_e)\ \Theta(\Omega_d-\omega)$	$\text{ES}\left[\text{Li}_2\left(\frac{\Omega_d-\omega_0}{\omega-\omega_0}\right)-\text{Li}_2\left(\frac{\omega_0-\omega_e}{\omega_0-\omega}\right)-\ln\frac{\omega_0-\omega_e}{\Omega_d-\omega_0}\ln\frac{\omega_e-\omega}{\omega_0-\omega}\right]$
3D	$\text{ES}\sqrt{\frac{2\omega-\omega_e}{\Omega_C}}\exp\left(-\frac{\omega-\omega_e}{\Omega_C}\right)\Theta(\omega-\omega_e)$	$\text{ES}\left[\pi\sqrt{\frac{\omega_e-\omega}{\Omega_C}}\exp\left(-\frac{2\omega-\omega_e}{\Omega_C}\right)\text{erfc}\sqrt{\frac{\omega_e-\omega}{\Omega_C}}-\sqrt{2\pi r}\right]$

$$\begin{aligned} \partial_\mu \to D_\mu = \partial_\mu = ieA_\mu \quad &[\text{covariant derivatives}] \\ A'_\mu = A_\mu + \frac{1}{e}\,\partial_\mu\alpha \quad &[A_\mu \text{ derivatives}] \end{aligned} \tag{2.47}$$

Here, the global $U(1)$ gauge denotes the invariant local Albanian for a complex scalar field which further can be expressed as:

$$\mathfrak{h} = (D^\mu)^\dagger (D_\mu \varnothing) - \frac{1}{4}F_{\mu\nu}F^{\mu\nu} - V(\phi) \tag{2.48}$$

The term $\frac{1}{4}F_{\mu\nu}F^{\mu\nu}$ is the dynamic term for the gauge field of the Earth surface and $V(\varnothing)$ denotes the extra term in the local Albanian which is $V(\phi^*\varnothing) = \mu^2(\varnothing^*\varnothing) + \lambda(\varnothing^*\varnothing)^2$.

Therefore, the generation of local Albanian ($\mathfrak{h}$) under the perturbational function of the quantum field of the Earth surface has been confirmed by the calculation of mass-scalar particles ϕ_1 and ϕ_2 along with a mass variable of μ. In this condition $\mu^2 < 0$ had an infinite number of quantum which is being clarified by $\phi_1^2 + \phi_2^2 = -\mu^2/\lambda = v^2$ and the $\mathfrak{h}$ through the variable derivatives using further shifted fields η and ξ defined the quantum field as $\phi_0 = \frac{1}{\sqrt{2}}[(v+\eta) + i\xi]$.

$$\begin{aligned} \text{Kinetic term: } \mathfrak{h}(\eta,\xi) &= (D^\mu\phi)^\dagger(D^\mu\phi) \\ &= (\partial^\mu + ieA^\mu)\phi^*(\partial_\mu - ieA_\mu)\phi \end{aligned} \tag{2.49}$$

Thus, this expanding term in the $\mathfrak{h}$ associated to the scalar field of the Earth surface is suggesting that the net Earth surface field is prepared to initiate the net

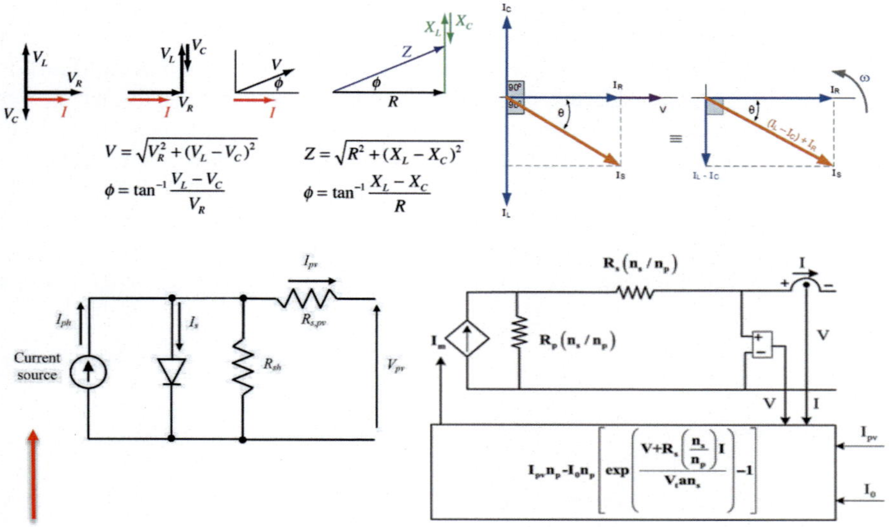

Fig. 2.8 (**a**) The scalar field of the Earth surface, (**b**) solar energy scalar field on Earth surface, (**c**) net electricity current energy generation on Earth from the total solar energy on Earth [19, 21]

electricity energy generation into its quantum field of induced photon energy, respectively, at the normal, normalized, and normal modes [19, 21].

To determine this electricity energy, hereby, a non-variable function of readily dynamics has been implemented for the calculation of $\overline{\varphi}[s_0]$ to confirm the expected value of s_0 considering the Earth surface which is expressed as follows:

$$\overline{\varphi}[s_0] = 2\, s_0(\ln 4s_0 - 2) + \ln 4s_0(\ln 4s_0 - 2) - \frac{(\pi^2 - 9)}{3} + s_0^{-1}\left(\ln 4s_0 + \frac{9}{8}\right) + \ldots (s_0 \gg 1); \tag{2.50}$$

$$\overline{\varphi}[s_0] = \left(\frac{2}{3}\right)(s_0 - 1)^{\frac{3}{2}} + \left(\frac{5}{3}\right)(s_0 - 1)^{\frac{5}{2}} - \left(\frac{1507}{420}\right)(s_0 - 1)^{\frac{7}{2}} \quad (1/2 \text{ instead of } 1) \tag{2.51}$$

Then the final equation can the rewritten as where s_0 is the areal value of electricity energy generation into the Earth surface (1 m^2).

$$\overline{\varphi}[s_0] = \left(\frac{2}{3}\right)(s_0 - 1)^{\frac{3}{2}} + \left(\frac{5}{3}\right)(s_0 - 1)^{\frac{5}{2}} - \left(\frac{1507}{420}\right)(s_0 - 1)^{\frac{7}{2}} \tag{2.52}$$

The function $\overline{\varphi}[s_0]$ thus determines the net electricity energy generation from the total solar energy into the atmosphere by calculation of Earth's cross-sectional area of 127,400,000 km^2, and the net solar energy intercepted by the surface of Earth is 1.740×10^{17} W. Considering the seasonal and climate variation the net power reaching the ground generally averages 200 Watts per square meter per day [40]. Thus, the average power reaching the Earth's surface at any time is calculated as $127.4 \times 10^6 \times 10^6 \times 200 = 25.4 \times 10^{15}$ W or 25,400 TW which is TW $\times$ 24 $\times$ 365 $=$ 222,504,000 Tera Watthours (TWh). Since the net annual electrical energy (not the total energy) consumed in the world from all sources in 2019 was 22,126 TWh, the available solar energy is over 10,056 times the world's consumption.

Conclusion

Since the fossil fuel utilization throughout the world is getting finite level and is the major contributor for climate change, usage of solar energy has been calculated in this research as the renewable clean energy source will indeed be an interesting source to mitigate the global energy and environmental perplexity. Simply, the energy from the solar irradiance surely can play a vital role to mitigate global energy crisis and green house reduction tremendously. Inevitably, Solar energy, here, the radiant energy from the Sun is thus estimated in order to propose to use as the primary source of renewable energy to capture to mitigate the global energy needs. Simply the utilization of solar energy in every sector of our daily life surely will have tremendous benefits for the mankind which will secure the global energy consumption naturally and subsequently will mitigate the global climate change dramatically.

Acknowledgments The Green Globe Technology (GGT) has supported this research with a green environment grant of RD-02019-02. Author wishes to thank sincerely the R&D group of GGT.

References

1. D.K. Armani, T.J. Kippenberg, S.M. Spillane, K.J. Vahala, Ultra-high-Q toroid microcavity on a chip. Nature **421**, 925 (2003)
2. G. Baur, K. Hencken, D. Trautmann, Revisiting unitarity corrections for electromagnetic processes in collisions of relativistic nuclei. Phys. Rep. **453**, 1 (2007)
3. F. Besharat, A.A. Dehghan, A.R. Faghih, Empirical models for estimating global solar radiation: A review and case study. Renew. Sust. Energ. Rev. **21**, 798–821 (2013)
4. K.M. Birnbaum et al., Photon blockade in an optical cavity with one trapped atom. Nature **436**, 87–90 (2005)
5. B. Bresar, Quasi-median graphs, their generalizations, and tree-like equalities. Eur. J. Comb. **24**, 557–572 (2003)
6. D.E. Chang, A.S. Sørensen, E.A. Demler, M.D. Lukin, A single-photon transistor using nanoscale surface plasmons. Nat. Phys. **3**, 807–812 (2007)

7. B. Dayan et al., A photon turnstile dynamically regulated by one atom. Science **319**, 1062–1065 (2008)
8. J.S. Douglas, H. Habibian, C.-L. Hung, A.V. Gorshkov, H.J. Kimble, D.E. Chang, Quantum many-body models with cold atoms coupled to photonic crystals. Nat. Photonics **9**, 326–331 (2015)
9. M. Faruque Hossain, Green science: Independent building technology to mitigate energy, environment, and climate change. Renew. Sustain. Energy Rev. **73**, 695–705 (2017)
10. M. Faruque Hossain, Photon energy amplification for the design of a micro PV panel. Int. J. Energy Res. (2018). https://doi.org/10.1002/er.4118
11. M. Faruque Hossain, Green science: Advanced building design technology to mitigate energy and environment. Renew. Sustain. Energy Rev. **81**, 3051–3060 (2018)
12. M. Faruque Hossain, Natural mechanism to console global water, energy, and climate change crisis. Sustain. Energy Technol. Assess. **35**, 347–353 (2019)
13. M. Faruque Hossain, *Green Building Complexes* (Elsevier BV, Amsterdam, 2019)
14. M. Faruque Hossain, *Applied Energy Technology* (Elsevier BV, Amsterdam, 2019)
15. M. Faruque Hossain, *Power Systems* (Elsevier BV, Amsterdam, 2019)
16. M. Faruque Hossain, *Best Management Practices* (Elsevier BV, Amsterdam, 2019)
17. P. Fratzl, Biomaterial systems for mechanosensing and actuation. Nature **462**, 442–448 (2009)
18. S. Gleyzes et al., Quantum jumps of light recording the birth and death of a photon in a cavity. Nature **446**, 297 (2007)
19. C. Guerlin et al., Progressive field-state collapse and quantum non-demolition photon counting. Nature **448**, 889 (2007)
20. Y. Guo, A. Al-Jubainawi, Z. Ma, Performance investigation and optimisation of electrodialysis regeneration for LiCl liquid desiccant cooling systems. Appl. Therm. Eng. **149**, 1023–1034 (2019)
21. M.F. Hossain, Solar energy integration into advanced building design for meeting energy demand and environment problem: Climate change, photoenergy, solar panel, and clean energy. Int. J. Energy Res. **40**, 1293–1300 (2016)
22. M.F. Hossain, Photonic thermal control to naturally cool and heat the building. Appl. Therm. Eng. (2018). https://doi.org/10.1016/j.applthermaleng.2017.12.041
23. M.F. Hossain, Transforming dark photons into sustainable energy. Int. J. Energy Environ. Eng. (2018). https://doi.org/10.1007/s40095-017-0257-1
24. F. Hossain, Photon application in the design of sustainable buildings to console global energy and environment. Appl. Therm. Eng. (2018). https://doi.org/10.1016/j.applthermaleng.2018.05.085
25. M.F. Hossain, Sustainable technology for energy and environmental benign building design. J. Build. Eng. **22**, 130–139 (2019)
26. M.F. Hossain, *Advanced Building Design* (Elsevier BV, Amsterdam, 2019)
27. M.F. Hossain, Theoretical mechanism to breakdown of photonic structure to design a micro PV panel. Energy Rep. **5**, 649–657 (2019)
28. M.F. Hossain, *Water* (Elsevier BV, Amsterdam, 2019)
29. M.F. Hossain, Green technology: Transformation of transpiration vapor to mitigate global water crisis. Polytechnica (2019). https://doi.org/10.1007/s41050-019-00009-y
30. M.F. Hossain, *Energy* (Elsevier BV, Amsterdam, 2019)
31. E. Jaivime, J.M.D. Scott, Global separation of plant transpiration from groundwater and streamflow. Nature **525**, 91–94 (2015)
32. J.D. Joannopoulos, P.R. Villeneuve, S. Fan, Photonic crystals: Putting a new twist on light. Nature **386**, 143 (1997)
33. C. Lang et al., Observation of resonant photon blockade at microwave frequencies using correlation function measurements. Phys. Rev. Lett. **106**, 243601 (2011)
34. L. Langer, S. Poltavtsev, M. Bayer, Access to long-term optical memories using photon echoes retrieved from semiconductor spins. Nat. Photonics **8**, 851–857 (2014)

35. R.M. Maxwell, L.E. Condon, Connections between groundwater flow and transpiration partitioning. Science **353**, 377–380 (2015)
36. R.B. Messaoud, Extraction of uncertain parameters of double-diode model of a photovoltaic panel using simulated annealing optimization. J. Phys. Chem. C (2019). https://doi.org/10.1021/acs.jpcc.9b07064
37. J.R. Newell, *A Story of Things Yet-to-Be: The Status of Geology in the United States in 1807* (Geological Society, London, Special Publications, 2009)
38. K. Park, P. Marek, R. Filip, Qubitmediated deterministic nonlinear gates for quantum oscillators. Sci. Rep. **7**, 11536 (2017)
39. T. Pregnolato, E. Lee, J. Song, D. Stobbe, P. Lodahl, Single-photon non-linear optics with a quantum dot in a waveguide. Nat. Commun. **6**, 8655 (2015)
40. A. Reinhard, Strongly correlated photons on a chip. Nat. Photonics **6**, 93–96 (2011)
41. C. Sayrin et al., Real-time quantum feedback prepares and stabilizes photon number states. Nature **477**, 73 (2011)
42. J. Scott, D. Zachary, Terrestrial water fluxes dominated by transpiration. Nature **496**, 347–350 (2013)
43. H.Z. Shen, X. Shuang, H.T. Cui, X.X. Yi, Non-Markovian dynamics of a system of two-level atoms coupled to a structured environment. Phys. Rev. A **99**, 032101 (2019)
44. M.S. Tame, K.R. McEnery, Ş.K. Özdemir, J. Lee, S.A. Maier, M.S. Kim, Quantum plasmonics. Nat. Phys. **9**, 329–340 (2013)
45. M.W.Y. Tu, W.M. Zhang, Non-Markovian decoherence theory for a double-dot charge qubit. Phys. Rev. B **78**, 235311 (2008)
46. R. Vargas, D. Carvajal, L. Madriz, B.R. Scharifker, Chemical kinetics in solar to chemical energy conversion: The photoelectrochemical oxygen transfer reaction. Energy Rep. **6**, 2–12 (2020)
47. T.D. Wheeler, A.D. Stroock, The transpiration of water at negative pressures in a synthetic tree. Nature **455**, 208–212 (2008)
48. Y.F. Xiao et al., Asymmetric Fano resonance analysis in indirectly coupled microresonators. Phys. Rev. A **82**, 065804 (2010)
49. W. Yan, H. Fan, Single-photon quantum router with multiple output ports. Sci. Rep. **4**, 4820 (2014)
50. W.-B. Yan, J.-F. Huang, H. Fan, Tunable single-photon frequency conversion in a Sagnac interferometer. Sci. Rep. **3**, 3555 (2013)
51. L. Yang, S. Wang, Q. Zeng, Z. Zhang, T. Pei, Y. Li, L.-M. Peng, Efficient photovoltage multiplication in carbon nanotubes. Nat. Photonics **5**, 672–676 (2011)
52. W.M. Zhang, P.Y. Lo, H.N. Xiong, M.W.Y. Tu, F. Nori, General non-Markovian dynamics of open quantum systems. Phys. Rev. Lett. **109**, 170402 (2012)
53. Y. Zhu, H. Xiaoyong, Y. Hong, G. Qihuang, On-chip plasmon-induced transparency based on plasmonic coupled nanocavities. Sci. Rep. **4**, 3752 (2014)

Chapter 3
Reconfiguration of Bose–Einstein Photonic Structure to Produce Clean Energy

Abstract The Bose–Einstein photonic structure has been deconstructed and modeled using the MATLAB software to design a *Modern Solar Photovoltaics Energy Systems* for trapping clean energy. Bose–Einstein photon distribution theory suggests that under low-temperature conditions, *photonic bandgap state* photons are induced locally and remain steady as long-lived equilibrium particles called *discrete energy state photons*. Thus, I assume that once a photon is in an extreme relativistic thermal condition, it will not obey Bose–Einstein discrete energy state theory. The photonic bandgap volume will be naturally increased within its vicinity as a result of the extreme relativistic thermal conditions, and the discrete energy state photon will be agitated by extreme relativistic thermal fluctuations. Consequently, the Bose–Einstein photonic dormant state will be broken down within its region and will create a multiple number of photons. Simply, a single discrete energy state photon will be transformed from the crossover phenomenon equilibrium state to a non-equilibrium state to exponentially create multiple photons, here named *Hossain nonequilibrium photons* (HnP^-). Calculations reveal that if only 0.00008% of a building's exterior skin curtain wall is used as a *Modern Solar Photovoltaics Energy* panel to transform *Bose–Einstein equilibrium photons* into HnP^-, it will produce enough clean energy to satisfy the total energy demand of a building.

Keywords Bose–Einstein photonic structure · Extreme relativistic condition · Thermal fluctuation · Non-equilibrium photon production · Clean energy conversion

Introduction

Clean energy is presently in demand to create a better planet for future generations. Conventional energy consumption by building sectors releases 8.01×10^{11} tons of CO_2 per year and causes nearly 40% of global warming. Therefore, in this study, I have proposed an advanced micro photovoltaic (PV) panel that utilizes Bose–Einstein photon distribution theory to convert a single photon into multiple photons to mitigate the total energy demand of a building and significantly reduce global warming. Photonic bandgap (PBG) structures have been studied in the last decade at the *nano*-scale to obtain better knowledge of the characteristic dispersion properties

M. F. Hossain, *Sustainable Design for Global Equilibrium*,
https://doi.org/10.1007/978-3-030-94818-4_3

of photons [2, 32]. Some interesting features of photonic structure have been identified; for example, when photons are induced by PBGs, spontaneous photonic inhibition occurs due to electron level energy state photons [3, 5]. For the last several decades, quantum optics has been extensively explored under zero-degree temperature and point break quantum electrodynamics (QED) considering photonic band edges (PBEs). Previous work has suggested [8, 12] that the emitted photons are at a slightly higher energy level in a dynamic equilibrium state [2, 7]. Although these findings are very interesting, the photonic non-equilibrium dynamics at the *nano*-point break in the PBG state have not yet been studied under extreme relativistic conditions. To activate this dormant photon into an activated energy state, optical circuit networks consisting of *nano*-point breaks and waveguides have been proposed for use under extreme relativistic conditions to confirm the creation of point defects. This would provide a mechanism for incorporation of PBG waveguide defect arrays into PV panels. Therefore, the quantum dynamics of photons will be activated by point defects and PBG waveguides under extreme relativistic conditions. Consequently, the point break photons in a PV semiconductor array will not obey Bose–Einstein photon distribution theory. The dormant state photon will be broken down in these regimes and will create what I term HnP^{-} in the PV panel. To calculate the energy conversion created by utilizing these HNEPs in the PV panel, I have developed a model of a PV module for a PV panel design using the MATLAB/Simulink software package. The model is explained using a detailed mathematical calculation considering a PV as a circuit including the PV current origin and a single diode to confirm the PV module behavior under extreme relativistic parameters for calculating the solar energy transformation rate. The calculations reveal that if only a mere 0.0008% of a building's curtain wall is designed as an extreme relativistic conditioned PV panel, it will capture sufficient solar irradiance to exponentially convert enough energy to satisfy the total energy demand of the building.

Methods and Simulations

Photon Dynamics Transformation

The calculation of the photon dynamics was conducted at the nano-scale via point break waveguides rooted within a photovoltaic semiconductor circuit, particularly at the extreme relativistic state. For this particular calculation, both point break and photovoltaic semiconductor have been treated as wave guides for photons reservoirs. Subsequently, within the photovoltaic panel, the *nano*-point break flaws, purely, fulfill electronic dynamics for unceasing conditions of photon manufacture at atomic spectra, bearing in mind contour maps; therefore, the expression of Hamiltonian is as under:

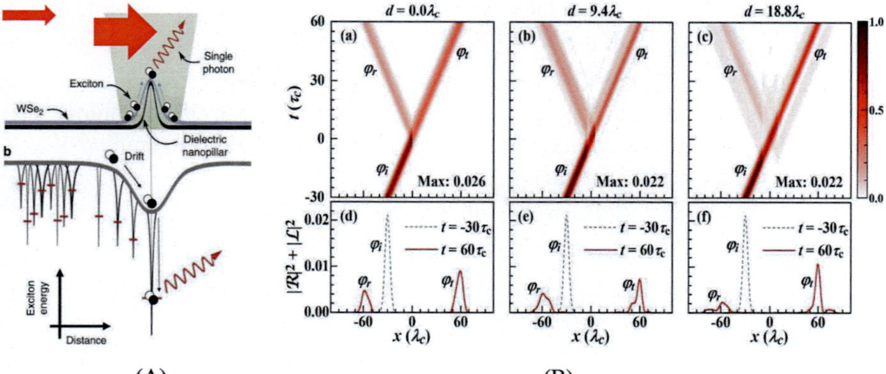

(A) (B)

Fig. 3.1 (**A**) The photonic mode for conversion of excited state energy in respect of distance of the photonic frequencies. (**B**) (a–c) Contour maps of the photon probability densities, normalized to their maximum values in the maps as functions of x and t. (d–f) Probability distributions of the incident (φ_i), reflected (φ_r), and transmitted (φ_t) pulses

$$H = \sum \omega_{ci} a_i^\dagger a_i + \sum_K \omega_k b_k^\dagger b_k + \sum_{ik} \left(V_{ik} a^\dagger b_k + V_{ik}^* b_k^\dagger a_i \right) \tag{3.1}$$

where $a_i \left(a_i^\dagger \right)$ is the nano-point break mode driver, $b_k \left(b_k^\dagger \right)$ is the driver of the photon nanostructure photodynamic modes, and V_{ik} is the photonic mode magnitude amid the photon *nano* structure and *nano*-breakpoints [1, 9, 11] (Fig. 3.1).

Considering into account the first photonic structure as within an equilibrium condition, the entire photonic reservoir structure is incorporated in view of excited coherent condition photons within the photovoltaic semiconductor; it is expressed as the below equation [4, 6]:

$$\begin{aligned} \rho(t) = -i\left[H_c'(t)\rho(t)\right] + \sum_{ij} \Big\{ k_{ij}(t) \left[2a_j \rho(t) a_i^\dagger - a_i^\dagger a_j \rho(t) - \rho(t) a_i^\dagger a_j \right] \\ + k_{ij}(t) \left[a_i^\dagger \rho(t) a_j + a_j \rho(t) a_i^\dagger - a_i^\dagger a_j \rho(t) - \rho(t) a_j a_i^\dagger \right] \Big\} \end{aligned} \tag{3.2}$$

In this case, $\rho(t)$ is the photons attenuated density within breakpoint conditions; $H_c'(t) = \sum_{ij} \omega_{cij}'(t) a_i^\dagger a_j$ is the point break re-standardized Hamiltonian with reference to the frequencies of point break $\omega_{cii}'(t) = \omega_{ci}'(t)$, as well as $\omega_{cij}'(t)$, which is the function instigated induced photons couplings amid the breakpoints. The factors $\widetilde{\kappa}_{ij}(t)$ and $\kappa_{ij}(t)$ are considered a photonic dynamics within the photovoltaic semiconductor beneath the maximum relativistic states. The non-perturbative principle is purely the one that resolves time-reliant factors $\widetilde{\kappa}_{ij}(t)$ and $\kappa_{ij}(t)$ and ω_{cij}'. In the case of photon reservoir, Hamiltonian is represented by $H_I = \sum_k \lambda_k x q_k$, with q_k and x being the secondary point break reservoir and the primary exclusive point break location, respectively. Taking into account a photon quantum dynamics, the entire Hamiltonian reservoir point break is revised as $H_{\mathrm{I}} = \sum_k V_k \left(a^\dagger b_k + b_k^\dagger a + a^\dagger b_k^\dagger + ab_k \right)$ to approve the photonic dynamics magnitude in the point break. As a result, one characterizes the photonic dynamics by the dissipated factors of photon: $\widetilde{\kappa}(t)$ and $\widetilde{\kappa}(t)$. These factors can therefore be determined as the below equations:

$$\omega_{\mathrm{c}}'(t) = -\mathrm{Im}[u\ (t, t_0)/u(t, t_0)] \tag{3.3}$$

$$k(t) = -\ \mathrm{Re}\left[u(t, t_0)/u(t, t_0)\right] \tag{3.4}$$

$$\widehat{k}(t) = \dot{v}(t, t) + 2v(t, t)\ k(t) \tag{3.5}$$

With reference to the above equations, $u(t, t_0)$ is the photonic region of the point break and $v(t, t)$ is the photon dynamics as a result of the induced reservoir. The function $v\ (t, t)$ is explained further by the use of non-equilibrium dynamics theory and the following integral-differential equation [18, 19]:

$$\dot{u}(t, t_0) = -i\omega_{\mathrm{c}} u(t, t_0) - \int_{t_0}^{t} dt' g(t - t')\ u(t', t_0) \tag{3.6}$$

$$v(t, t) = \int_{t_0}^{t} dt \int_{0}^{t} dt_2 u^*(t_1, t_0) \widehat{g}(t_1 - t_2)\ u(t_2, t_0) \tag{3.7}$$

The primary frequency at the point break is represented by v_{c}. Consequently, in Eqs. (3.6) and (3.7) above, the integral functions can be utilized to calculate the back-up function in the point breaks. The number of photons generated by the non-stable condition is exceptionally expressed per *unit* area of the photonic structure $J(\varepsilon)$ via the following connections;

$$g(t - t') = \int d\omega J(\omega) e^{-i\omega(t-t')} \quad \text{and} \quad \widetilde{g}(t - t') = \int d\omega J(\omega) \overline{n}(\omega, T) e^{-i\omega(t-t')},$$

where, in this case, $\overline{n}(\omega, T) = 1/\left[e^{\hbar\omega/k_{\mathrm{B}}T} - 1\right]$ is the prime photon dynamics within the photovoltaic panel at a temperature (T). The clarification the unit area $J(\varepsilon)$ is done in connection to the density of states $\varrho(\omega)$ production of photon within the photovoltaic at the V_k magnitude amid the photovoltaic and point break circuits:

$$J\ (\omega) = \sum_k |V_k|^2\ \delta\ (\omega - \omega_k) = \varrho\ (\omega) \mid V\ (\omega)|^2 = [n * e(1 + 2n)]^4 \tag{3.8}$$

Lastly, the summary of the proliferation of the photon production is done with regard to the condition of photon dynamic within the photovoltaic panel, so that

$V_k \rightarrow V(\omega)$ and i of V_{ik} may be calculated at the one-diode mode breakpoint. Hence the production of the non-equilibrium photon ($J(\omega)$, given in Eq. (3.8)) can be determined exactly through the use of the simplified equation in Eq. (3.9) below:

$$J\,(\omega) = [n * e(1 + 2n)]^4 \tag{3.9}$$

Photovoltaic (PV) Modeling

The use of the exterior curtain wall skin of a building as a micro photovoltaic panel to carry out ultra-relativistic reaction within the cell has been recommended to deform Bose–Einstein equilibrium photon to numerous HnP^- to transform it into electricity by using a single-diode model PV panel (Fig. 3.2).

Therefore, the PV cells *I–V* equation within the single-diode model describes the photovoltaic model. Within the photovoltaic panel the equation of *V–I* relationship can be as under:

$$I = I_{\mathrm{L}} - I_{\mathrm{o}} \left\{ \exp^{\left[\frac{q(V+I_{\mathrm{RS}})}{AkT_{\mathrm{c}}}\right]} - 1 \right\} - \frac{(V + I_{\mathrm{RS}})}{R_{\mathrm{Sh}}} \tag{3.10}$$

where I_{L} is the photon creating energy, I_{o} is the saturated energy within the diode, R_{s} is the resistance within the series, A is the diode inactive function, k ($=1.38 \times 10^{-23}$ W/m^2 K) is the Boltzmann's constant, q ($=1.6 \times 10^{-19}$ C) is the electron charge amplitude, and T_{c} is the functional cell temperature. Consequently, within the photovoltaic cells, the *I–q* connection varies because of the diode energy as well as/or saturation energy, which may be expressed as follows [16, 17, 30]:

$$I_{\mathrm{o}} = I_{\mathrm{Rs}} \left(\frac{T_{\mathrm{c}}}{T_{\mathrm{ref}}} \right)^3 \exp \left[\frac{qEG\left(\frac{1}{T_{\mathrm{ref}}} - \frac{1}{T_{\mathrm{c}}} \right)}{KA} \right] \tag{3.11}$$

I_{RS} in the above equation is the saturated current given that the solar irradiance and qEG and functional temperature denote the bandgap energy into the graphene and silicon photovoltaic cell taking into account the perfect, normalized, and normal modes (Fig. 3.3).

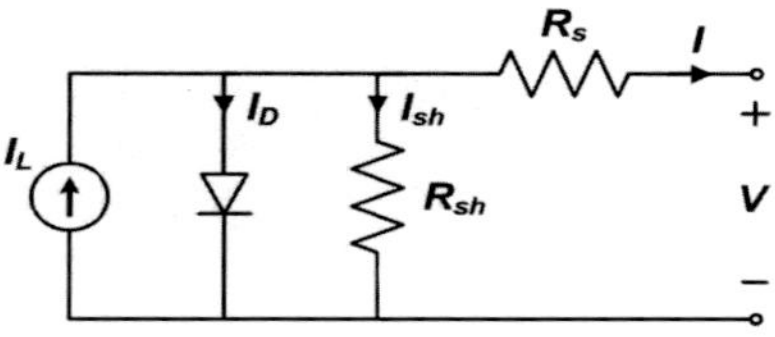

Fig. 3.2 The above model integrates parallel and series circuit connections using two resistors, a diode for PV cell

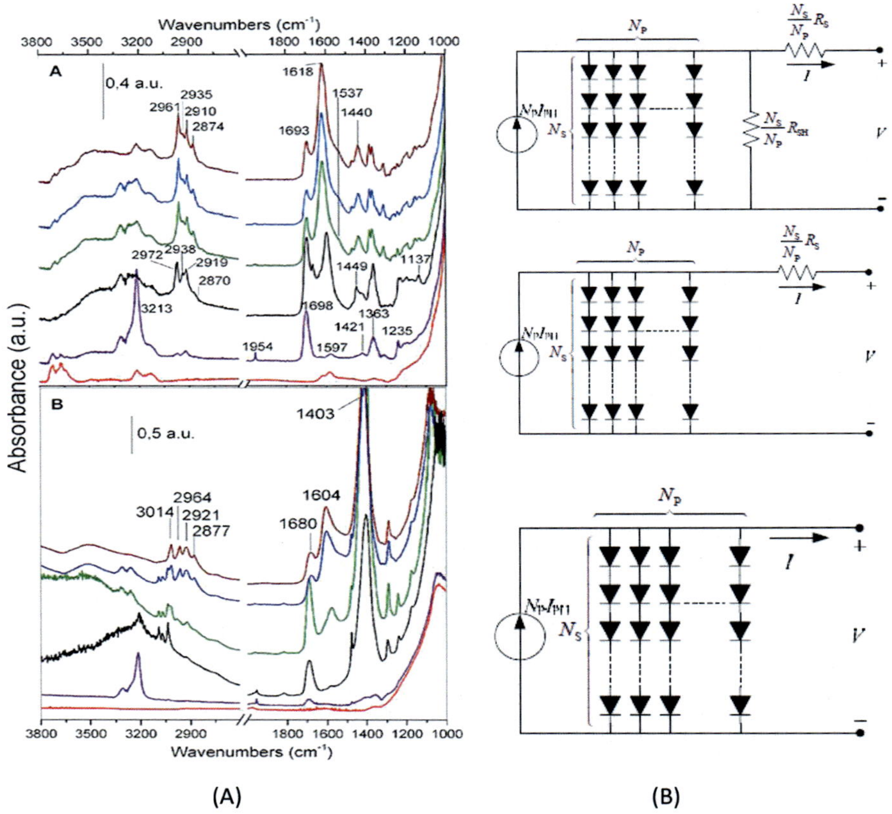

Fig. 3.3 (**A**) The absorption of solar irradiance (a.u.) in relation to the photonic wave per unit area, (**B**) Equivalent circuit models of the PV module (a) normal, (b) normalized, and (c) perfect modes

In consideration of a photovoltaic module, not including the *I–V* curve, are *I–V* curve conjunctions amid all the photovoltaic panel cells. As a result, the *I–V* equation can be changed as under, to find out the *V–R* connection:

$$V = -IR_s + K \log \frac{I_L - I + I_0}{I_0}. \tag{3.12}$$

From the above Eq. (3.12), V_{mo} and I_{mo} represent the voltage and current within the photovoltaic panel and K stands for a constant $(= \frac{AKT}{q})$. Thus, the connection amid V_{mo} and I_{mo} will be similar to that of photovoltaic cell *I–V* relation:

$$V_{mo} = -I_{mo}R_{Smo} + K_{mo} \log \left(\frac{I_{Lmo} - I_{mo} + I_{Omo}}{I_{Omo}} \right) \tag{3.13}$$

I_{Lmo} is the photon-generated current, I_{Omo} is the saturated current going into the diode, R_{Smo} is the resistant within the series, and K_{mo} is the fractional constant. As soon as the interconnection of all the non-series (N_s) cells within the series is done, the series resistant will be regarded as the sum of each cell series resistant $R_{Smo} = N_s \times R_s$. The constant factor can be written as follows:

$$V_{mo} = -I_{mo} N_s R_s + N_s K \log \left(\frac{I_L - I_{mo} + I_o}{I_o} \right) \tag{3.14}$$

In the same way, after all N_p cells are linked within a parallel mode, the calculation of the current–voltage can be expressed as below:

$$V_{mo} = -I_{mo} \frac{R_s}{N_p} + K \log \left(\frac{I_{Sh} I_L - I_{mo} + N_p I_o}{N_P I_o} \right) \tag{3.15}$$

Mostly, since the photon-generated current shall rely on the PV panel relativistic temperature states and solar irradiance, the following equation can be utilized to determine the current:

$$I_L = G[I_{SC} + K_I (T_c - T_{ref})] * V_{mo} \tag{3.16}$$

where G signifies the solar energy in kW/m^2, T_{ref} signifies the functional temperature of photovoltaic panel, K_I denotes the PV panel relativistic coefficient factor, and I_{SC} is the photovoltaic current at kW/m^2 and 25 °C.

Results and Discussion

Photon Production Proliferation

Determining arithmetically, the HnP^- production in the photovoltaic panel requires one to first solve the dynamic photon proliferation by incorporating Eqs. (3.7) and (3.8). It is well known that, due to the state of areal unit variable, different dynamics photon proliferation are produced by the photovoltaic panels. $J(\omega)$, unit area, possesses an insistent weak-coupling perimeter, and the approximation rule of Weisskopf–Winger and/or Markovian master equation is equivalent to the proliferation of photon production. As a result, all proliferation of photon production will take a dynamic photon condition mode (1D, 2D, 3D) within the photovoltaic cell. Finally, minus following the Bose–Einstein photon distribution theory, the production will transform from the current non-equilibrium state. This can clearly be expressed as shown in Table 3.1 [10, 46].

Consequently, to avoid the DOS bifurcation within a 3D photovoltaic cell, a fine cutoff at Ω_C which is of high-level frequency is employed. Correspondingly, a

Table 3.1 The photonic structures in different DOS dimensional modes in the PV cell. The photonic structures in the table (1) correspond to different $J(\omega)$, as well as self-energy induction at $\Sigma(\omega)$, reservoir. The variables η, χ, and C work like joined forces amid the point break, and photovoltaic of 3D, 2D, and 1D into the photovoltaic cell

Photovoltaic (PV)	Unit area $J(\omega)$ for different DOS	Reservoir-induced self-energy correction $\Sigma(\omega)$
1D	$\frac{C}{\pi}\frac{1}{\sqrt{\omega-\omega_e}}\Theta(\omega-\omega_e)$	$-\frac{C}{\sqrt{\omega_e-\omega}}$
2D	$-\eta\left[\ln\left\|\frac{\omega-\omega_0}{\omega_0}\right\|-1\right]\Theta(\omega-\omega_e)$ $\Theta(\Omega_d-\omega)$	$\eta\left[\mathrm{Li}_2\left(\frac{\Omega_d-\omega_0}{\omega-\omega_0}\right)-\mathrm{Li}_2\left(\frac{\omega_0-\omega_e}{\omega_0-\omega}\right)\right.$ $\left.-\ln\frac{\omega_0-\omega_e}{\Omega_d-\omega_0}\ln\frac{\omega_e-\omega}{\omega_0-\omega}\right]$
3D	$\chi\sqrt{\frac{\omega-\omega_e}{\Omega_C}}\exp\left(-\frac{\omega-\omega_e}{\Omega_C}\right)$ $\Theta(\omega-\omega_e)$	$\chi\left[\pi\sqrt{\frac{\omega_e-\omega}{\Omega_C}}\exp\left(-\frac{\omega-\omega_e}{\Omega_C}\right)\mathrm{erfc}\sqrt{\frac{\omega_e-\omega}{\Omega_C}}-\sqrt{\pi}\right]$

sharpened cutoff at Ω_C, which is also of high-level frequency, maintains a positive DOS within 2D and 1D photovoltaic cells. Therefore, while erfc(x) functions as an addictive variable, $\mathrm{Li}_2(x)$ acts as a dilogarithm variable. Successively, DOS of different photovoltaic cells represented by $Q_{PC}(\omega)$ can be determined through calculating Eigen functions and Eigen frequencies of Maxwell's rules in consideration of the photovoltaic *nano* structure [19, 20, 23]. The equivalent DOS for a 1D photovoltaic cell is represented by $Q_{PC}(\omega)\alpha\,\frac{1}{\sqrt{\omega-\omega_e}}\,\theta(\omega-\omega_e)$. The variable ω_e denotes the existing frequency within the PBE in consideration of DOS, and θ-$(\omega-\omega_e)$ denotes the Heaviside step function.

As a result, this density of state is determined to carry out 3D isotropic analysis within photovoltaic cells to project the mistake free qualitative condition of the mode of Weisskopf–Winger, as well as the photon-photon collision condition within the photovoltaic cell [13, 20, 21]. Consequently, within a 3D photovoltaic cell, the density of state near the photon band energy is effected by anisotropic density of state: $Q_{PC}(\omega)\alpha\,\frac{1}{\sqrt{\omega-\omega_e}}\,\theta(\omega-\omega_e)$. In the case of the 1D and 2D photovoltaic cells, the DOS of photon shows a pure logarithm divergence near the photon band energy, which is estimated as $Q_{PC}(\omega)\alpha-\left[\ln\left|\frac{\omega-\omega_0}{\omega_0}\right|-1\right]\theta(\omega-\omega_e)$, with $\omega_e$being the central peal logarithm point. $J(\omega)$, unit area, is explained as the DOS production field within the photovoltaic cell by $V(\omega)$ in the photon band and photon voltage cell as under [15, 47]:

$$J(\omega)=Q(\omega)|V(\omega)|^2 \tag{3.17}$$

Henceforth, it was assumed through the proliferative photon dynamics and photon band frequency ω_c using $u(t,t_0)$ for the structure of photon in the connection $\langle a(t)\rangle=u(t,t_0)\langle a(t_0)\rangle$.

Therefore,

$$(t, t_0) = \frac{1}{1 - \Sigma'(\omega_b)} e^{-i\omega(t - t_0)} + \int_{\omega_e}^{\infty} d\omega \frac{J(\omega) e^{-i\omega(t - t_0)}}{[\omega - \omega_c - \Delta(\omega)]^2 + \pi^2 J^2(\omega)} \tag{3.18}$$

$\Sigma'(\omega_b) = [\partial\Sigma(\omega)/\partial\omega]_{\omega=\omega_b}$, and $\Sigma(\omega)$ is the reservoir-induced photonic band photon self-energy correction:

$$\Sigma(\omega) = \int_{\omega_e}^{\infty} d\omega' \frac{J(\omega')}{\omega - \omega'} \tag{3.19}$$

The function ω_b in Eq. (3.2) denotes the mode of photonic frequency in the photonic bandgap (PBG). PBG ($0 < \omega_b < \omega_e$) can be determined through the use of the pole state; $\omega_b - \omega_c - \Delta(\omega_b) = 0$, with a principal-value integral being $\lesssim \Delta(\omega) = \mathcal{P}\left[\int d\omega' \frac{J(\omega')}{\omega - \omega'}\right]$.

Moreover, in consideration of the proliferation magnitude, the comprehensive photonic dynamics have been determined and indicated in Fig. 3.4a for 3D, 2D, and 1D photovoltaic cells with regard to different detuning δ incorporated from the area of photonic bandgap to the area of photonic band [14, 23, 41], where *k*(*t*), photonic dynamic rate, is indicated in Fig. 3.4b disregarding $\delta = 0.1\ \omega_e$. Based on the

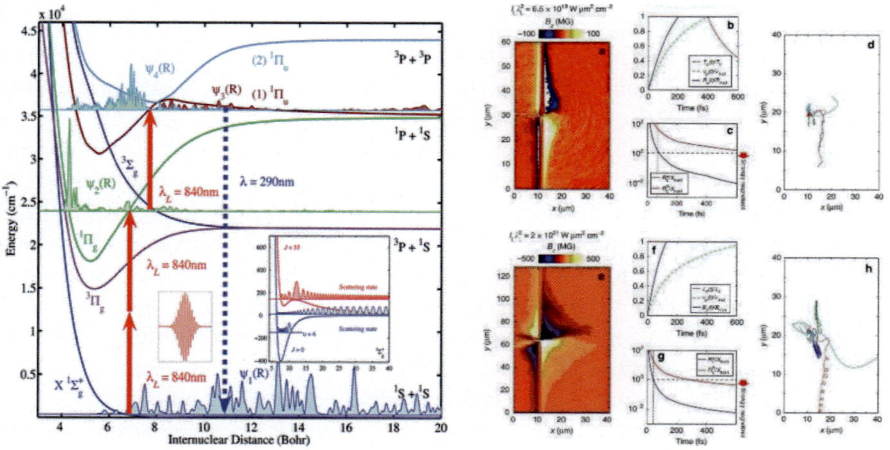

A B

Fig. 3.4 (**a**) Heating photon energy formation against the internuclear distance (Bohr) at different variable frequencies where the photon particle carries out the energy measurements (DQD) by integrating Higgs boson quantum field. (**b**) The photon particle dynamics into the particle in simulation of cell (PIC). (Source: M. Nakatsutsumi et al. *Nature Communications*; volume 9, article number: 280, 2018)

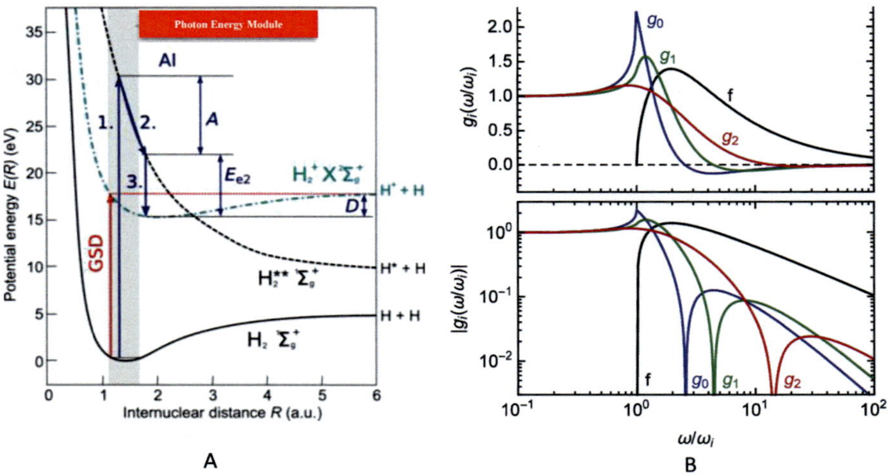

Fig. 3.5 (**a**) *I–V* clarification of the single diode where curtain wall is in the standard condition to deform energy considering photonic irradiance. (**b**) Photonic energy factor irradiance where black line is the photon absorptive region. Blue line is the dispersive region for related photon irradiance with static energy and static direction. Green line is the angle average for photonic irradiance isotropic dispersal with static energy. Red line represents photonic thermal isotropic average. (Source: Alexandra Dobrynina et al. Demidov Yaroslavl State University, Sovietskaya; 15 December 2014)

calculated result, the rate of producing dynamic photons is high as soon as ω_c crosses to PB area from the photonic bandgap area. Since the range within $u(t, t_0)$ is $1 \geq |u(t, t_0)| \geq 0$, the crossing area can be described as connected to the condition $0.9 \gtrsim |u(t \rightarrow \infty, t_0)| \geq 0$. This relates to $-0.025\omega_e \lesssim \delta \lesssim 0.025\omega_e$, with $k(t)$ as the rate of production in the photonic bandgap ($\delta < -0.025\omega_e$), as well as within the PBG ($0.025\omega_e \lesssim \delta \lesssim 0.025\omega_e$) vicinity.

Figure 3.5—Dynamic photons proliferation in photovoltaic cells; (a) in consideration of the photon band area, $\langle a(t)\rangle = 5u(t, t_0)\langle a(t_0)\rangle$, (b) is the rate of dynamic photon ($k(t)$), which is plotted for 1D, 2D, and 3D photovoltaic cells [27, 28, 36].

For $\delta \gg 0.025\omega_e$, the production of dynamic photon is nearly exponential, which represents a Markov factor. Since the photonic band frequency in the crossover area is small within the PBE vicinity, this sharply raises the dynamic photon production mode [22, 42, 43]. As a result, this dynamic photon production proliferation proves the ionic photonic condition counts within the PBG vicinity in the photovoltaic cell, where the existing photons are in a non-equilibrium photonic condition.

Dynamics photon proliferation within the photovoltaic panel are afterwards explained in consideration of thermal variations in regard to $v(t, t)$, through the determination of the non-equilibrium photon scattering theorem.

$$v(t,t) = \int_{t_0}^{t} dt_1 \int_{t_0}^{t} dt_2 u^*(t_1,t_0)\widetilde{g}(t_1,t_2)u(t_2,t_0) \tag{3.20}$$

From the above equation, $\widetilde{g}(t_1, t_2) = \int d\omega J(\omega)\widetilde{n}(\omega, T)e^{-i\omega(t-t')}$ discloses the variations of photonic dynamic induced particularly by the thermal relativistic state, with $\overline{n}(\omega, T) = 1/\left[e^{\hbar\omega/k_BT} - 1\right]$ being the photon production proliferation within the photovoltaic cell at relativistic temperature *T*, and it is expressed as follows:

$$v(t, t \to \infty) = \int_{\omega_e}^{\infty} d\omega \mathcal{V}(\omega), \tag{3.21}$$

with

$$\mathcal{V}(\omega) = \overline{n}(\omega, T)[\mathcal{D}_l(\omega) + \mathcal{D}_d(\omega)]$$

The above Eq. (3.21) is simplified so as to determine $V(\omega) = \overline{n}(\omega, T)\mathcal{D}_d(\omega)$, which represents the non-equilibrium state. Dissipation of Einstein's photon fluctuation under low-temperature states is non-dynamically viable at the photonic band, although the linking photonic structures are measurable; $n(t) = \langle a^{\dagger}(t)a(t)\rangle = |u(t, t_0)|^2 n(t_0)v(t, t)$, with $n(t_0)$ being the prime PB. For this reason, in Fig. 3.3b, the dynamic photon numbers have been plotted versus temperature so as to confirm the non-equilibrium production of proliferated photon, as indicated by the curve that is solid-blue. Basically, the distribution of Bose–Einstein does not necessarily apply for non-equilibrium photon generation proliferation within the photonic bandgap religion, because of the extreme relativistic states. More precisely, one has to initially consider the PB as the Fock condition photon number n_0, i. e. , $\rho(t_0) = |n_0\rangle\langle n_0|$, in order convert the distributed Bose–Einstein photons into HnP^- proliferated photons. The photon number n_0 is determined as follows:

$$\rho(t) = \sum\nolimits_{n=0}^{\infty} p_n^{(n_0)}(t)|n_0\rangle\langle n_0| \tag{3.22}$$

$$p_n^{(n_0)}(t) = \frac{[v(t, t)]^n}{[1 + v(t, t)]^{n+1}}[1 - \Omega(\mathrm{t})]^{n_0} \times \sum_{k=0}^{\min\{n_0 n\}} \binom{n_0}{k} \times \binom{n}{k}\left[\frac{1}{v(t, t)}\frac{\Omega(\mathrm{t})}{1 - \Omega(\mathrm{t})}\right]^k \tag{3.23}$$

In this case, $\Omega(\mathrm{t}) = \frac{|u(t, t_0)|^2}{1 + v(t, t)}$ suggesting that a Fock condition photon will transform into different Fock conditions of $|n_0\rangle$ is $p_n^{(n_0)}(t)$. The dissipation of photon proliferation $p_n^{(n_0)}(t)$ within the prime condition $|n_0 = 5\rangle$ as well as the steady-condition $p_n^{(n_0)}(t \to \infty)$ is therefore indicated in Fig. 3.6. The photon generation proliferation, therefore, will finally reach the thermal non-equilibrium condition with the photonic structure. The distribution of Bose–Einstein photon can be overlooked:

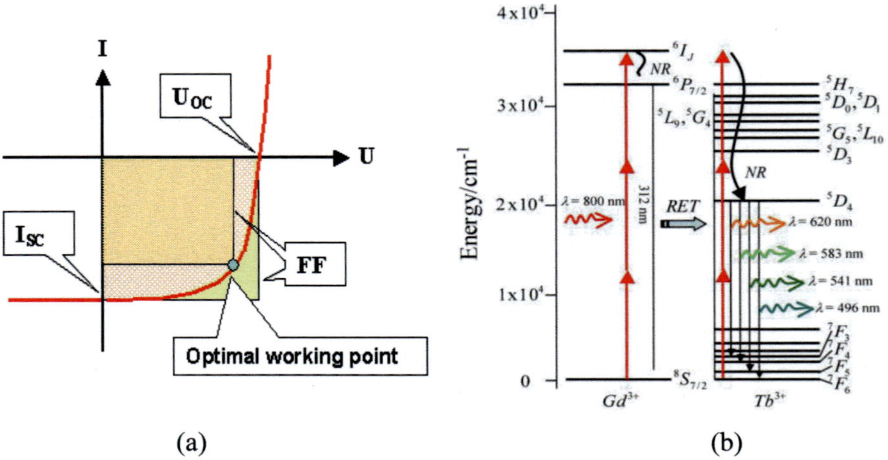

Fig. 3.6 Above represents the solar cell current–voltage distinctive features for conceptual function of (**a**) current production optimum working point, (**b**) the current source current–voltage module nearby the short circuit point, and as a production of energy per unit area within the open circuit point vicinity

$$p_n^{(n_0)}(t \to \infty) = \frac{[\overline{n}(\omega_c, T)]^n}{[1 + \overline{n}(\omega_c, T)]^{n+1}} \tag{3.24}$$

To confirm this generation of HnP^-, it is suggestively essential to breakdown the Bose–Einstein distribution theory via extreme relativistic comprehensible conditions and work out Eq. (3.2) in consideration of the photons proliferation conditions, which can as well be rewritten as below:

$$\rho(t) = \mathcal{D}[\alpha(t)]\rho_T[v(t,t)]\mathcal{D}^{-1}[\alpha(t)] \tag{3.25}$$

where $\mathcal{D}[\alpha(t)] = \exp\{\alpha(t)\alpha^\dagger - \alpha^*(t)\alpha\}$ represents the displacement driver with $\alpha(t) = u(t, t_0)\alpha_0$ and

$$\rho_T[v(t,t)] = \sum\nolimits_{n=0}^{\infty} \frac{[v(t,t)^n]}{[1 + v(t,t)]^{n+1}} \mid n\rangle\langle n| \tag{3.26}$$

From the above equation, ρ_T denotes the thermal condition with mean practical number $v(t, t)$, with Eq. (3.11) disclosing that the prime point break cavity condition will transform into a displaced thermal condition [24, 25, 37], which represents the combination of displaced number conditions $\mathcal{D}[\alpha(t)] \mid n\rangle$ [35]. As a result, Eq. (3.25) can as well be rewritten as follows:

$$\langle m|\rho(t)|n\rangle = J(\omega) = e^{-\Omega(t)|\alpha_0|^2} \frac{[\alpha(t)]^m[\alpha^*(t)]^n}{[1+v(t,t)]^{m+n+1}}$$

$$= \sum_{k=0}^{\min\{m,n\}} \frac{\sqrt{m!n!}}{(m-k)!(n-k)!k!} \left[\frac{v(t,t)}{\Omega(t)|\alpha_0|^2}\right]^k \quad (3.27)$$

Here, the *HnP*-production within the photovoltaic panel ($\langle m|\rho(t)|n\rangle$) will certainly transform into a non-equilibrium state $[\alpha(t)]^m[\alpha^*(t)]^n$ and an extreme relativistic thermal condition $[1 + v(t, t)]^{m+n+1}$. This clearly recommends that the photonic band photon number shall not comply with the Bose–Einstein distribution, instead it produces the photon exponentially $J(\omega) = \sum_{k=0}^{\min\{m,n\}} \frac{\sqrt{m!n!}}{(m-k)!(n-k)!k!} \left[\frac{v(t,t)}{\Omega(t)|\alpha_0|^2}\right]^k = [n * e(1+2n)]^4$. Basically, Bose–Einstein photon dissemination breakdown and transformation into numerous photons within the photovoltaic panels is relatively viable through implementing extreme relativistic states, and clearly explaining dynamic photon, thus transforming the equilibrium photons into non-equilibrium stable-condition photons in regard to DOS photon activation [34, 38].

Electricity Conversion by a Photovoltaic Panel

Modeling a one-diode photovoltaic panel, comprising a small semiconductor disk attached by wire to a circuit with a negative and a positive silicon film located beneath a thin glass slice, and attached to graphene, by the use of an exterior curtain wall skin of a building, has been suggested in this study as a way of converting HnP^- into electricity beneath extreme relativistic states. Essentially, the photovoltaic panel will require to have an open *circuit point*, with maximum voltage and current, open circuit voltage V_{oc}, and also the extreme power point being explained through the use of maximum current as well as calculation of voltage instantaneously upon capturing the non-equilibrium photons [26, 44, 45]. Power that will be supplied by a photovoltaic panel shall therefore have the ability to acquire maximum values at I_{mp} and V_{mp} [29, 40]. Analyzing the optimum working point as well as the dark functionality under revealing the photovoltaic panel status will help in confirming the photovoltaic panels' ability of such kind of voltage–current flow.

Finally, the total photon production from a single photon is calculated as, $J(\omega) = [n * e(1 + 2n)]^4$ (Eq. 3.9). Therefore, the total number of photons would be 81 photons, which originate from a single photon. Then, I performed an estimate to convert these photons into electricity. The light quanta of a certain type of polarization has a frequency range of ν_r to $\nu_r + d\nu_r$; thus, the maximum solar radiation is achieved from 81 photons at 1.4 eV × 81 = 113.4 eV (one photon is 1.4 eV). Therefore, the total energy would be (1.4 eV = 27.77 mW/m^2 × 81) 2249.37 mW/m^2 eV/h [29, 31]. The electricity produced at this stage is DC (direct current) that must be converted to AC (alternating current) for use in the building sector and battery system storage. The estimate reveals that in a year with an average of 5 h a day maximum for 365 days, this total energy production is equivalent to 4,105,100.25 kW/year [33, 39, 40]. This means that 11,246.85 kW of energy can be supplied by a 1 m^2 solar panel per day. If an office and commercial building consumption is approximately 3800 kWh/day for a building with a 32 m $\times$ 31 m footprint and height of 30 m (10 floors), 0.33 m^2 of the building's exterior skin must be used as solar panels to meet the total energy demand of the building, which is equivalent to 0.00008% of the exterior curtain wall.

Conclusions

Bose–Einstein photon distribution theory is broken down theoretically under ultra-relativistic conditions to create HnP^- and produce clean energy. A series of mathematical calculations have been performed to confirm the production of multiple HnP^- from a single Bose–Einstein dormant photon under extreme relativistic conditions in the PV panel. Mathematical analyses suggest that extreme relativistic conditions break down the Bose–Einstein dormant photons and produce nearly exponential rate photons from an equilibrium to a non-equilibrium stage. The results suggest that the proliferation of these photons is agitated by the extreme relativistic condition to conduct photon–photon interactions in the PV panel to exponentially produce photons. Naturally, these created photons can be captured by the PV circuit system to produce DC current and then convert it into AC current to satisfy the energy demand of a building. The energy production from the PV panel has also been calculated, which revealed that if only a mere 0.00008% of a building exterior curtain wall is used as a *Modern Solar Photovoltaics Energy* panel, it will meet the total energy demand of the building using 100% clean energy. Naturally, implementation of this technology would introduce a new era of science to meet the total energy demand of the building sector and dramatically reduce climate change.

Acknowledgments This research was supported by Green Globe Technology under grant RD-02017-03. Any findings, predictions, and conclusions described in this article are solely those of the author. The author confirms that he has no conflicts of interest.

References

1. D.K. Armani, T.J. Kippenberg, S.M. Spillane, K.J. Vahala, Ultra-high-Q toroid microcavity on a chip. Nature **421**, 925 (2003)
2. K.M. Birnbaum et al., Photon blockade in an optical cavity with one trapped atom. Nature **436**, 87–90 (2005)
3. K. Busch, G. von Freymann, S. Linden, S.F. Mingaleev, L. Tkeshelashvili, M. Wegener, Periodic nanostructures for photonics. Phys. Rep. **444**, 101 (2007)

4. D.E. Chang, A.S. Sørensen, E.A. Demler, M.D. Lukin, A single-photon transistor using nanoscale surface plasmons. Nat. Phys. **3**, 807–812 (2007)
5. J. Chen, C. Wang, R. Zhang, J. Xiao, Multiple plasmon-induced transparencies in coupled-resonator systems. Opt. Lett. **37**, 5133–5135 (2012)
6. M.T. Cheng, Y.Y. Song, Fano resonance analysis in a pair of semiconductor quantum dots coupling to a metal nanowire. Opt. Lett. **37**, 978–980 (2012)
7. B. Dayan et al., A photon turnstile dynamically regulated by one atom. Science **319**, 1062–1065 (2008)
8. D. Englund et al., Resonant excitation of a quantum dot strongly coupled to a photonic crystal nanocavity. Phys. Rev. Lett. **104**, 073904 (2010)
9. S. Gleyzes et al., Quantum jumps of light recording the birth and death of a photon in a cavity. Nature **446**, 297 (2007)
10. C. Guerlin et al., Progressive field-state collapse and quantum non-demolition photon counting. Nature **448**, 889 (2007)
11. Z. Han, S.I. Bozhevolnyi, Plasmon-induced transparency with detuned ultracompact Fabry-Pérot resonators in integrated plasmonic devices. Opt. Express **19**, 3251–3257 (2011)
12. M.F. Hossain, Green science: Independent building technology to mitigate energy, environment, and climate change. Renew. Sustain. Energ. Rev. (2017a). https://doi.org/10.1016/j.rser.2017.01.136
13. M.F. Hossain, Design and construction of ultra-relativistic collision PV panel and its application into building sector to mitigate total energy demand. J. Build. Eng. (2017b). https://doi.org/10.1016/j.jobe.2016.12.005
14. M. Hossain, Faruque., Solar energy integration into advanced building design for meeting energy demand and environment problem. Int. J. Energy Res. (2016). https://doi.org/10.1002/er.3525
15. J.F. Huang, T. Shi, C.P. Sun, F. Nori, Controlling single-photon transport in waveguides with finite cross section. Phys. Rev. A **88**, 013836 (2013)
16. J.D. Joannopoulos, P.R. Villeneuve, S. Fan, Photonic crystals: Putting a new twist on light. Nature **386**, 143 (1997)
17. S. John, J. Wang, Quantum optics of localized light in a photonic band gap. Phys. Rev. B **43**, 12772 (1991)
18. A.G. Kofman, G. Kurizki, B. Sherman, Spontaneous and induced atomic decay in photonic band structures. J. Mod. Opt. **41**, 353 (1994)
19. P. Kolchin, R.F. Oulton, X. Zhang, Nonlinear quantum optics in a waveguide: Distinct single photons strongly interacting at the single atom level. Phys. Rev. Lett. **106**, 113601 (2011)
20. C. Lang et al., Observation of resonant photon blockade at microwave frequencies using correlation function measurements. Phys. Rev. Lett. **106**, 243601 (2011)
21. C.U. Lei, W.M. Zhang, A quantum photonic dissipative transport theory. Ann. Phys. **327**, 1408 (2012)
22. Q. Li, D.Z. Xu, C.Y. Cai, C.P. Sun, Recoil effects of a motional scatterer on single-photon scattering in one dimension. Sci. Rep. **3**, 3144 (2013)
23. J.Q. Liao, C.K. Law, Correlated two-photon transport in a one-dimensional waveguide side-coupled to a nonlinear cavity. Phys. Rev. A **82**, 053836 (2010)
24. J.Q. Liao, C.K. Law, Correlated two-photon scattering in cavity optomechanics. Phys. Rev. A **87**, 043809 (2013)
25. P.-Y. Lo, H.-N. Xiong, W.-M. Zhang, Breakdown of Bose-Einstein distribution in photonic crystals. Sci. Rep. **5**, 9423 (2015)
26. P. Longo, P. Schmitteckert, K. Busch, Few-photon transport in low-dimensional systems. Phys. Rev. A **83**, 063828 (2011)
27. X.-Y. Lü, W.-M. Zhang, S. Ashhab, Y. Wu, F. Nori, Quantum-criticality-induced strong Kerr nonlinearities in optomechanical systems. Sci. Rep. **3**, 2943 (2013)
28. G.D. Mahan, *Many-Body Physics*, 3rd edn. (Kluwer Academic/Plenum Publishers, New York, 2000)

29. C. Martens, P. Longo, K. Busch, Photon transport in one-dimensional systems coupled to three-level quantum impurities. N. J. Phys. (2013). https://doi.org/10.1063/1.4750125
30. S. Noda, T. Baba, *Roadmap on Photonic Crystals* (Kluwer Academic Publishers Groups, Dordrecht, 2003)
31. D. O'Shea, C. Junge, J. Volz, A. Rauschenbeutel, Fiber-optical switch controlled by a single atom. Phys. Rev. Lett. **111**, 193601 (2013)
32. A. Reinhard, Strongly correlated photons on a chip. Nat. Photonics **6**, 93–96 (2012)
33. D. Roy, Two-photon scattering of a tightly focused weak light beam from a small atomic ensemble: An optical probe to detect atomic level structures. Phys. Rev. A **87**, 063819 (2013)
34. E. Saloux, Explicit model of photovoltaic panels to determine voltages and currents at the maximum power point. Sol. Energy **85**, 713–722 (2011)
35. C. Sayrin et al., Real-time quantum feedback prepares and stabilizes photon number states. Nature **477**, 73 (2011)
36. J.T. Shen, S. Fan, Strongly correlated two-photon transport in a one-dimensional waveguide coupled to a two-level system. Phys. Rev. Lett. **98**, 153003 (2007)
37. T. Shi, S. Fan, C.P. Sun, Two-photon transport in a waveguide coupled to a cavity in a two-level system. Phys. Rev. A **84**, 063803 (2011)
38. S. Sreekumar, A. Benny, Maximum power point tracking of photovoltaic system using Fuzzy Logic Controller based boost converter, in *2013 International Conference on Current Trends in Engineering and Technology (ICCTET)*, 2013
39. M.S. Tame, K.R. McEnery, Ş.K. Özdemir, J. Lee, S.A. Maier, M.S. Kim, Quantum plasmonics. Nat. Phys. **9**, 329–340 (2013)
40. M.W.Y. Tu, W.M. Zhang, Non-Markovian decoherence theory for a double-dot charge qubit. Phys. Rev. B **78**, 235311 (2008)
41. M.W.-Y. Tu, W.-M. Zhang, J. Jin, O. Entin-Wohlman, A. Aharony, Transient quantum transport in double-dot Aharonov-Bohm interferometers. Phys. Rev. B **86**, 115453 (2012)
42. X.H. Wang, B.Y. Gu, R. Wang, H.Q. Xu, Decay kinetic properties of atoms in photonic crystals with absolute gaps. Phys. Rev. Lett. **91**, 113904 (2003)
43. Y.F. Xiao et al., Asymmetric Fano resonance analysis in indirectly coupled microresonators. Phys. Rev. A **82**, 065804 (2010)
44. W.-B. Yan, H. Fan, Single-photon quantum router with multiple output ports. Sci. Rep. **4**, 4820 (2014)
45. W.-B. Yan, J.-F. Huang, H. Fan, Tunable single-photon frequency conversion in a Sagnac interferometer. Sci. Rep. **3**, 3555 (2013)
46. Z. Yu, X. Hu, H. Yang, Q. Gong, On-chip plasmon-induced transparency based on plasmonic coupled *nanos* cavities. Sci. Rep. **4**, 3752 (2014)
47. W.M. Zhang, P.Y. Lo, H.N. Xiong, M.W.Y. Tu, F. Nori, General non-Markovian dynamics of open quantum systems. Phys. Rev. Lett. **109**, 170402 (2012)

Chapter 4
Transformation of Building's Biowaste into Electricity Energy to Mitigate the Global Energy Vulnerability

Abstract *Green Science*, a sustainable mechanism, is being described to fulfill the complete need of energy for a building that can be created by the building itself. To meet the complete energy demand for a building, the domestic biowaste including human empathetic excrement (stool) is suggested to be executed into converting process in situ where separated sludge is to be collected into an anaerobic shut tank bioreactor (BR) into the basement, facilitating to form biogas (CH_4) by *methanogenesis* in order to convert biogas into electricity energy to power the entire building. Thereafter, the discharged wastewater is to be stored into another detention tank in situ in order to conduct a complete treatment process of primary, secondary, tertiary and UV application to utilize the treated wastewater for gardening. Implementation of this technology indeed shall be an inventive field of science where a building can form electricity by itself to complete its total energy need without any connection with the utility authorities which is benevolent to environment.

Keywords Domestic biowaste · Bioreactor · In site biowaste treatment technology · Methanogenesis · Bioenergy · Environmental sustainability

Introduction

Environmental vulnerability correlated much on building sector since 40% of global fossil energy is consumed by building sector throughout the world [11, 29]. In 2018, the net energy consumption globally accounted for 5.59×10^{20} J = 559 EJ, where 2.236×10^{20} EJ energy is alone engulfed by the building sector [21, 43]. Consequently, building sector triggered to release nearly 8.01×10^{11} ton CO_2 (218 gtC by building sector of worldwide total carbon production of 545 gtC; 1 gtC = 10^9 ton C = 3.67 gt CO_2) into the atmosphere in year 2018. The quickening of fossil fuel consumption by building sector is getting higher and higher globally and the situation shall remain unchanged until an innovative technology is used to power the building sector globally. At present, the atmospheric CO_2 level is 400 ppm where building sector is the major player for creating this high level of CO_2 concentration into the atmosphere and it is accelerating by 2.11% per year which is the clear and present danger to the survival of all living beings in this planet in the near future [1,

M. F. Hossain, *Sustainable Design for Global Equilibrium*,
https://doi.org/10.1007/978-3-030-94818-4_4

3, 59]. Necessarily, the atmospheric CO_2 level must be lowered to a clean breathable level of 300 ppm CO_2. Therefore, a sustainable energy mechanism in building sector is an urgent demand to confirm a clean and green environment on earth.

Though there are some recent interesting studies showing that a person can produce an average of 0.4 kg/day of feces that can form 0.4 m^3 biogas/day, and this amount of biogas (0.4 m^3/day) production is good enough to cook three meals for a family of four persons in a day [6, 30, 35]. However, no one has shown that the mechanism of using cellar of a building as an acting bioreactor to transform biowaste into electricity energy can satisfy the total energy demand of a building.

Therefore, in this research, a net zero carbon release by a building has been proposed by producing bioenergy by the building itself and transforming it into electricity energy to meet its net energy need. Simply, the domestic biowaste including human stool and wastewater of the building is being chosen to collect it into the sealed separation chamber into the basement. Thereafter, this biowaste is being isolated into (a) wastewater, and (b) sludge and transferred into two separation tanks into the cellar. Then the wastewater is being conducted for treatment process in site by integrating required all chemical and physical process in order to use for landscaping. Consequently, the solid biowaste has been permitted to undergo the *methanogenesis* process into the bioreactor to form bioenergy and then convert it into electricity energy. Implementation of this innovative mechanism shall indeed be a promising technology to fulfill the net need for a building which is delivered by the building itself.

Materials and Methods

For the conversion of domestic biowaste into bioenergy, structurally sound long-lasting bioreactor (BR) needs to be designed. Thus, load resistant factor design (LRFD) bioreactor must be constructed for a structurally sound bioreactor to operate regularly under high water velocity pressure considering the mathematical calculation of water velocity (379 mile/h), water density (1.2 kg/m^3), and the friction loss cofactor 1.00/m^2, respectively [9, 35, 36]. As the water dynamic force is 0.5 of half of the density of the water, the equation for water force into the bioreactor can be expressed as $p_w = 0.5\rho C_p v_r^2$, where p_w represents the water force (Pa), ρ considers water density (kg/m^3), C_p denotes water force gradient which is 1, and v_r^2 is the water velocity (m/s) into the bioreactor. Thus, the net resultant force of $P_w = 0.5 \times 1.2$ kg/m^3 $\times$ 379^2 m/s is 86,185 Pa is the water pressure resistance capacity of the bioreactor. It can be simplified as force of $F =$ area $\times$ drag coefficient (constant = 1.00) $\times$ water dynamic force by following equation $F = 1$ m^2 $\times$ 1.0 $\times$ 86,185 = 86,185 N (8788 kgf) = 19,375 ibf to confirm that the bioreactor is structurally sound which flows the water velocity must be less than 19,000 ibf to operate bioreactor normally throughout the year.

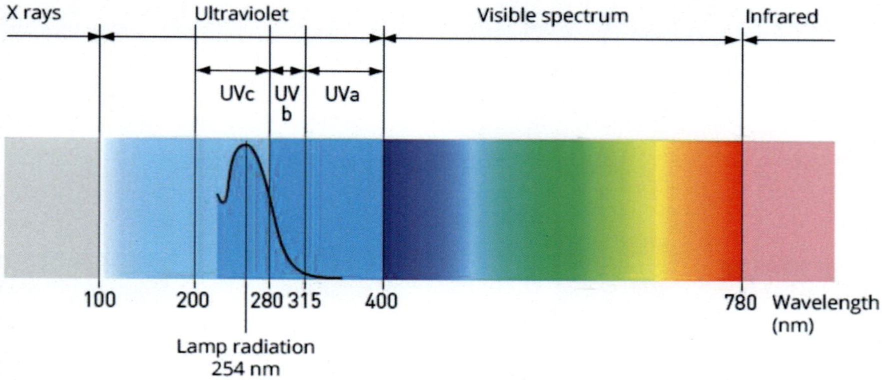

Fig. 4.1 The application of photophysics radiation in purifying water that illustrates that once one applies UV light of 320 nm into the wastewater, it begins to kill the microorganisms once the temperature momentum hits at 50 °C

Once the sophisticated water force resistance two chambers bioreactor has been constructed, the bioreactor is to be connected into biowaste chamber in order to collect the biowaste into the shut separation chamber into the basement. The other chamber is to be connected with the separated wastewater for the process of treatment of primary, secondary, and tertiary mechanism and then implemented into UV application to disinfect the wastewater. The UV application and filtration constitutes the simplest way of treating wastewater involving *Disinfection* (DIS) system in which one fills a detention chamber with water and exposes it to full UV light for a few hours. Once the wastewater temperature hits 50 °C due to the subject of UV light of approximately 320 nm, it functions immediately to kill all bacteria, viruses, and molds and disinfect the water completely thorough bacteriological disinfection process (Fig. 4.1).

This treatment mechanism removes nearly 100% microorganism and other contaminants from the wastewater effluent which could be used for local gardening (Fig. 4.2).

Then, the other product sludge (humane feces including domestic waste) in another chamber of the bioreactor is being conducted for disinfection process in site into an anaerobic chamber. This is the conversion mechanism performed by electrochemical filters of activated carbon nanotubes (CNT), which has the capability to electrolyze and oxidize pollutant in the anode actively from the sludge [7, 22, 44]. It is an advanced mechanism of biowaste disinfection mechanism in combining both electrolyze and oxidation process into the anode of carbon nanotubes and catalyzed by the process of oxidation by H_2O_2 into the cathode of carbon nanotubes. The function here accelerates the rate of sludge treatment, and its active oxidation process into the tank is being calculated and demonstrated a pathway that H_2O_2 flow is very much effective to disinfect the biowaste by the electrode and the cathode potential in order to achieve the content of biowaste pH, flow rate, and oxygen dissolved into a normal clean biomaterial form [26, 40, 46]. Hence, the maximum

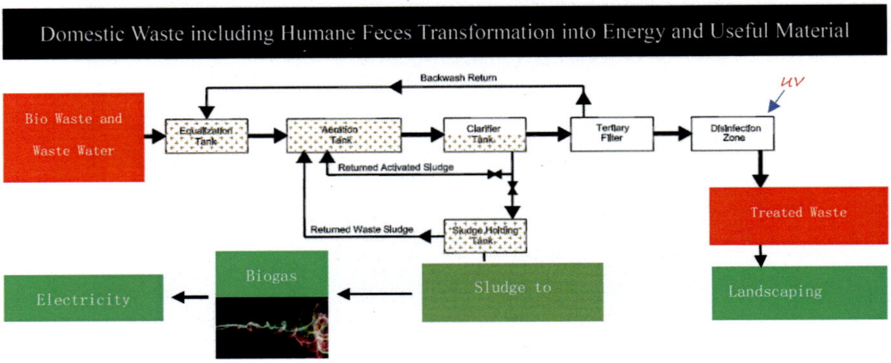

Fig. 4.2 The schematic diagram of wastewater treatment mechanism where treated water could be utilized for landscaping and the sludge is to be used for transforming process to produce the energy

flow of H_2O_2 is being accounted for 1.38 mol L^{-1} m^{-2} C by achieving CNT L^{-1} m^{-2} with the implementation of cathode potential V −0.4 (vs Ag/AgCl), a pH of 6.46, with the flowing rate of 1.5 mL min^{-1} and the dissolved oxygen (DO) content of 1.95 mol L^{-1} m^{-2}. Additionally, phenol (C_6H_5OH) is being induced as an aromatic element for addressing the removal efficiency by clarifying the oxidation rate directed to the H_2O_2 flow [2, 31, 32]. Consequently, the electrochemical carbon nanotube filters activated the H_2O_2 generation tremendously as carbon nanotubes can work most effectively to remove the organic contaminants from the biowaste nearly 100%.

Once the sludge is disinfected, the product is being placed into the closed bioreactor tank to allow for anaerobic co-digestion process [4, 10, 47]. Thereafter, the product is being heated for 95 °F for 15 days which will stimulate the growth of anaerobic bacteria, *Desulfovibrio* and *Methanococcus,* which engulf organic material of the sludge and produce biogas through biosynthesis process (Fig. 4.3).

Then the biogas is to be conducted for transforming process to generate electricity energy through the semi-conductor diodes of the circuit panel. Hence, the electricity production from biogas into the circuit panel is being examined by detailed mathematical computations [21, 23, 41] (Fig. 4.4).

Hence, to achieve a successful conversion of biogas into electricity energy, the first order perturbation theory has been implemented considering the production of biogas [26, 37, 40]. The first order mechanism of the transformation of the biogas into the electricity energy needs the adequate surface into the bioreactor to separate the electrons into the semiconductor to produce the electric charge by the given term below [15, 42, 58]:

$$I = I_{\mathrm{ph}} - I_{\mathrm{ph}}\left[\exp\left(\frac{V + R_{\mathrm{s}}I}{V_{\mathrm{r}}}\right) - 1\right] - \frac{V + R_{\mathrm{s}}I}{R_{\mathrm{p}}} \tag{4.1}$$

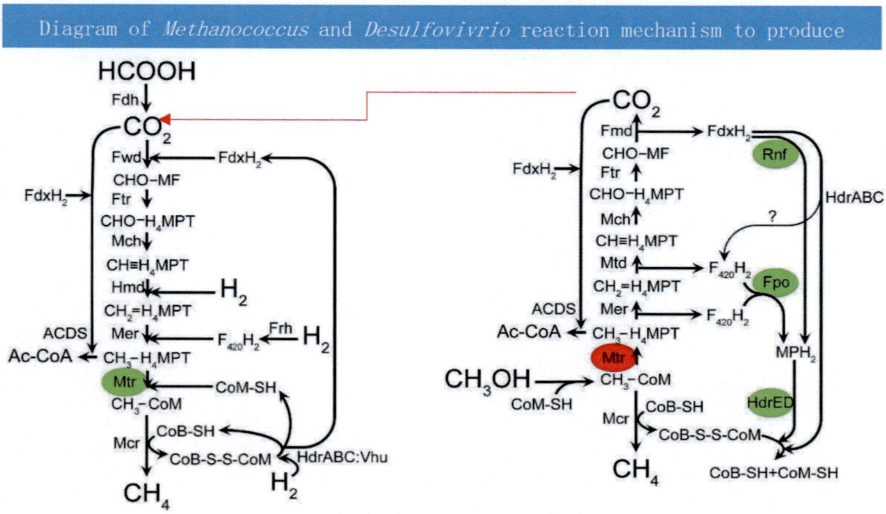

(a)

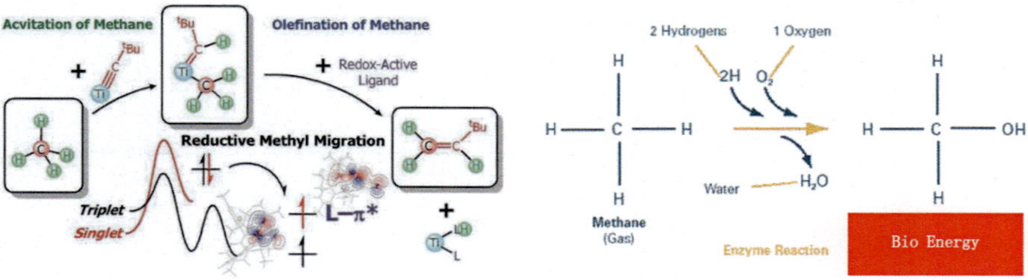

(b)

Fig. 4.3 (**a**) Biosynthesis mechanism of methanogenesis which shows the conduction of chain reaction to form methane from sludge where two bacteria, Methanococcus and Desulfovibrio, are the primary inhibitors to conduct this reaction, (**b**) the process showing the transformation of methane into bioenergy

Here I is the current and V is the voltage into the circuit panel. I_{ph} $(=N_p I_{ph,cell})$ is the electricity energy-created current running inside the circuit module which consists of N_p cells that are connected in parallel. I_0 $(=N_{pI0,cell})$ is called the reverse current passing through N_p cells that are connected in parallel, wherein the reverse saturation current $I_{0,cell}$ passes through each cell. Subsequently, V_T $(=aN_s{\cdot}kT/q)$ is represented as a matrix of thermal stress of N_s cells that are connected in series where (~1.5 = 1.0) keeping in mind the diode ideality factor, k $(=1.38e^{-23}$ J/K) is a constant, q $(=1.602e^{-19}$ C) is the charge on an electron, and T is the temperature in kelvin. Here R_p is the equivalent resistance in parallel while R_s is the equivalent resistance in series for circuit generator. Depending on the operational point, the circuit device, in practice, operates as a mixed performance of the current source or

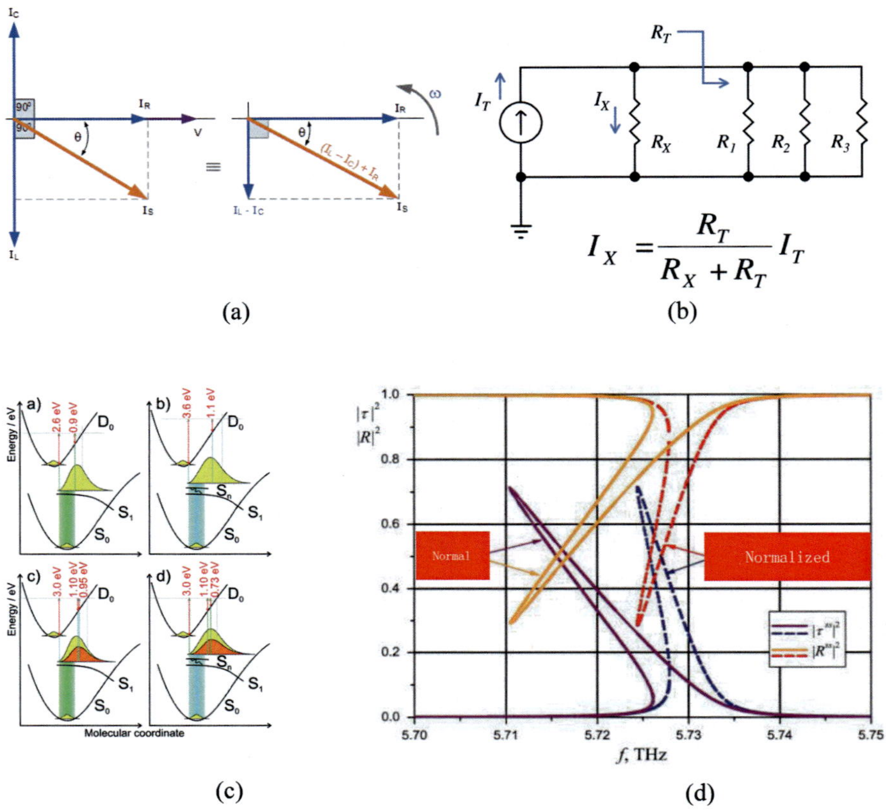

Fig. 4.4 The conversion mechanism of bioenergy into electricity energy, (**a**) mode of electricity production dynamics with respect to power factor (pf), (**b**) flow of electricity current generation, (**c**) net electric energy production (eV) rate at molecular rate, (**d**) the rate of electricity energy generation at normal and normalized circuit parameters, respectively

the voltage source [39, 52]. Practically, for the circuit panel, the effect of R_p parallel resistance will be greater in the operating area having a current source, while the R_s series resistance has a bigger effect on the functioning of the photovoltaic modules when the device works in the area having a voltage source [11, 19, 56]. Based on studies of various researchers, it can be concluded that for simplifying the model, the value of R_p can be ignored as it is very high [16, 24, 38]. Likewise, the value of R_s being very low, can be neglected too [12, 13, 49], thus the temperature of the circuit panel can be shown as follows [17, 18].

$$T = 3.12 + 0.25\frac{S}{S_n} + 0.899T_a - 1.3v_a + 273 \tag{4.2}$$

Here S and S_n (=1000 W/m^2) are the electricity energy available in working condition, respectively, and T_a is the surrounding temperature and v_a is the surrounding energy flow. The I–V features of photovoltaic panel are based on the internal qualities of the device, i.e., R_s and R_p; consequently, electricity energy as well as surrounding temperature affect outer features. The electricity energy that is responsible for producing the electric current is linked linearly to the electricity energy and temperature and can be stated as follows [5, 14, 55]:

$$I_{ph} = \left(I_{ph,n} + \alpha_I \Delta T\right) \frac{S}{S_n} \tag{4.3}$$

Here I_{ph} is the current that is produced because of biogas at STC and $\Delta T = T - T_n$, T is the temperature of the circuit panel because of the electricity energy whereas T_n is the supposed temperature. For preventing any problems faced by the electricity energy current in deciding the series resistance (very low) as well as the parallel resistance (very high), it has been presumed that $I_{sc} \approx I_{ph}$ so that an explanation can be given for the complex circuit modeling and the open circuit voltage that is dependent on the temperature can be confirmed [8, 19, 50]. This can be shown by (Fig. 4.1):

$$V_{oc} = V_{oc,n}(1 + \alpha_v \Delta T) + V_T \ln\left(\frac{S}{S_n}\right) \tag{4.4}$$

Here $V_{oc,n}$ is the open circuit voltage that is calculated at the given conditions and α_v is the voltage-temperature coefficient. The electrical and thermal features of the electricity energy panels can be achieved from these characteristics which are integrated to achieve the I–V curve to produce much electricity energy, Eq. (4.1). The characteristics of the suggested electricity energy panel should consist of the following: the short-circuit current/temperature coefficient (α_I), the open-circuit voltage/temperature coefficient (α_v), the experimental peak power (P_{max}), the insignificant short-circuit current ($I_{sc,n}$), the Maximum Power Point (MPP) voltage (V_{mp}), the MPP current (I_{mpp}), and the insignificant open-circuit voltage ($V_{oc,n}$), to calculate at the supposed conditions or standard test conditions (STC) of temperature $T = 298$ K and electricity energy of $S = 1000$ W [52, 53]. The simple equation at STC can be expressed as follows

$$I = I_{ph,n} - I_{0,n}\left[\exp\left(\frac{V + R_s I}{V_{T,n}}\right) - 1\right] - \frac{V + R_s I}{R_p} \tag{4.5}$$

Here "n" is evaluated at STC and the values are expected to show that the resistance in series and the resistance in parallel are not dependent on each other. Hence, the modeling in Eq. (4.5) can be simplified as below

$$I = I_{\mathrm{ph},n} - I_{0,n}\left[\exp\left(\frac{V + R_{\mathrm{s}}I}{V_{\mathrm{T},n}}\right) - 1\right] \tag{4.6}$$

There are three significant points on the I–V curve of electricity energy: maximum power point (V_{mp}, I_{mpp}), open circuit (V_{oc}, 0) and short circuit (0, I_{sc}) that can be shown as:

$$I_{\mathrm{sc},n} = I_{\mathrm{ph},n} - I_{0,n}\left[\exp\left(\frac{R_{\mathrm{s}}I_{\mathrm{sc},n}}{V_{\mathrm{T},n}}\right) - 1\right] \tag{4.7}$$

$$0 = I_{\mathrm{ph},n} - I_{0,n}\left[\exp\left(\frac{V_{\mathrm{oc},n}}{V_{\mathrm{T},n}}\right) - 1\right] \tag{4.8}$$

$$I_{\mathrm{mpp},n} = I_{\mathrm{ph},n} - I_{0,n}\left[\exp\left(\frac{V_{\mathrm{mpp},n} + R_{\mathrm{s}}I_{\mathrm{mpp},n}}{V_{\mathrm{T},n}}\right) - 1\right] \tag{4.9}$$

The diode saturation current can thus be shown by its dependence on the temperature of the bioreactor [15],

$$I_0 = I_{0,n}\left(\frac{T_n}{T}\right)^3 \exp\left[\frac{qE_{\mathrm{G}}}{ak}\left(\frac{1}{T_n} - \frac{1}{T}\right)\right] \tag{4.10}$$

Here EG represents the bandgap energy of the electricity energy. Equation (4.8) shows that the diode saturation current at the STC and the photocurrent at STC are linked,

$$I_{0,n} = \frac{I_{\mathrm{ph},n}}{\left[\exp\left(\frac{V_{\mathrm{oc},n}}{V_{\mathrm{T},n}}\right) - 1\right]} \tag{4.11}$$

The electricity generation model can be further enhanced if Eq. (4.8) is substituted by

$$I_0 = \frac{I_{\mathrm{sc},n} + \alpha_{\mathrm{I}}\Delta T}{\exp\left(\frac{V_{\mathrm{oc},n+\alpha_{\mathrm{V}}\Delta T}}{V_{\mathrm{T}}}\right) - 1} \tag{4.12}$$

By assuming $V_{\mathrm{oc},n}/V_{\mathrm{T},n} \gg 1$, $I_{0,n}$ can be shown as:

$$I_{0,n} = I_{\mathrm{ph},n} \exp\left(-\frac{V_{\mathrm{oc},n}}{V_{\mathrm{T},n}}\right) \tag{4.13}$$

Using Eqs. (4.13) and (4.6), it can be shown that

$$V = V_{\mathrm{oc},n} + V_{\mathrm{T},n} \ln\left(1 + \frac{I_{\mathrm{ph},n} - I}{I_{0,n}}\right) - R_{\mathrm{s}} I \tag{4.14}$$

Thus, Eq. (4.14) is considered as modest electricity energy generation model that is transformed from the biogas from bioreactor and it can be explained as simply as the following equations.

$$V = V_{\mathrm{oc},n} + V_{\mathrm{T},n} \ln\left(1 + \frac{I}{I_{\mathrm{ph},n}}\right) - R_{\mathrm{s}} I \tag{4.15}$$

Results and Discussion

Since the anaerobic *Co-digestions* of domestic biowaste including human feces lead into an anaerobic bioreactor, the *methanogenesis* process began to produce biogas into the bioreactor right way. Naturally, the formation of biogas from the biowaste is being examined by computerized gas chromatograph [20, 33, 45] (Figs. 4.5 and 4.6).

Therefore, a model of bioreactor module described the generation of maximum bioenergy from domestic waste considering protective anerobic detention chamber (Fig. 4.1). Naturally, the model of the bioreactor module is being simplified by the determination of accurate form of the current–voltage (*I*–*V*) curb considering the mode of single-diode electricity circuit [28, 29, 54].

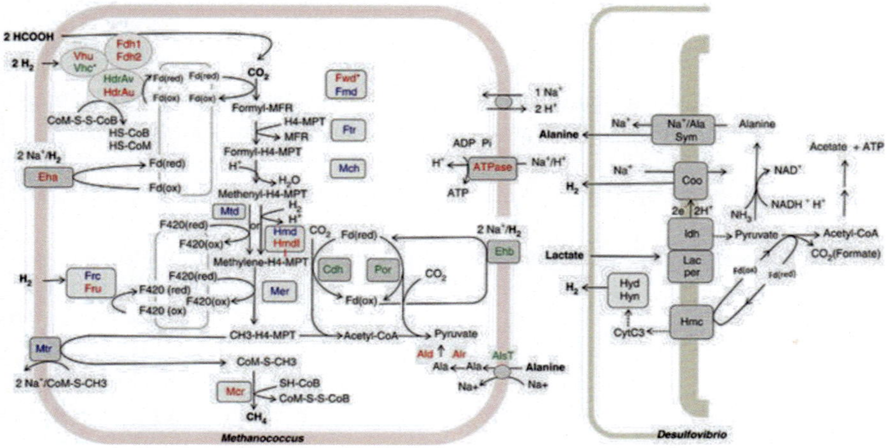

Fig. 4.5 The pathway of the methanogenesis mechanism depicts the biosynthesis of Methanococcus maripaludis and Desulfovibrio vulgaris to conduct bioenergy generation by consuming sludge

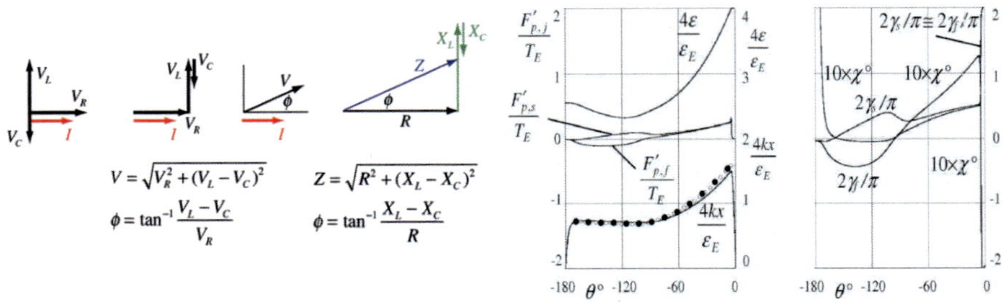

(a)

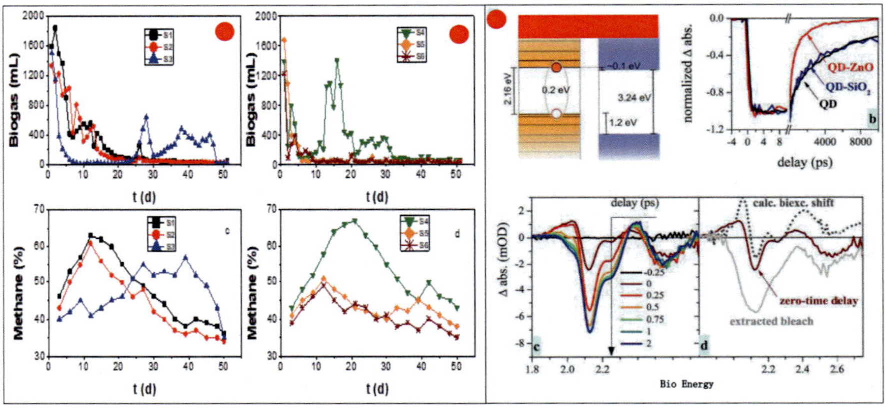

(b)

Fig. 4.6 (**a**) The biowaste transformation rate into the bioreactor in different direction and angles, (**b**) the production rate of biogas and the bioenergy considering bioreactor methane content of the biowaste

The next step is to calculate the electricity energy generation I_{pv} form biogas production by the calculation from the mode of current flow into the diode panel (Fig. 4.7a), accounting *I*–*V*–*R* relationship (Fig. 4.7b), and biogas received by the diode to convert to alternating current (AC) for using domestic energy demand (Fig. 4.7c).

The below equation represents the electricity energy output from biogas (CH_4):

$$P_{pv} = \eta_{pvg} A_{pvg} G_t \tag{4.16}$$

where η_{pvg} represents the methane-generation efficiency, A_{pvg} represents to the electricity energy generation, and G_t represents the current flow in the circuit cell. Thus, η_{pvg} can be rewritten as follows:

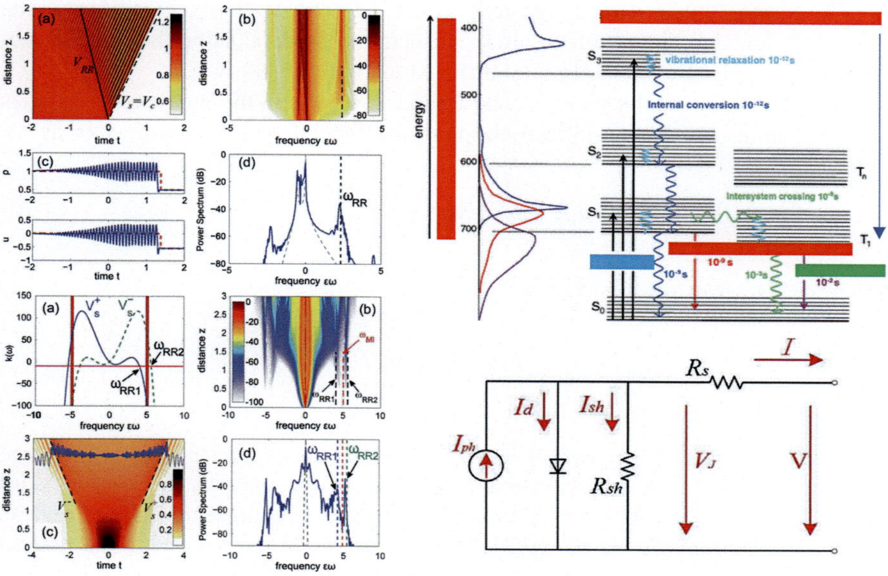

(A) (B)

Fig. 4.7 (**A**) MATLAB simulation calculated that the generating of electricity energy from bioenergy shows at various frequencies and the distances of the electric charges into the single-diode circuit cell, (**B**) shows the conversion mechanism of electricity energy DC into AC for the use as the prime source of power supply for a building

$$\eta_{\text{pvg}} = \eta_{\text{r}}\eta_{\text{pc}}[1 - \beta(T_{\text{c}} - T_{\text{cref}})] \tag{4.17}$$

η_{pc} represents the power factor effectiveness once it is equal to 1; β represents the energy cofactor (0.004–0.006 per °C); η_{r} represents the mode of energy production; and T_{cref} is the cell temperature in °C which can be obtained from the equation follow:

$$T_{\text{c}} = T_{\text{a}} + \left(\frac{\text{NOCT} - 20}{800}\right)G_{\text{t}} \tag{4.18}$$

Here, T_{a} represents the ambient temperature in °C, G_{t} represents the current flow in a circuit cell (W/s), and NOCT represents the standard operating cell temperature in Celsius (°C) degree. The total electricity energy production in the circuit panel is estimated by the following equation:

$$I_{\text{t}} = I_{\text{b}}R_{\text{b}} + I_{\text{d}}R_{\text{d}} + (I_{\text{b}} + I_{\text{d}})R_{\text{r}} \tag{4.19}$$

The current flow into the circuit cells which is determined by the functional mode of its P-N junction that is able to produce electricity by conducting the interconnection of series-parallel configuration of the circuit cell [25, 34, 48].

Implementation of the standard single-diode circuit cell, the function of N_s series and N_p parallel connection in relation to current generation can be expressed as

$$I = N_p \left[I_{ph} - I_{rs} \left[\exp\left(\frac{q(V + IR_s)}{AKTN_s} - 1 \right) \right] \right] \tag{4.20}$$

where

$$I_{rs} = I_{rr} \left(\frac{T}{T_r} \right)^3 \exp\left[\frac{E_G}{AK} \left(\frac{1}{T_r} - \frac{1}{T} \right) \right] \tag{4.21}$$

Here, in Eqs. (4.5) and (4.6), q represents the generation of electron charge (1.6×10^{-19} C), K is the Boltzmann's constant, A represents the cell standard cofactor, and T represents the cell temperature (K). I_{rs} represents the cell reverse current at T, T_r represents the cell referred temperature, I_{rr} represents the reverse current at T_r, and E_G represents the bandgap energy flow into the circuit cell. The electric current I_{ph} formation conforming the circuit cell's temperature can be simplified as follows:

$$I_{ph} = \left[I_{SCR} + k_i (T - T_r) \frac{S}{100} \right] \tag{4.22}$$

I_{SCR} represents the cell short-circuit current and electricity energy generation, k_i represents the short-circuit current temperature coefficient, and S represents the electricity energy (kW). Thus, the I–V relationship into the circuit cell can be expressed simply as:

$$I = I_{ph} - I_D \tag{4.23}$$

$$I = I_{ph} - I_0 \left[\exp\left(\frac{q(V + R_s I)}{AKT} - 1 \right) - \frac{V + R_s I}{R_{sh}} \right] \tag{4.24}$$

I_{ph} represents the electricity current (A), I_D represents the functional current (A), I_0 represents the inverse current (A), A represents the functional constant, q represents the charge of the electron (1.6×10^{-19} C), K is the Boltzmann's constant, T represents the cell temperature (°C), R_s represents the series resistance (ohm), R_{sh} represents to the shunt resistance (Ohm), I represents the cell current (A), and V represents the circuit cell voltage (V). Thus, the output electricity current into the circuit panel is thus described as follows:

$$I = I_{PV} - I_{D1} - \left(\frac{V + IR_s}{R_{SH}}\right) \tag{4.25}$$

where

$$I_{D1} = I_{01}\left[\exp\left(\frac{V + IR_s}{a_1 V_{T1}}\right) - 1\right] \tag{4.26}$$

$$I_{D2} = I_{01}\left[\exp\left(\frac{V + IR_s}{a_2 V_{T2}}\right) - 1\right] \tag{4.27}$$

I_{01} and I_{02} represent the reverse currents of cell, respectively, and V_{T1} and V_{T2} represent the thermal voltages of the respective cell. The cell idealist constants are denoted as a_1 and a_2. Then the simplified equation of the cell mode is described as:

$$v_{oc} = \frac{V_{oc}}{cKT/q} \tag{4.28}$$

$$P_{max} = \frac{\frac{V_{oc}}{cKT/q} - \ln\left(\frac{V_{oc}}{cKT/q} + 0.72\right)}{\left(1 + \frac{V_{oc}}{KT/q}\right)}\left(1 - \frac{V_{oc}}{\frac{V_{oc}}{I_{SC}}}\right) \times \left(\frac{V_{oc0}}{1 + \beta \ln \frac{G_0}{G}}\right)\left(\frac{T_o}{T}\right)^y I_{sc0}\left(\frac{G}{G_o}\right)^a \tag{4.29}$$

where v_{oc} represents the normal value of the open-circuit voltage V_{oc} represents the thermal voltage $V_t = nkT/q$, c represents constant current flow, K is the Boltzmann's constant, T represents to the temperature in Kelvin, α represents the nonlinear cofactor, q represents the electron charge, γ represents the factor representing all the nonlinear temperature-voltage function, while β represents the cell module coefficient. Since Eq. (4.29) depicted the tip energy generation by the circuit cell, the equation of total power output for an array with N_s cells connected in series and N_p cells connected in parallel with power P_M for each mode can be expressed as

$$P_{array} = N_s N_p P_M \tag{4.30}$$

Conversely, the derivative of the power with respect to current will equate to peak electricity energy production

$$\left.\frac{dP}{dI}\right|_{mpp} = \left.\frac{d(VI)}{dI}\right|_{mpp} = V_{mpp} + I_{mpp}\left.\frac{dV}{dI}\right|_{mpp} \tag{4.31}$$

and

$$V = V_{\mathrm{oc},n} + V_{\mathrm{T},n} \ln\left(1 - \frac{I}{I_{\mathrm{ph,n}}}\right) - R_{\mathrm{s}} I \tag{4.32}$$

Thus, the net electricity energy production volt (V) from biogas is finally computed as, $V = V_{\mathrm{oc},n} + V_{\mathrm{T},n} \ln\left(1 - \frac{I}{I_{\mathrm{ph},n}}\right) - R_{\mathrm{s}} I$, where the total amount of power has been determined using the equation $P_{\mathrm{max}} = \frac{\frac{V_{\mathrm{oc}}}{cKT/q} - \ln\left(\frac{V_{\mathrm{oc}}}{cKT/q} + 0.72\right)}{\left(1 + \frac{V_{\mathrm{oc}}}{KT/q}\right)} \left(1 - \frac{V_{\mathrm{oc}}}{\frac{V_{\mathrm{oc}}}{I_{\mathrm{SC}}}}\right) \times \left(\frac{V_{\mathrm{oc0}}}{1 + \beta \ln \frac{G_0}{G}}\right) \left(\frac{T_{\mathrm{o}}}{T}\right)^{y} I_{\mathrm{sc0}} \left(\frac{G}{G_{\mathrm{o}}}\right)^{a}$ (Eq. 4.29) considering the parameter of $P_{\mathrm{array}} = N_{\mathrm{s}} N_{\mathrm{p}} P_{\mathrm{M}}$ (Eq. 4.30). The electricity energy production is therefore accomplished per mole biogas production which is equivalent to 1.4 eV/mol. Since 0.4 kg biowaste can produce 81 mol biogas, the total electricity generation from 0.4 kg biowaste is equivalent to $1.4 \times 81 = 113.4$ eV (cc). Since 1.4 eV is equal to 27.77 kWT, the total electricity energy production would be $(27.77 \text{ kW} \times 81) = 2249.37$ kW eV/day [27, 37, 51]. If a commercial building and an office consumption is roughly 2200 kWh/day for a building with a 20 m × 20 m footprint and height of 20 m, 0.4 kg/day biowaste is sufficient enough to meet the total energy demand for this building which is environmentally friendly.

Conclusion

The advancement of building construction in both urban and sub-urban regions around the globe are quickening tremendously for the past 50 years. Thus, environmental change is expanding quickly because of the traditional utilization of fossil fuel by building sector throughout the world. Likewise, conventional domestic waste and wastewater treatment process are causing/executing the serious ecological contamination, making harm to human well-being, hampering the animal and plant kingdom. Here, the "Green Science," an inventive technology, could be the front-line science to mitigate the complete energy need for a building without using any utility service connection. This innovation "Green Science" could deliver sustainable energy by utilizing building's cellar as the acting bioreactor to create biogas from the household biowaste and then convert it into electricity energy to meet the total indispensable energy need for a building which is environmentally friendly.

Acknowledgments This research was conducted by the support of the grant RD-02018-01 provided by Green Globe Technology to build a better environment. It does not have any financial interest by any means. Any discoveries, conclusions, and recommendations expressed in this paper are exclusively those of the author, who affirm that the article has no conflict of interest for publication in a suitable journal.

References

1. F. Achard et al., Determination of tropical deforestation rates and related carbon losses from 1990 to 2010. Glob. Change Biol. **20**, 2540–2554 (2014)
2. A. Alessandro, P. Friedlingstein, et al., Spatio-temporal patters of terrestrial gross primary production: A review: GPP spatio-temporal patters. Rev. Geophys. (2015). https://doi.org/10.1002/2015RG000483
3. M. Alfredo et al., In focus: Biotechnology and chemical technology for biorefineries and biofuel production. J. Chem. Technol. Biotechnol. **92**, 897–898 (2017)
4. R. Andres, T. Boden, D. Higdon, A new evaluation of the uncertainty associated with CDIAC estimates of fossil fuel carbon dioxide emission. Tellus B **66**, 23616 (2014). https://doi.org/10.3402/tellusb.v66.23616
5. N. Arnell et al., in *Climate Change 1995: Impacts, Adaptations, and Mitigation of Climate Change*, ed. by R. T. Watson et al., (Cambridge University Press, Cambridge, 1996), pp. 325–363
6. B. Ashley et al., Accelerating net terrestrial carbon uptake during the warming hiatus due to reduced respiration. Nat. Clim. Change **7**, 148–152 (2017)
7. A.P. Ballantyne, C.B. Alden, J.B. Miller, P.P. Tans, J.W.C. White, Increase in observed net carbon dioxide uptake by land and oceans during the last 50 years. Nature **488**, 70–72 (2012)
8. J.E. Bauer, W.-J. Cai, P.A.G. Regnier, The changing carbon cycle of the coastal ocean. Nature **504**, 61–70 (2013)
9. R.A. Betts, C.D. Jones, J.R. Knight, R.F. Keeling, J.J. Kennedy, El Nino and a record CO_2 rise. Nat. Clim. Change **6**, 806–810 (2016)
10. T.A. Boden, R.J. Andres, *Global Regional, and National Fossil-Fuel CO_2 Emissions* (Oak Ridge National Laboratory, US Department of Energy, Oak Ridge, TN, 2016). Available at: http://cdiac.ornl.gov/trends/emis/overview_2013.html
11. J.G. Canadell et al., Contributions to accelerating atmospheric CO_2 growth from economic activity, carbon intensity, and efficiency of natural sinks. Proc. Natl. Acad. Sci. USA **104**, 18866–18870 (2007)
12. J. Chen, D. Dimitrov, T. Dimitrova, P. Timans, et al., Carrier density profiling of ultra-shallow junction layer through corrected C-V plotting, in *Extended Abstracts—2008 8th International Workshop on Junction Technology (IWJT'08)*, 2008
13. Y. Chen et al., A pan-tropical cascade of fire driven by El Niño/Southern Oscillation. Nat. Clim. Change **7**, 906–911 (2017)
14. F. Chevallier, On the statistical optimality of CO_2 atmospheric inversions assimilating CO_2 column retrievals. Atmos. Chem. Phys. **15**, 11133–11145 (2015). https://doi.org/10.5194/acp-15-11133-2015
15. P. Ciais et al., Chapter 6: Carbon and other biogeochemical cycles, in *Climate Change 2013. The Physical Science Basis*, ed. by T. Stocker, D. Qin, G.-K. Platner, (Cambridge University Press, Cambridge, 2013)
16. P. Colonna, E. Casati, C. Trapp, T. Mathijssen, J. Larjola, T. Turunen-Saaresti, A. Uusitalo, Organic Rankine cycle power systems: From the concept to current technology, applications and an outlook to the future. J. Eng. Gas Turb. Power **137**, 100801 (2015)
17. J. C. van Dam (ed.), *Impacts of Climate Change and Climate Variability on Hydrological Regimes* (Cambridge University Press, Cambridge, 1999)
18. S.J. Davis, K. Calderia, Consumption-based accounting of CO_2 emissions. Proc. Natl. Acad. Sci. USA **107**, 5687–5692 (2010)
19. E. Dietzenbacher, J.S. Pei, C.H. Yang, Trade, production fragmentation, and China's carbon dioxide emissions. J. Environ. Econ. Manag. **64**, 88–101 (2012)
20. R.A. Duce et al., Impacts of atmospheric anthropogenic nitrogen on the open ocean. Science **320**, 893–897 (2008)
21. J.M. Earles, S. Yeh, K.E. Skog, Timing of carbon emissions from global forest clearance. Nat. Clim. Change **2**, 682–685 (2012)

22. K.-H. Erb et al., Bias in the attribution of forest carbon sinks. Nat. Clim. Change **3**, 854–856 (2013)
23. R.A. Feely, Global nitrogen deposition and carbon sinks. Nat. Geosci. **1**, 430–437 (2008)
24. I. Gelfand, R. Sahajpal, X. Zhang, R. Izaurralde, K. Gross, G. Robertson, Sustainable bioenergy production from marginal lands in the US Midwest. Nature **493**(7433), 514–517 (2013)
25. M. Grätzel, Photoelectrochemical cells. Nature **414**(6861), 338–344 (2001)
26. M.F. Hossain, Production of clean energy from cyanobacterial biochemical products. Strateg. Plan. Energy Environ. **3**, 6–23 (2016a)
27. M.F. Hossain, Solar energy integration into advanced building design for meeting energy demand. Int. J. Energy Res. **40**, 1293–1300 (2016b)
28. M.F. Hossain, Green science: Independent building technology to mitigate energy, environment, and climate change. Renew. Sustain. Energy Rev. **73**, 695–705 (2017)
29. M.F. Hossain, Bose-Einstein (B-E) photonic energy structure reformation for cooling and heating the premises naturally. Adv. Therm. Eng. **142**, 100–109 (2018a). https://doi.org/10.1016/j.applthermaleng.2018.06.057
30. M.F. Hossain, Green science: Decoding dark photon structure to produce clean energy. Energy Rep. **4**, 41–48 (2018b)
31. M.F. Hossain, Global environmental vulnerability and the survival period of all living beings on earth. Int. J. Environ. Sci. Technol. (2018c). https://doi.org/10.1007/s13762-018-1722-y
32. M.F. Hossain, Photonic thermal energy control to naturally cool and heat the building. Appl. Therm. Eng. **131**, 576–586 (2018d)
33. M.F. Hossain, Green science: Advanced building design technology to mitigate energy and environment. Renew. Sustain. Energy Rev. **81**(2), 3051–3060 (2018e)
34. M.F. Hossain, Photon energy amplification for the design of a micro PV panel. Int. J. Energy Res. (2018f). https://doi.org/10.1002/er.4118.(Wiley).2017
35. M.F. Hossain, Breakdown of Bose-Einstein photonic structure to produce sustainable energy. Energy Rep. **5**, 202–209 (2019a)
36. M.F. Hossain, Sustainable technology for energy and environmental benign building design. J. Build. Eng. **22**, 130–139 (2019b)
37. R.A. Houghton, Balancing the global carbon budget. Annu. Rev. Earth Planet. Sci. **35**, 313–347 (2017)
38. N. Izadyar, H. Ong, W. Chong, K. Leong, Resource assessment of the renewable energy potential for a remote area: A review. Renew. Sustain. Energy Rev. **62**, 908–923 (2016)
39. M. Kane, Small hybrid solar power system. Energy **28**(14), 1427–1443 (2003)
40. D.M. Karl, M.J. Church, Microbial oceanography and the Hawaii Ocean time-series programme. Nat. Rev. Microbiol. **12**, 699–713 (2014)
41. G. Kenneth et al., Unravelling the link between global rubber price and tropical deforestation in Cambodia. Nat. Plants **5**, 47–53 (2019)
42. A.J. Landig, J.V. Koski, P. Scarlino, U.C. Mendes, A. Blais, C. Reichl, W. Wegscheider, A. Wallraff, K. Ensslin, T. Ihn, Coherent spin–photon coupling using a resonant exchange qubit. Nature **560**, 179–184 (2018)
43. C. Le Querel, R.M. Andrew, et al., Global carbon budget 2016. Earth Syst. Sci. Data **8**, 605–649 (2016). https://doi.org/10.5194/essd-8-605-2016
44. W. Li, P. Ciais, Y. Wang, S. Peng, et al., Reducing uncertainties in decadal variability of the global car-bon budget with multiple datasets. Proc. Natl. Acad. Sci. USA (in press) (2016). https://doi.org/10.1073/pnas.1603956113
45. Z. Liu, D. Guan, et al., Reduced carbon emission estimates from fossil fuel combustion and cement production in China. Nature **524**, 335–338 (2015a)
46. Y. Liu, J. Xie, C. Ong, C. Vecitis, Z. Zhou, Electrochemical wastewater treatment with carbon nanotube filters coupled with in situ generated H_2O_2. Environ. Sci. Water Res. Technol. **1**(6), 769–778 (2015b)

47. Y. Liu, W. Peng, F. Liu, F. Li, X. An, J. Liu, Z. Wang, C. Shen, W. Sand, Electroactive modified carbon nanotube filter for simultaneous detoxification and sequestration of Sb (III). Environ. Sci. Technol. **53**, 1527–1535 (2019)
48. P.P. Miller et al., Audit of the global carbon budget: Estimate errors and their impact on uptake uncertainty. Biogeosciences **12**, 2565–2584 (2015). https://doi.org/10.5194/bg-12-2565-2015
49. J. Milliman, R. Mei-e, in *Climate Change: Impact on Coastal Habitation*, ed. by D. Eisma, (CRC Press, Boca Raton, FL, 1995), pp. 57–83
50. R. Pierre, R. Lauerwald, P. Ciais, Carbon leakage through the terrestrial-aquatic interface: Implications for the anthropogenic CO_2 budget. Proc. Earth Planet. Sci. **10**, 319–324 (2014)
51. S.L. Postel et al., Science **271**, 785 (1996)
52. J. Prietzel, L. Zimmermann, A. Schubert, D. Christophel, Organic matter losses in German Alps forest soils since the 1970s most likely caused by warming. Nat. Geosci. **9**, 543–548 (2016)
53. J. Romero-García, A. Sanchez, G. Rendón-Acosta, J. Martínez-Patiño, E. Ruiz, G. Magaña, E. Castro, An olive tree pruning biorefinery for co-producing high value-added bioproducts and biofuels: Economic and energy efficiency analysis. Bioenergy Res. **9**(4), 1070–1086 (2016)
54. H. Ruiz, A. Martínez, W. Vermerris, Bioenergy potential, energy crops, and biofuel production in Mexico. Bioenergy Res. **9**(4), 981–984 (2016)
55. S. Schwietzke et al., Upward revision of global fossil fuel methane emissions based on isotope database. Nature **538**, 88–91 (2016)
56. G.R. Van der Werf, J. Dempewolf, et al., Climate regulation of fire emissions and deforestation in equatorial Asia. Proc. Natl. Acad. Sci. USA **15**, 20350–20355 (2008)
57. C.J. Vorosmarty, B. Fekete, M. Meybeck, R. Lammers, Glob. Biogeochem. Cycles **14**, 599 (2000)
58. P. Weiland, Biogas production: Current state and perspectives. Appl. Microbiol. Biotechnol. **85** (4), 849–860 (2009)
59. Y. Yin, P. Ciais, F. Chevallier, et al., Variability of fire carbon emissions in equatorial Asia and its non-linear sensitivity to El, Nino. Geophys. Res. Lett. **43**, 10472–10479 (2016)

Part III
Sustainable Building, Water, and Transportation Technology

Chapter 5
Photonic Thermal Control to Cool and Heat the Housing Naturally

Abstract Photon particle has been modeled to be decoded by the curtain wall of the building to create a natural cooling and heating system for a building by implementing the Bose–Einstein (*B–E*) photon distribution mechanism into helium-assisted building curtain wall where the *photonic bandgap state* photon will be locally induced to cool the building naturally by the cooling state of the photons. This cooling-state photon is denominated as *Hossain Cooling Photon* (*HcP*‾). Once needed, this (*HcP*‾) can be transformed into thermal state photon by implementing the quantum Higgs boson BR ($H \rightarrow \gamma\gamma^{-}$) to create the electromagnetic field (*EmF*) using two-diode thermal semiconductors. Just because Higgs boson BR ($H \rightarrow \gamma\gamma$) quantum field has extremely short range of weak force that initiates Higgs boson BR ($H \rightarrow \gamma\gamma$), the quantum gets excited. Therefore, electrically charged cooling-state photon (*HcP*‾) will be transformed into a thermal stage photon, referred to here as the Hossain Thermal Photon (*HtP*‾). To confirm this *HcP*‾ formation and its transformation into *HtP*‾, a series of mathematical tests has been performed which revealed that formation and transformation of *HcP*‾ and *HtP*‾ are indeed doable to decode the photons into the curtain wall to cool and heat the building naturally.

Keywords Decoding photon particle · Bose–Einstein photon distribution · Helium · Higgs boson BR ($H \rightarrow \gamma\gamma^{-}$) quantum field · Natural cooling and heating technology

Introduction

The conventional heating and cooling of a building is causing serious environmental and atmospheric impacts. Traditional heating technology consumes fossil fuels and releases CO_2, causing climate change, which triggers a deadly natural disaster on the earth and makes its environment vulnerable. On the other hand, conventional cooling technology releases CFCs, creating holes in the ozone layers. The ozone layer lies between 9.3 and 18.6 miles (15 and 30 km) above the Earth's surface and acts as a blanket to block most of the sun's high-frequency ultraviolet rays. Due to the creation of holes in the ozone layer, UV rays are easily penetrating the surface of the earth causing deadly skin cancer to human and causing serious reproductive problems in all mammals [21, 23]. Although there have been numerous studies in the

M. F. Hossain, *Sustainable Design for Global Equilibrium*,
https://doi.org/10.1007/978-3-030-94818-4_5

past on clean energy technology, climate change, and conventional heating and cooling [9, 16, 19, 22], no study has been done on the natural cooling and heating system in the building sectors to avoid greenhouse gases and CFC emissions. Therefore, in this study, I proposed a natural cooling and heating technology by using Bose–Einstein photon distribution mechanism and Higgs boson quantum activation to decode photons (solar energy) in the states of cooling and heating. To decode this photon, cooling photon emission panel consisting of *nano*-point breaks and waveguides has been proposed using helium in a portion of the exterior curtain wall [14, 15]. Therefore, the quantum dynamics of photons will be cooled by the quantum electrodynamics (QED) waveguides using photon band edges (PBEs) by photon emission from the sun to cool the building naturally [2, 24]. Then this cooling-state photon can be transformed into a heating-state photon by employing quantum Higgs boson BR ($H \rightarrow \gamma\gamma^{-}$) to create an electromagnetic field with the help of two-diode semiconductors to naturally heat the building [7, 12, 30]. This cooling and heating transformation process will in indeed be a new field of science to mitigate energy usage and damage to the environment and protect the ozone layer.

Methods and Simulation

Cooling Mechanism

To decode activated photon into cooling-state one, photon emission networks of *nano*-point breaks, waveguides, and helium-assisted curtain wall will create point defects into the photon emission panel [1, 4]. This would provide a mechanism for incorporation of photonic bandgap (PBG) waveguide defect arrays into the curtain wall [10, 31]. Therefore, the quantum dynamics of photons will be decoded by point defects and PBG waveguides under helium cooling conditions. Consequently, the solar state photon will be cooled down within this regime. To calculate this detailed formation of cooling state of photon conversion from sun, I have used MATLAB software to calculate detailed mathematical analysis. Since the *nano*-point break through helium waveguides embedded in a curtain wall, thus I have treated it as waveguides for reservoirs of photons to satisfy purely electron dynamics for cooling states of photon by considering contour maps (Fig. 5.1) which can be expressed by Hamiltonian as [11, 24, 28]

$$H = \sum \omega_{ci} a_i^{\dagger} a_i + \sum_{K} \omega_k b_k^{\dagger} b_k + \sum_{ik} \left(V_{ik} a^{\dagger} b_k + V_{ik}^{*} b_k^{\dagger} a_i \right) \tag{5.1}$$

where $a_i \left(a_i^{\dagger} \right)$ represents the driver of the *nano*-point break mode, $b_k \left(b_k^{\dagger} \right)$ represents the driver of the photodynamic modes of the photon *nano* structure, and the coefficient V_{ik} represents the magnitude of the photonic mode among the *nano*-breakpoints and photon *nano* structure.

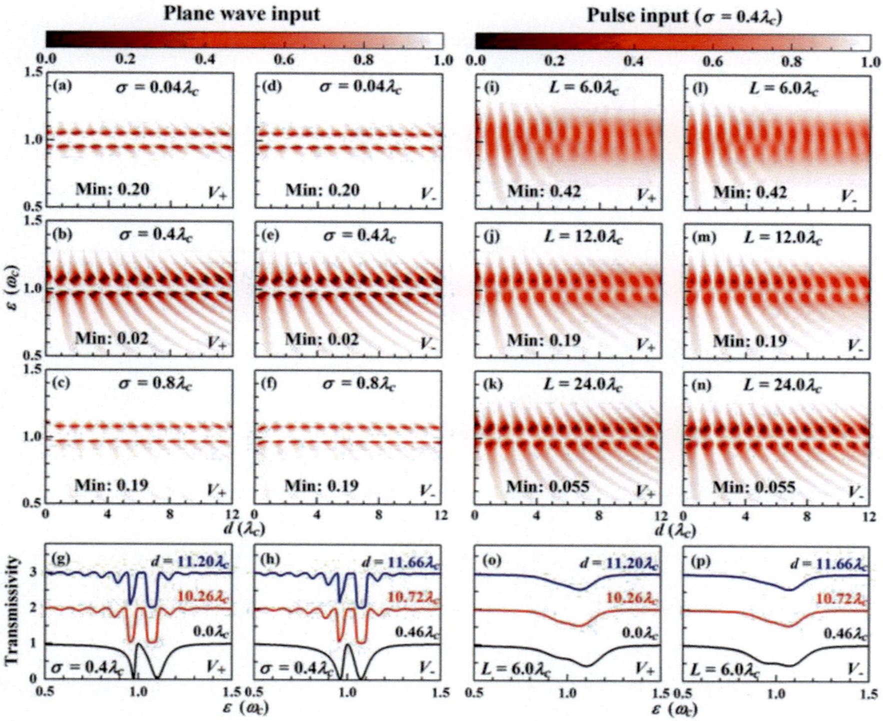

Fig. 5.1 Contour maps of the transmissivity of the single photon (**a–f**) plane wave and (**i–n**) pulse as functions of d and ε. Transmission spectra for the (**g**, **h**) plane wave and (**o**, **p**) pulse for different values of d, offset in steps of 1 denoted in the figures

where $a_i\left(a_i^\dagger\right)$ represents the driver of the nanoscale point break mode, $b_k\left(b_k^\dagger\right)$ represents the driver of the photodynamic modes of the nanoscale structures of the cooling photons, and the coefficient V_{ik} represents the magnitude of the cool photonic mode among the nanoscale break points and photon band structures.

Thus, the induced solar photon will be formed as HcP^{-} where point break photon module is built by helium by utilizing photon energy (current), two diodes, and two resistors to cooling the building (Fig. 5.2).

More detail can be explained by the I–V equation of photon cells for the single-diode mode by expressing as

$$I = I_{\mathrm{L}} - I_{\mathrm{o}}\left\{\exp^{\left[\frac{q(V+I_{\mathrm{RS}})}{AkT_{\mathrm{c}}}\right]} - 1\right\} - \frac{(V + I_{\mathrm{RS}})}{R_{\mathrm{s}}}, \tag{5.2}$$

I_{L} represents the photon generating current, I_{o} represents the saturated current in the diode, R_{s} represents the resistance in the series, A represents the passive function

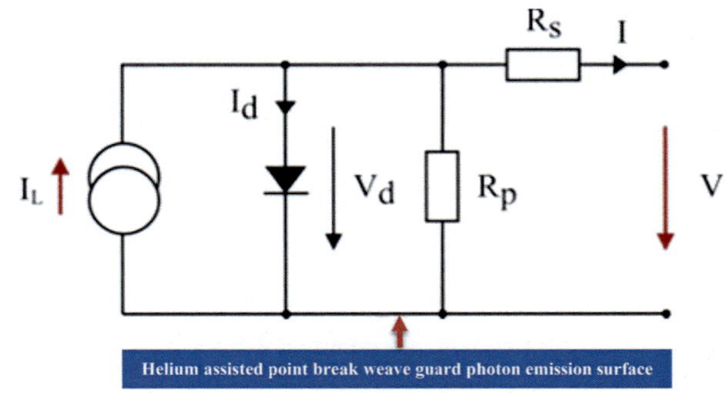

Fig. 5.2 The mechanism of two-diode model of solar irradiance receptor to cooling down electron into the curtain wall skin by photon induction assisted by helium-assisted point break

of the diode, k ($=1.38 \times 10^{-23}$ W/m^2 K) represents Boltzmann's constant, q ($=1.6 \times 10^{-19}$ C) represents the magnitude of the charge of an electron, and T_c represents the functional cell temperature. Subsequently, the I–q relationship in the photon cells varies owing to the diode current and/or saturation current, which can be expressed as [3, 25].

$$I_o = I_{RS}\left(\frac{T_c}{T_{ref}}\right)^3 \exp\left[\frac{qEG\left(\frac{1}{T_{ref}} - \frac{1}{T_c}\right)}{KA}\right] \quad (5.3)$$

where I_{RS} represents the saturation current, considering the functional temperature and solar irradiance speed, and qEG represents the bandgap energy moving into the photon cell respectively per unit area, and thus considering the I–V curbs it can be further explained in Fig. 5.3 as the following.

Considering the photon module, the I–V equation, apart from the I–V curve, is a conjunction of the I–V curves of all the cells in the photon emission panel, and thus the equation can be rewritten as follows to determine the V–R relationship:

$$V = -IR_s + K \log\left[\frac{I_L - I + I_o}{I_o}\right] \quad (5.4)$$

Here, K is a constant $\left(= \frac{AkT}{q}\right)$, and I_{mo} and V_{mo} are the current and voltage in the PV panel. Therefore, the relationship between I_{mo} and V_{mo} is the same as the PV cell I–V relationship:

$$V_{mo} = -I_{mo}R_{Smo} + K_{mo} \log\left(\frac{I_{Lmo} - I_{mo} + I_{Omo}}{I_{Omo}}\right) \quad (5.5)$$

where I_{Lmo} represents the photon-generated current, I_{Omo} represents the saturated current into the diode, R_{Smo} represents the resistance in series, and K_{mo} represents the factorial constant. Once all non-series (NS) cells are connected in series, the series resistance is counted as the summation of the resistance of each cell in the series

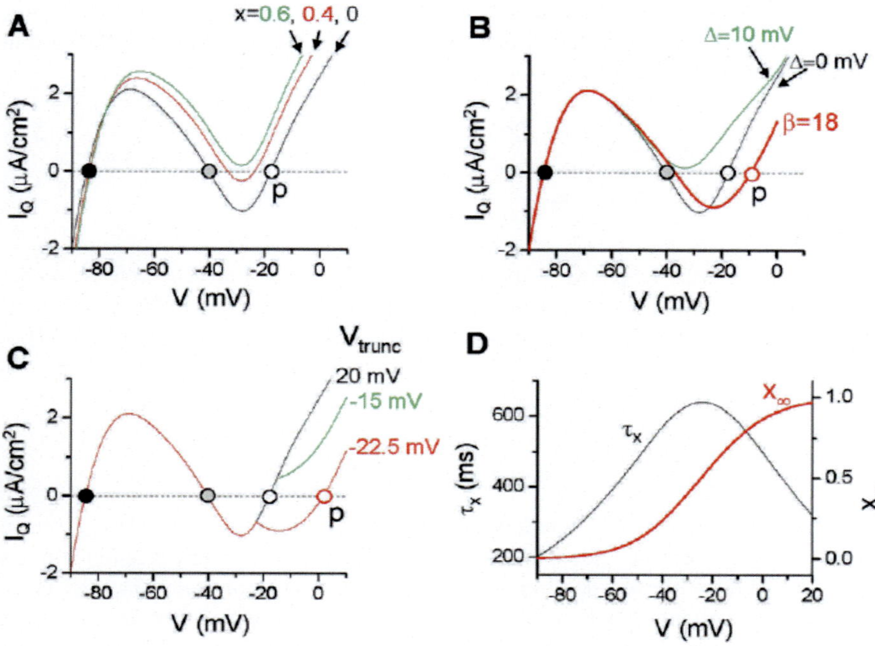

Fig. 5.3 The calculative model of *I–V* characteristics described the cooling mechanism on respectively light speed and unit of area, (**a**) two-diode model for curtain wall skin cell at normal state, (**b**) double-diode model for curtain wall skin cell at normalized condition, (**c**) two-diode model for curtain wall skin cell module normal condition, and (**d**) double-diode model for curtain wall skin cell module at normalized condition

$R_{Smo} = N_s \times R_s$, and the constant factor can be expressed as $K_{mo} = N_s \times K$. There is a certain amount of current flowing into the series of connected cells; thus, the current flow in Eq. (5.4) remains the same for each component, i.e., $I_{Omo} = I_o$ and $I_{Lmo} = I_L$. Thus, the module I_{mo}–V_{mo} equation for the N_s series of connected cells is written as

$$V_{mo} = -I_{mo}N_sR_s + N_sK\log\left(\frac{I_L - I_{mo} + I_o}{I_o}\right) \tag{5.6}$$

Similarly, the current–voltage calculation can be rewritten for the parallel connections once all N_p cells are connected in parallel and can be expressed as follows [18, 26]:

$$V_{mo} = -I_{mo}\frac{R_s}{N_p} + K\log\left(\frac{N_{sh}I_L - I_{mo} + N_pI_o}{N_pI_o}\right) \tag{5.7}$$

Because the photon-generated current will depend primarily on the solar irradiance and relativistic temperature conditions of the photon emission panel, the current can be calculated using the following equation:

$$I_{\mathrm{L}} = G[I_{\mathrm{sc}} + K_{\mathrm{I}}\ (T_{\mathrm{cool}})]V_{\mathrm{mo}} \tag{5.8}$$

$$T_{\mathrm{cool}} = \left(\frac{I_{\mathrm{L}}}{(G * V_{\mathrm{mo}}) \times (I_{\mathrm{sc}} + K_{\mathrm{I}})}\right),$$

where I_{sc} represents the photon current at 25 °C and per unit area, K_I represents the relativistic photon panel coefficient, T_{cool} represents the photon cell's cooling temperature determination, and G represents the solar energy in per unit area [5, 13].

Heating Mechanism

Essentially, a detailed mathematical calculation associated with the Higgs boson electromagnetic field, accuracy, and parameters of a photon-heating relationship has been performed to convert cooling photon into heating-state photons [34, 42]. Thus, to create Higgs quantum field locally (into the curtain wall skin), I have implemented abelian local symmetries using MATLAB software. Therefore, the gauge field symmetry will be broken down due to the penetration of solar irradiance, and the Goldstone scalar will become longitudinal mode of the vector [26, 32]. In the abelian case, for each spontaneously broken particle T^{α} of the local symmetry will be the corresponding gauge field of $A^{\alpha}_{\mu}(x)$ where the Higgs quantum field started to work at a local $U(1)$ phase symmetry [8, 18, 20]. Thus, the model can comprise a complex scalar field $\Phi(x)$ of electric charge q coupled to the EM field $A^{\mu}(x)$ which can be expressed by *Lagrangian* as:

$$\mathcal{L} = \frac{1}{4}\ F_{\mu\nu}F^{\mu\nu} + D_{\mu}\Phi^*\ D^{\mu}\Phi - V(\Phi^*\Phi) \tag{5.9}$$

where

$$\begin{aligned} D_{\mu}\Phi(x) &= \partial_{\mu}\Phi(x) + iqA_{\mu}(x)\Phi(x) \\ D_{\mu}\Phi^*(x) &= \partial_{\mu}\Phi^*(x) - iqA_{\mu}(x)\Phi^*(x) \end{aligned} \tag{5.10}$$

and

$$V(\Phi^*\Phi) = \frac{\lambda}{2}\ (\Phi^*\Phi)^2 + m^2(\Phi^*\Phi) \tag{5.11}$$

Suppose $\lambda > 0$ but $m^2 < 0$, so that $\Phi = 0$ is a local maximum of the scalar potential, the minima form a degenerate circle $\Phi = \frac{v}{\sqrt{2}} * e^{i\theta}$,

$$v = \sqrt{\frac{-2m^2}{\lambda}} \quad \text{for any real } \theta \tag{5.12}$$

Consequently, the scalar field Φ develops a non-zero vacuum expectation value $\langle \Phi \rangle \neq 0$, which spontaneously creates the $U(1)$ symmetry of the magnetic field. The breakdown would lead to a massless Goldstone scalar stemming from the phase of the complex field $\Phi(x)$. But for the local $U(1)$ symmetry, the phase of $\Phi(x)$—not just the phase of the expectation value $\langle \Phi \rangle$ but the x-dependent phase of the dynamical $\Phi(x)$ field.

To confirm this mechanism, I have used polar coordinates in the scalar field space; thus

$$\Phi\ (x) = \frac{1}{\sqrt{2}}\ \Phi_{\mathrm{r}}(x) * e^{i\Theta(x)}, \quad \text{real } \Phi_{\mathrm{r}}(x) > 0, \text{real } \Phi(x) \tag{5.13}$$

This field redefinition is singular when $\Phi(x) = 0$, so I never used it for theories with $\langle \Phi \rangle \neq 0$, but it is adequate for spontaneously broken theories where we expect $\Phi\langle x \rangle \neq 0$ almost everywhere. In terms of the real fields $\phi_{\mathrm{r}}(x)$ and $\Theta(x)$, the scalar potential depends only on the radial field ϕ_{r},

$$V(\phi) = \frac{\lambda}{8}\left(\phi_{\mathrm{r}}^2 - v^2\right)^2 + \text{const} \tag{5.14}$$

or in terms of the radial field shifted by its VEV, $\Phi_{\mathrm{r}}(x) = v + \sigma(x)$,

$$\phi_{\mathrm{r}}^2 - v^2 = (v+\sigma)^2 - v^2 = 2v\sigma + \sigma^2 \tag{5.15}$$

$$V = \frac{\lambda}{8}\ \left(2v\sigma - \sigma^2\right)^2 = \frac{\lambda v^2}{2} * \sigma^2 + \frac{\lambda v}{2} * \sigma^3 + \frac{\lambda}{8} * \sigma^4 \tag{5.16}$$

At the same time, the covariant derivative $D_\mu \phi$ becomes

$$\begin{aligned} D_\mu \phi &= \frac{1}{\sqrt{2}}\ \left(\partial_\mu \left(\phi_{\mathrm{r}} e^{i\Theta}\right) + iqA_\mu * \phi_{\mathrm{r}} e^{i\Theta}\ \right) \\ &= \frac{e^{i\Theta}}{\sqrt{2}}\ \left(\partial_\mu \phi_{\mathrm{r}} + \phi_{\mathrm{r}} * i\partial_\mu \Theta + \phi_{\mathrm{r}} * iqA_\mu\right) \end{aligned} \tag{5.17}$$

$$\begin{aligned} \left|D_\mu \phi\right|^2 &= \frac{1}{2}\ \left|\partial_\mu \phi_{\mathrm{r}} + \phi_{\mathrm{r}} * i\partial_\mu \Theta + \phi_{\mathrm{r}} * iqA_\mu\right|^2 \\ &= \frac{1}{2}\left(\partial_\mu \phi_{\mathrm{r}}\right) + \frac{\phi_{\mathrm{r}}^2}{2} * \left(\partial_\mu \Theta qA_\mu\right)^2 \\ &= \frac{1}{2}\ \left(\partial_\mu\ \sigma\right)^2 + \frac{(v+\sigma)^2}{2} * \left(\partial_\mu \Theta + qA_\mu\right)^2 \end{aligned} \tag{5.18}$$

Altogether,

$$\mathcal{L} = \frac{1}{2}\left(\partial_\mu \sigma\right)^2 - v(\sigma) - \frac{1}{4} F_{\mu\nu}F^{\mu\nu} + \frac{(v+\sigma)^2}{2} * \left(\partial_\mu \Theta + qA_\mu\right)^2 \tag{5.19}$$

To determine the heating ($\mathcal{L}_{\text{heat}}$) into magnetic field properties of this *Lagrangian*, it has been expanded in powers of the fields (and their derivatives) and focus on the quadratic part describing the free particles,

$$\mathcal{L}_{\text{heat}} = \frac{1}{2}\left(\partial_\mu \sigma\right)^2 - \frac{\lambda v^2}{2} * \sigma^2 - \frac{1}{4} F_{\mu\nu}F^{\mu\nu} + \frac{v^2}{2} * \left(qA_\mu + \partial_\mu \Theta\right)^2 \tag{5.20}$$

Here this *Lagrangian* ($\mathcal{L}_{\text{free}}$) function obviously will suggest a real scalar particle of positive mass$^2 = \lambda v^2$ involving the $A_\mu(x)$ and the $\Theta(x)$ fields to initiate to create high heating within the quantum field of the curtain wall (Fig. 5.4).

Results and Discussion

Cooling Mechanism

To mathematically determine the formation of cooling photon by the helium-assisted curtain wall skin, I have initially solved the dynamic photon proliferation by integrating Eqs. (5.15) and (5.16). It is assumed that owing to the cooling unit areal condition $J(\omega)$, the curtain wall skin produces photon proliferation dynamics [18, 24] since it has a persistent weak-coupling limit, and the Weisskopf-Winger approximation rule and/or Markovian master equation which is equal to the photon proliferation production. Consequently, all *HcP* photon proliferation induction will have a dynamic photon state mode (1D, 2D, 3D) in the curtain wall skin, which can be expressed as described in Table 5.1 [23, 35].

Subsequently, a fine frequency cutoff Ω_C is employed to avoid bifurcation of the DOS in a 3D curtain wall skin. Similarly, a sharpened frequency cutoff at Ω_d maintains a positive DOS in 2D and 1D curtain wall skin (Fig. 5.5). Hence $\text{Li}_2(x)$ acts as a dilogarithm variable and erfc(x) acts as an additive variable. Subsequently, the DOS of various curtain wall skin, denoted as $\varrho_{PC}(\omega)$, is determined by calculating photon eigenfrequencies and eigenfunctions of Maxwell's rules considering the *nano* structure [10, 18, 25]. For a 1D curtain wall skin, the corresponding DOS is denoted as $\varrho_{PC}(\omega) \propto \frac{1}{\sqrt{\omega-\omega_e}}\Theta(\omega-\omega_e)$, where $\Theta(\omega-\omega_e)$ represents the Heaviside step function and ω_e represents the frequency in the PBE considering the DOS.

This DOS is therefore calculated to conduct 3D isotropic analysis in curtain wall skin to predict the error-free qualitative state of the non-Weisskopf-Winger mode and the photon-cooling state in the photon cell considering density of states (DOS) and projected density of states (PDOS) (Fig. 5.6). Therefore, for a 3D curtain wall skin, the DOS close to the PBE is implemented by anisotropic DOS: $\varrho_{PC}(\omega) \propto \frac{1}{\sqrt{\omega-\omega_e}}\Theta(\omega-\omega_e)$, which is then clarified with respect to the electromagnetic field

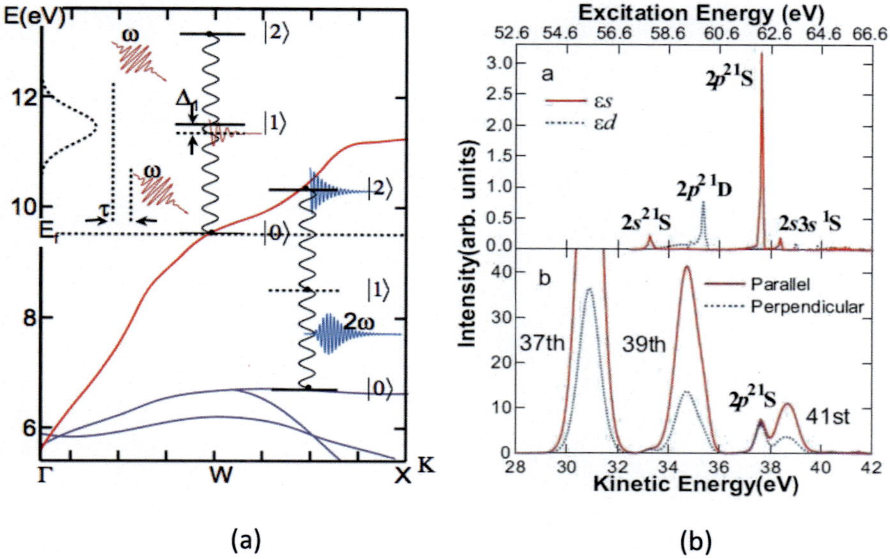

Fig. 5.4 (**a**) Shows the mechanism of photon (2ω) transformation from (ω) at energy level (eV) into the two-diode feed semiconductors. (**b**) Shows the electron spectra through the photon excitation energy (eV) into the quantum field intensity (arb. units) that transform into kinetic energy (eV) by the conversion of heating state of photon

(EMF) vector [8, 18, 21, 24]. For 2D and 1Dcurtain wall skin, the cooling photon DOS exhibits a pure logarithm divergence close to the PBE, which is approximated as $\varrho_{PC}(\omega) \propto -[\ln|(\omega - \omega_0)/\omega_0| - 1]\Theta(\omega - \omega_e)$, where ω_e represents the central point of peak logarithm.

The quantum field area $J(\omega)$ is clarified as the production field of the DOS in the PV cell by the fine cooling photonic magnitude $V(\omega)$ within the PB and PV cell [3, 24, 26],

$$J(\omega) = \varrho(\omega)|V(\omega)|^2 \tag{5.21}$$

Hereafter, I consider the PB frequency ω_c and proliferative photon dynamics using the function $u(t, t_0)$ for photon structure in the relation $\langle a(t)\rangle = u(t, t_0)\langle a(t_0)\rangle$. It is calculated using the dissipative integro-differential equation given in Eq. (5.18) and expressed as

$$u(t, t_0) = \frac{1}{1 - \Sigma'(\omega_b)} e^{-i\omega(t-t_0)} + \int_{\omega_e}^{\infty} d\omega \frac{J(\omega) e^{-i\omega(t-t_0)}}{[\omega - \omega_c - \Delta(\omega)]^2 + \pi^2 J^2(\omega)} \tag{5.22}$$

Table 5.1 The photonic structures in different DOS dimensional modes in the curtain wall skin. They correspond with different unit area $J(\omega)$ and self-energy induction at reservoir $\Sigma(\omega)$, which is determined by the photon dynamics into the extreme relativistic curtain wall skin. The variables C, η, and χ function like coupled forces between the point break and PV of 1D, 2D, and 3D into the curtain wall skin [18, 24]

Photon	Unit area $J(\omega)$ for different DOS	Reservoir-induced self-energy correction $\Sigma(\omega)$
1D	$\frac{C}{\pi}\frac{1}{\sqrt{\omega-\omega_e}}\Theta(\omega-\omega_e)$	$-\frac{C}{\sqrt{\omega_e-\omega}}$
2D	$-\eta\left[\ln\left\lvert\frac{\omega-\omega_0}{\omega_0}\right\rvert-1\right]\Theta(\omega-\omega_e)\Theta(\Omega_d-\omega)$	$\eta\left[\mathrm{Li}_2\left(\frac{\Omega_d-\omega_0}{\omega-\omega_0}\right)-\mathrm{Li}_2\left(\frac{\omega_0-\omega_e}{\omega_0-\omega}\right)-\ln\frac{\omega_0-\omega_e}{\Omega_d-\omega_0}\ln\frac{\omega_e-\omega}{\omega_0-\omega}\right]$
3D	$\chi\sqrt{\frac{\omega-\omega_e}{\Omega_C}}\exp\left(-\frac{\omega-\omega_e}{\Omega_C}\right)\Theta(\omega-\omega_e)$	$\chi\left[\pi\sqrt{\frac{\omega_e-\omega}{\Omega_C}}\exp\left(-\frac{\omega-\omega_e}{\Omega_C}\right)\mathrm{erfc}\sqrt{\frac{\omega_e-\omega}{\Omega_C}}-\sqrt{\pi}\right]$

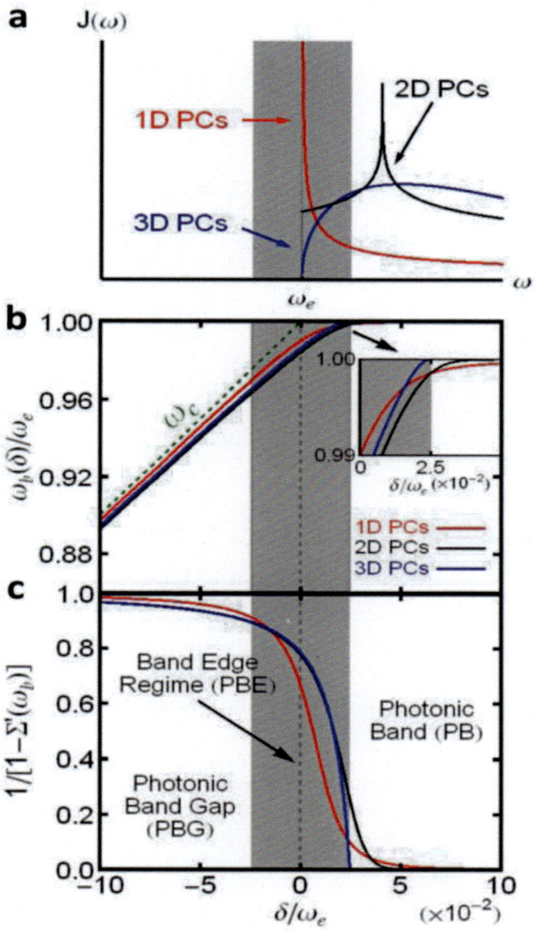

Fig. 5.5 The photonic band structure and mode for energy conversion. (**a**) Unit area at various DOS values for 1D, 2D, and 3D curtain wall skin. (**b**) Photonic mode of frequencies for functional tuning. (**c**) Photonic mode of magnitude to release energy calculated using Eq. (5.2). The photonic modes depict the crossover at 1D and 2D into curtain wall skin, and a complex transitional state at 3D into the PV cell once the point break frequency v_c transforms from a PBG area to a photonic band (PB) area [18]

where $\Sigma'(\omega_b) = [\partial\Sigma(\omega)/\partial\omega]_{\omega=\omega_b}$ and $\Sigma(\omega)$ represents the reservoir-induced PB photon self-energy correction,

$$\Sigma(\omega) = \int_{\omega_e}^{\infty} d\omega' \frac{J(\omega')}{\omega - \omega'} \tag{5.23}$$

Here, the frequency ω_b in Eq. (5.1) represents the cooling photonic frequency mode in the PBG ($0 < \omega_b < \omega_e$) and is calculated using the pole condition: $\omega_b - \omega_c - \Delta(\omega_b) = 0$, where $\lesssim\Delta(\omega) = \mathcal{P}\left[\int d\omega' \frac{J(\omega')}{\omega-\omega'}\right]$ is a principal-value integral.

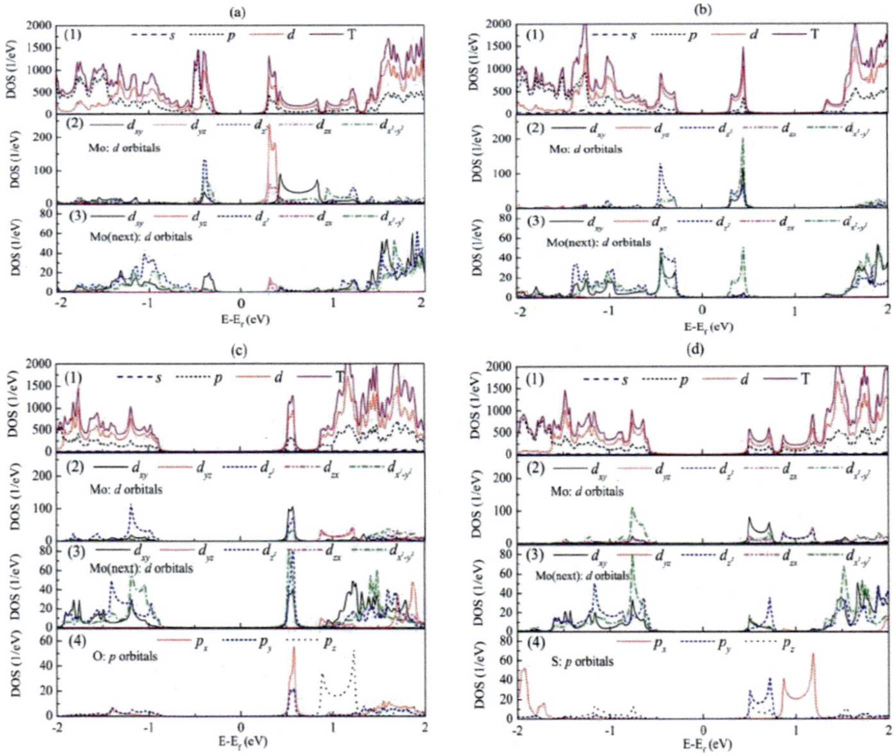

Fig. 5.6 (**a**) Shows the total density of states (DOS) and projected density of states (PDOS) of decoded photon to transform into cooling state: where (1) total DOS and DOS projected on orbitals. Notations are *s* for *s* orbital DOS, *p* for *p* orbital DOS, *d* for *d* orbital DOS, and *T* for total DOS, and (2) PDOS of *d* orbitals on the very edge Mo atoms, and (3) PDOS of *d* orbitals on the next edge Mo atoms. (**b**) The symbols are the same as in (**a**). (**c**) From (1) to (3), the symbols are the same as in (**a**), and (4) is for PDOS of *p* orbitals of O atoms. (**d**) From (1) to (3), the symbols are the same as in (**a**), and (4) is for PDOS of *p* orbitals of external S atoms

Furthermore, the detailed cooling photonic dynamics, considering the proliferation magnitude $|u(t, t_0)|$, have been calculated and are shown in Fig. 5.5a for 1D, 2D, and 3D photon cells with respect to various detuning δ integrated from the PBG area to the PB area [17, 38, 44]. The cooling photonic dynamic rate $\kappa(t)$ is shown in Fig. 5.5b, neglecting the function $\delta = 0.1\omega_e$. The calculated result indicates that dynamic photons are produced at a high rate once ω_c crosses from the PBG to PB area. Because the range in $u(t, t_0)$ is $1 \geq |u(t, t_0)| \geq 0$, I have defined the crossover area as related to the condition $0.9 \gtrsim |u(t \to \infty, t_0)| \geq 0$. This corresponds to $-0.025\omega_e \lesssim \delta \lesssim 0.025\omega_e$, with a cooling photon induction rate $\kappa(t)$ within the PBG ($\delta < -0.025\omega_e$) and near the PBE($-0.025\omega_e \lesssim \delta \lesssim 0.025\omega_e$).

To be more specific, I have considered first the PB as the Fock cooling determination n_0, i. e. , $\rho(t_0) = |n_0\rangle\langle n_0|$, which is obtained theoretically through the real-time quantum feedback control [18, 27, 33] and then by solving Eq. (5.1), considering the state of cooling photon induction at time t:

$$\rho(t) = \sum_{n=0}^{\infty} \mathcal{P}_n^{(n_0)}(t)|n_0\rangle\langle n_0| \tag{5.24}$$

$$\mathcal{P}_n^{(n_0)}(t) = \frac{[v(t,t)]^n}{[1+v(t,t)]^{n+1}}[1-\Omega(\text{t})]^{n_0} \times \sum_{k=0}^{\min\{n_0,n\}} \binom{n_0}{k} \times \binom{n}{k}\left[\frac{1}{v(t,t)}\frac{\Omega(t)}{1-\Omega(t)}\right]^k \tag{5.25}$$

where $\Omega(t) = \frac{|u(t,t_0)|^2}{1+v(t,t)}$. Therefore, the result suggests that a Fock state cooling photon will be induced into dynamic states of $|n_0\rangle$ is $\mathcal{P}_n^{(n_0)}(t)$. The proliferation of photon dissipation $\mathcal{P}_n^{(n_0)}(t)$ in the primary state $|n_0 = 5\rangle$ and steady-state limit $\mathcal{P}_n^{(n_0)}(t \to \infty)$ is thus shown in Fig. 5.7. Therefore, the proliferation of cooling photon production will ultimately reach non-equilibrium cooling state with the photonic structure to cool the building.

Heating Mechanism

Since the Higgs boson quantum field creates electromagnetic field employed by two-diode semiconductors, the local $U(1)$ gauge invariant (QED) did allow to add a mass-term for the gauge particle under $\varnothing' \to e^{i\alpha(x)}\varnothing$ to transform cooling photons into heating photons. In detail, it can be explained by a covariant derivative with a special transformation rule for the scalar field expressing by Refs. [6, 36, 41]:

$$\partial_\mu \to D_\mu = \partial_\mu = ieA_\mu \quad [\text{covariant derivatives}],$$
$$A'_\mu = A_\mu + \frac{1}{e}\,\partial_\mu\alpha \quad [A_\mu \text{ derivatives}], \tag{5.26}$$

where the local $U(1)$ gauge invariant *Lagrangian* for a complex scalar field is given by:

$$\mathcal{L} = (D^\mu)^\dagger\,(D_\mu\varnothing) - \frac{1}{4}\,F_{\mu\nu}F^{\mu\nu} - V(\varnothing) \tag{5.27}$$

The term $\frac{1}{4}\,F_{\mu\nu}F^{\mu\nu}$ is the kinetic term for the gauge field (heating photon) and $V(\varnothing)$ is the extra term in the *Lagrangian* that can be seen as: $V(\varnothing^*\varnothing) = \mu^2(\varnothing^*\varnothing) + \lambda\,(\varnothing^*\varnothing)^2$.

Therefore, the *Lagrangian* ($\mathcal{L}$) under perturbations into the quantum field have initiated with the massive scalar particles ϕ_1 and ϕ_2 along with a mass μ. In this

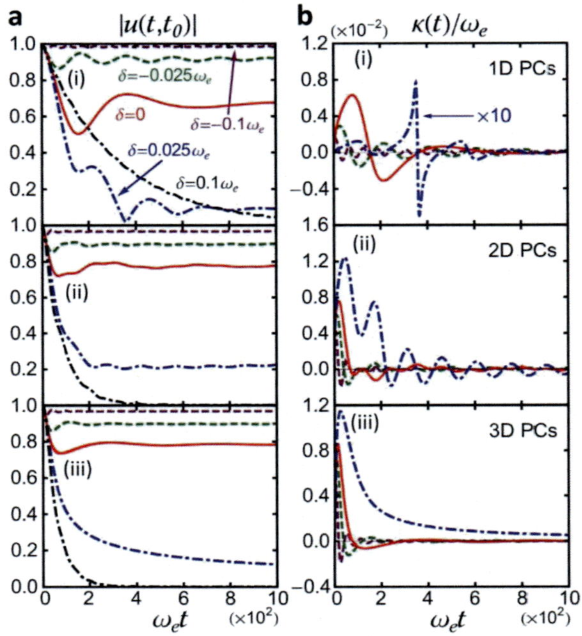

Fig. 5.7 Proliferation of dynamic photons in PV cells. (**a**) Considering the PB area, $\langle a(t)\rangle = 5u(t, t_0)\langle a(t_O)\rangle$, and (**b**) the dynamic photonic rate $k(t)$, plotted for (i) 1D, (ii) 2D, and (iii) 3D PV cells [18]

situation $\mu^2 < 0$ had an infinite number of quantum, each has been satisfied by $\phi_1^2 + \phi_2^2 = -\mu^2/\lambda = v^2$ and the *Lagrangian* through the covariant derivatives using again the shifted fields η and ξ defined the quantum field as $\phi_0 = \frac{1}{\sqrt{2}}\,[(v+\eta) + i\xi]$.

$$\text{Kinetic term}: \quad \mathcal{L}_{\text{kin}}(\eta,\xi) = (D^{\mu}\phi)^{\dagger}(D^{\mu}\phi) \\ = (\partial^{\mu} + ieA^{\mu})\phi^{*}\left(\partial_{\mu} - ieA_{\mu}\right)\phi \tag{5.28}$$

Potential term: $V(\eta, \xi) = \lambda\, v^2\eta^2$, up to second order in the fields, and thus the full Lagrangian can be written as:

$$\mathcal{L}_{\text{kin}}(\eta,\xi) = \frac{1}{2}\left(\partial_{\mu}\eta\right)^2 - \lambda v^2\eta^2 + \frac{1}{2}\left(\partial_{\mu}\xi\right)^2 - \frac{1}{4}F_{\mu\nu}F^{\mu\nu} + \frac{1}{2}e^2v^2A_{\mu}^2 \\ - evA_{\mu}(\partial^{\mu}\xi) + \text{int.terms} \tag{5.29}$$

Here massive η, massless ξ (as before) and also a mass term for the quantum and A_{μ} is fixed up to a term $\partial_{\mu}\alpha$ as can be seen from Eq. (5.27). In general, A_{μ} and ϕ change simultaneously and thus it can be redefined to accommodate the heating photon particle spectrum within the quantum field by expressing:

$$\mathcal{L}_{\text{scalar}} = (D^{\mu}\phi)^{\dagger}(D^{\mu}\phi) - V(\phi^{\dagger}\phi)$$

$$= (\partial^{\mu} + ieA^{\mu})\frac{1}{\sqrt{2}}(v+h)\left(\partial_{\mu} - ieA_{\mu}\right)\frac{1}{\sqrt{2}}(v+h) - V\left(\phi^{\dagger}\phi\right) \tag{5.30}$$

$$= \frac{1}{2}\left(\partial_{\mu}h\right)^{2} + \frac{1}{2}e^{2}A_{\mu}^{2}(v+h)^{2} - \lambda v^{2}h^{2} - \lambda v h^{3} - \frac{1}{4}\lambda h^{4} + \frac{1}{4}\lambda h^{4} \tag{5.31}$$

Thus, this expanding term in the *Lagrangian* associated to the scalar field is suggesting that Higgs boson quantum field is prepared to initiate heating photon into its quantum field.

To confirm this heating photon transformation, the isotropic distribution of movement on the differential cone has to be calculated considering the angle θ and within θ and $\theta + d\theta$ is $\frac{1}{2}\sin\theta d\theta$. Then, the differential photon density at energy $\in$ and angle θ is

$$dn = \frac{1}{2}n(\in)\sin\theta d \in d\theta \tag{5.32}$$

Consequently, the functional speeds of the high-energy photons were calculated considering the directional form of $c\ (1 - \cos\theta)$, where the absorption per unit path length would be

$$\frac{d\tau_{\text{abs}}}{dx} = \int\int \frac{1}{2}\sigma n\ (\in)(1 - \cos\theta)\sin\theta d \in d\theta. \tag{5.33}$$

Modifying the functions to an integration over s instead of θ by (5.2) and (5.4), I have calculated

$$\frac{d\tau_{\text{abs}}}{dx} = \pi r_0^2\left(\frac{m^2c^4}{E}\right)^2 \int\limits_{\frac{m^2c^4}{E}}^{\infty} \in^{-2} n(\in)\,\overline{\varphi}[s_0(\in)]de, \tag{5.34}$$

where

$$\overline{\varphi}[s_0(\in)] = \int\limits_{1}^{s_0(\in)} s\overline{\sigma}\ (s)ds, \quad \overline{\sigma}(s) = \frac{2\sigma(s)}{\pi r_0^2}. \tag{5.35}$$

This result has been identified as the dimensional variable $\overline{\varphi}$ and dimensionless cross section $\overline{\sigma}$. The variable $\overline{\varphi}[s_0]$ is calculated based on a detailed graphical frame for $1 < s_0 < 10$. I gave a reliable functional asymptotic calculation for $\overline{\varphi}$ where $s_0 - 1 \ll 1$ and $s_0 \gg 1$, which I have expressed as

$$\overline{\varphi}[s_0] = \frac{1+\beta_0^2}{1-\beta_0^2}\ln\omega_0 - \beta_0^2\ln\omega_0 - \ln^2\omega_0 - \frac{4\beta_0}{1-\beta_0^2} + 2\beta_0$$
$$+ 4\ln\omega_0\ln(\omega_0+1) - L(\omega_0), \tag{5.36}$$

where

$$\beta_0^2 = \frac{1-1}{s_0}, \quad \omega_0 = \frac{(1+\beta_0)}{(1-\beta_0)}, \tag{5.37}$$

$$L(\omega_0) = \int_1^{\omega_0} \omega^{-1}\ln(\omega+1)d\omega.$$

The last integral can be written as

$$(\omega+1) = \omega\left(\frac{1+1}{\omega}\right), \quad L(\omega_0) = \frac{1}{2}\ln^2\omega_0 + L'(\omega_0),$$

where

$$L'(\omega_0) = \int_1^{\omega_0} \omega^{-1}\ln\left(1+\frac{1}{\omega}\right)d\omega,$$
$$= \frac{\pi^2}{12} - \sum_{n=1}^{\infty}(-1)n^{-1}n^{-2}\omega_0^{-n}. \tag{5.38}$$

The accurate representation heating photon here readily allows the calculation of $\overline{\varphi}[s_0]$ to any needed accuracy for the expected value of s_0. Thus, the corrective functional asymptotic formulas are being used as follows:

$$\overline{\varphi}[s_0] = 2s_0(\ln 4s_0 - 2) + \ln 4s_0(\ln 4s_0 - 2) - \frac{(\pi^2-9)}{3}$$
$$+ s_0^{-1}\left(\ln 4s_0 + \frac{9}{8}\right) + \cdots \quad (s_0 \gg 1), \tag{5.39}$$

$$\overline{\varphi}[s_0] = \left(\frac{2}{3}\right)(s_0-1)^{\frac{3}{2}} + \left(\frac{5}{3}\right)(s_0-1)^{\frac{5}{2}} - \left(\frac{1507}{420}\right)(s_0-1)^{\frac{7}{2}}$$
$$+ \cdots \quad (s_0 - 1 \ll 1). \tag{5.40}$$

The function $\frac{\overline{\varphi}[s_0]}{(s_0-1)}$ is shown in Fig. 5.5 for $1 < s_0 < 10$; for larger s_0, it contains natural logarithmic dependence on s_0. The heating photon spectrum absorption by the power law has been calculated with respect to the form $n(\in) \propto \in_m$ considering two system is in a pristine, (b) system with BN in sheet of graphene.

Thus, I have considered that the function of light absorption for the spectrum will follow the features of a high-energy cutoff with $m > 0$.

Here, the heating photon spectra with a high-energy cutoff consider now a spectrum of the form

$$n(\in) = D\in^{\beta}, \in < \in_m, \quad \beta \leq 0 \tag{5.41}$$

$$= 0, \in > \in_m \tag{5.42}$$

For this spectrum, I have found

$$\frac{d\tau_{\text{abs}}}{dx} = \pi r_0^2 D \left(\frac{m^2c^4}{E}\right)^{1+\beta} \times \begin{cases} 0, & E < E_m, \\ F_\beta(\sigma_m), & E > E_m, \end{cases} \tag{5.43}$$

where

$$\sigma_m = \frac{E}{E_m} = \frac{\in_m E}{m^2c^4}, \tag{5.44}$$

$$F_\beta(\sigma_m) = \int_1^{\sigma_m} s_0^{\beta-2}\overline{\varphi}[s_0]ds_0. \tag{5.45}$$

Again, by Eqs. (5.40) and (5.41), we can obtain the asymptotic forms

$$\begin{aligned} \beta = 0: \quad & F_\beta(\sigma_m) \to A_\beta + \ln^2\sigma_m - 4\ln\sigma_m + \cdots, \\ \beta \neq 0: \quad & F_\beta(\sigma_m) \to A_\beta + 2\beta^{-1}\sigma_m^{\beta}\left(\ln 4\sigma_m - \beta^{-1} - 2\right) + \cdots, \quad \sigma_m > 10 \end{aligned} \tag{5.46}$$

$$\begin{aligned} \text{All } \beta: \quad & F_\beta(\sigma_m) \\ & \to \left(\frac{4}{15}\right)(\sigma_m - 1)^{\frac{5}{2}} + \left[\frac{2(2\beta+1)}{21}\right](\sigma_m - 1)^{\frac{7}{2}} + \cdots, \quad \sigma_m - 1 \ll 1. \end{aligned} \tag{5.47}$$

$\sigma_m^{-\beta}F_\beta(\sigma_m)$ is shown in Fig. 5.6 for $\beta = 0, 0.5, 1.0, 1.5, 2.0, 2.5$, and $3.0\ A_\beta$, which contribute to the integral of the region [1, 14]. The values here have been calculated as $A_\beta = 8.111, 13.53, 9.489, 15.675, 34.54, 85.29$, and 222.9 for $\beta = 0, 0.5, 1.0, 1.5, 2.0, 2.5$, and 3.0, respectively. Subsequently, I have considered heating photon terms corresponding to the spectra for both a negative and positive index:

$$\begin{aligned} n(\in) &= 0; \quad \in < \in_0 \\ &= C_{\in}^{-\alpha} \text{ or } D_{\in}^{\beta}, \ \in_0 < \in < \in_m \\ &= 0, \quad \in > \in_m. \end{aligned} \tag{5.48}$$

Then, I have calculated

$$\left(\frac{d\tau_{\rm abs}}{dx}\right)_\alpha = \pi r_0^2 C\left(\frac{m^2c^4}{E}\right)^{1-\alpha} \times \begin{cases} 0, & E < E_m, \\ [F_\alpha(1) - F_\alpha(\sigma_m)], & E_m < E < E_0, \\ [F_\alpha(\sigma_0) - F_\alpha(\sigma_m)], & E > E_0, \end{cases} \tag{5.49}$$

$$\left(\frac{d\tau_{\rm abs}}{dx}\right)_\beta = \pi r_0^2 D\left(\frac{m^2c^4}{E}\right)^{1+\beta} \times \begin{cases} 0, & E < E_m, \\ [F_\beta(\sigma_m)], & E_m < E < E_0, \\ [F_\beta(\sigma_m) - F_\beta(\sigma_0)], & E > E_0, \end{cases} \tag{5.50}$$

Asymptotic formulas for this case are quite valid for analysis of the heating photon spectrum, where I can further clarify $\Gamma_\gamma^{\rm LPM}$ as denoting the photon's contribution to emitting irradiance with a rate per unit volume according to bremsstrahlung processes [37, 39].

$$\Gamma_\gamma \equiv \frac{dn_\gamma}{dVdt} \tag{5.51}$$

The contribution $\Gamma_\gamma^{\rm LPM}$ was summed in order to confirm the rate of $O(\alpha_{\rm EM}\alpha_{\rm s})$. Here, I have assumed that the result of polarized emission rate $\Gamma_\gamma^{\rm LPM}$ at the thermodynamically controlled equilibrium of the plasma surface at temperature T and the photophysical reaction is μ; thus, the equation is expressed as

$$\frac{d\Gamma_\gamma^{\rm LPM}}{d^3k} = \frac{d_{\rm F} q_{\rm s}^2 \alpha_{\rm EM}}{4\pi^2 k} \int_{-\infty}^{\infty} \frac{dp_\parallel}{2\pi} \int \frac{d^2 p_\perp}{(2\pi)^2} A\left(p_\parallel, k\right) \ {\rm Re}\left\{2p_\perp \cdot f\left(p_\perp; p_\parallel, k\right)\right\}, \tag{5.52}$$

where $d_{\rm F}$ represents the functional strategy of a quark particle [$N_{\rm c}$ in SU($N_{\rm c}$)], $q_{\rm s}$ represents the abelian charge of a quark, $k \equiv |k|$, and the kinetic function $A(p_\parallel, k)$ of the emitted particle

$$A\left(p_\parallel, k\right) \equiv \begin{cases} \dfrac{n_{\rm b}\left(k+p_\parallel\right)\left[1+n_{\rm b}\left(p_\parallel\right)\right]}{2p_\parallel\left(p_\parallel+k\right)}, & \text{scalars} \\ \dfrac{n_{\rm f}\left(k+p_\parallel\right)\left[1-n_{\rm f}\left(p_\parallel\right)\right]}{2\left[p_\parallel\left(p_\parallel+k\right)\right]^2}\left[{p_\parallel}^2+\left(p_\parallel+k\right)^2\right], & \text{fermions} \end{cases} \tag{5.53}$$

with

$$n_{\mathrm{b}}(p) \equiv \frac{1}{\exp[\beta(p-\mu)]-1}, \quad n_{\mathrm{f}}(p) \equiv \frac{1}{\exp[\beta(p-\mu)]+1} \tag{5.54}$$

where the calculation $f(p_\perp; p_\|, k)$ has been integrated into Eq. (5.52) to solve the following linear integral equation to confirm that three-diode scattering occurred for multiple photon production [29, 40, 43]:

$$\begin{aligned} 2p_\perp = i\delta E f\left(p_\perp; p_\|, k\right) &+ \frac{\pi}{2} C_{\mathrm{F}} g_{\mathrm{s}}^2 m_{\mathrm{D}}^2 \int \frac{d^2 q_\perp}{(2\pi)^2} \frac{dq_\|}{2\pi} \frac{dq^0}{2\pi} 2\pi\delta\left(q^0 - q_\|\right) \\ &\times \frac{T}{|q|} \left[\frac{2}{\left|q^2 - \Pi_{\mathrm{L}}(Q)\right|^2} + \frac{\left[1 - \left(q^0/\left|q_\|\right|\right)^2\right]^2}{\left|(\mathrm{q}^0)^2 - q^2 - \Pi_{\mathrm{T}}(Q)\right|^2} \right] \\ &\times \left[f\left(p_\perp; p_\|, k\right) - f\left(q + p_\perp; p_\|, k\right) \right] \end{aligned} \tag{5.55}$$

C_{F} is a quadratic quark [$C_{\mathrm{F}} = (N_{\mathrm{c}}^2 - 1)/2N_{\mathrm{c}} = 4/3$ in QCD], m_{D} is the leading-order Debye mass, and δE is the difference in quasi-particle energies, which has also been calculated considering the photon emission and the state of thermodynamic temperature equilibrium,

$$\delta E \equiv k^0 + E_p \mathrm{sign}\left(p_\|\right) - E_{p+k} \mathrm{sign}\left(p_\| + k\right) \tag{5.56}$$

For an SU(N) gauge theory with N_{s} complex scalars and N_{f} Dirac fermions in the fundamental representation, the Debye mass is given by [38]

$$m_{\mathrm{D}}^2 = \frac{1}{6}(2N + N_{\mathrm{s}} + N_{\mathrm{f}}) g^2 T^2 + \frac{N_{\mathrm{f}}}{2\pi^2} g^2 \mu^2 \tag{5.57}$$

At the last integration (5.52), to confirm the photon energy emission rate accurately in the region $p_\| > 0$, I have calculated the distribution of $n(k + p_\|)[1 \pm n(p_\|)]$ that contains $A(p_\|, k)$, which confirms the production of pair annihilation by using the following equation:

$$n_{\mathrm{b}}(-p) = -[1 + \bar{n}_{\mathrm{b}}(p)], \quad n_{\mathrm{f}}(-p) = [1 - \bar{n}_{\mathrm{f}}(p)], \tag{5.58}$$

with $n(p) \equiv 1/[e\beta(p + \mu) \mp 1]$ as the appropriate anti-particle distribution function, the factor $A(p'', k)$ in this interval may be rewritten in the form

$$A\left(p_{\|}, k\right) \equiv \begin{cases} \dfrac{n_{\mathrm{b}}\left(k-\left|p_{\|}\right|\right)\overline{n}_{\mathrm{b}}\left(\left|p_{\|}\right|\right)}{2\left|p_{\|}\right|\left(k-\left|p_{\|}\right|\right)}, & \text{scalars,} \\ \dfrac{n_{\mathrm{f}}\left(k-\left|p_{\|}\right|\right)\overline{n}_{\mathrm{f}}\left(\left|p_{\|}\right|\right)}{2\left[\left|p_{\|}\right|\left(k-\left|p_{\|}\right|\right)\right]^{2}}\left[p_{\|}{}^{2}+\left(k-\left|p_{\|}\right|\right)^{2}\right], & \text{fermions.} \end{cases} \tag{5.59}$$

Thus, the explicit form of the energy E_p of a hard quark with momentum $|p|$ is given by

$$E_p = \sqrt{p^2 + m_\infty^2} \simeq |p| + \frac{m_\infty^2}{2|p|} \simeq \left|p_{\|}\right| + \frac{p_\perp^2 + m_\infty^2}{2\left|p_{\|}\right|} \tag{5.60}$$

where the asymptotic thermal "mass" is

$$m_\infty^2 = \frac{C_{\mathrm{f}} g^2 T^2}{4} \tag{5.61}$$

Substituting the explicit form of E_p into the definition (5.61) gives the following equation:

$$\delta E = \left[\frac{p_\perp^2 + m_\infty^2}{2}\right]\left[\frac{k}{p_{\|}\left(k + p_{\|}\right)}\right] \tag{5.62}$$

Thus, the results here present a derivation of Eqs. (5.52) and (5.55) and are shown in Fig. 5.8 considering the leading-order heating photoemission rate with respect to power counting analysis and electron time of flight for the photon into the curtain wall skin (Fig. 5.9).

Conclusions

The traditional cooling and heating system for the building sectors is certainly problematic, as it is not only causing climate change but also destroying our ozone layer. To avoid these two deadly effects, a series of mathematical tests have been performed using MATLAB software to transform solar irradiation into the cooling state of photons by implementing the Bose–Einstein photon distribution mechanism on helium-assisted curtain walls to cool the building. In addition, the Higgs boson [BR ($H \rightarrow \gamma\gamma^{-}$] quantum field has been created by two thermal semiconductor diodes through the helium-assisted curtain wall to transform the cooling photon into

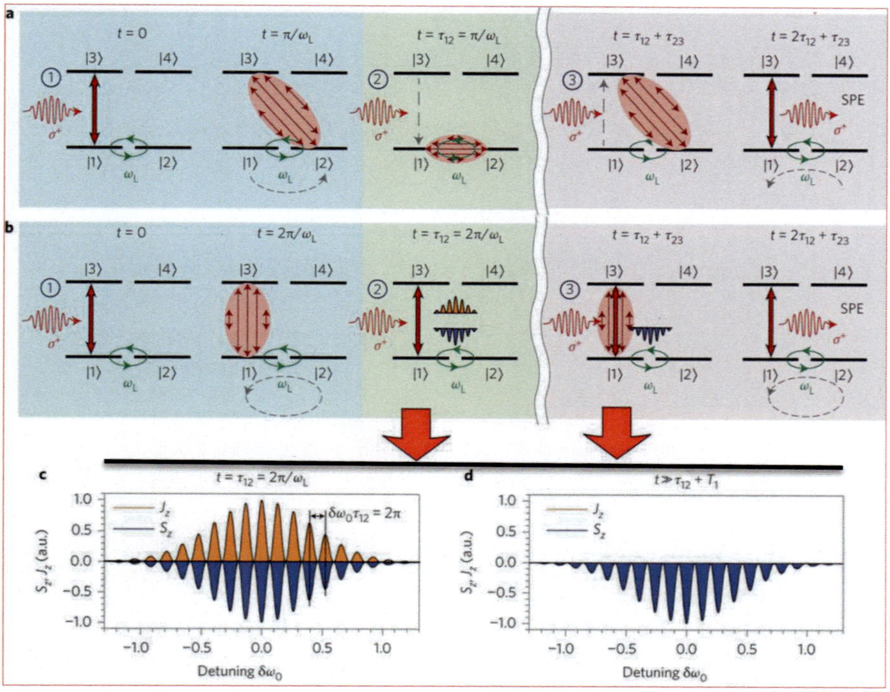

Fig. 5.8 Schematic presentation of the main mechanisms responsible for magnetic-field-induced photon. (**a**) Shows the heating of photons are simultaneously coupled into the fundamental mode and higher-order mode of the quantum field. (**b**) Coincidence rate of the fundamental mode output. (**c**) Coincidence rate of the higher-order mode output respectively detuning. (**d**) Classical coincidence rates are shown as a function of heating photons with respect to detuning [24]

heating-state photon to naturally heat the building. All mathematical calculation suggested that formation of cooling photon from sunlight and transformation heating photon from cooling photon is quite feasible into the building exterior curtain wall skin to naturally cool and heat the building. Simply, it can be said that naturally the cooling and heating process is but a New field of science to decode the Photon Thermodynamics to modify the solar irradiance into cooling-state photon (*HcP*⁻) and transform the cooling-state (*HcP*⁻) photon into the heating-state photon (*HtP*⁻) to mitigate the global energy, environmental, and ozone layer crisis.

Acknowledgments This research was supported by Green Globe Technology under grant RD-02017-05 for building a better environment. Any findings, predictions, and conclusions described in this article are solely performed by the authors and we confirm that there is no conflict of interest for publishing in a suitable journal.

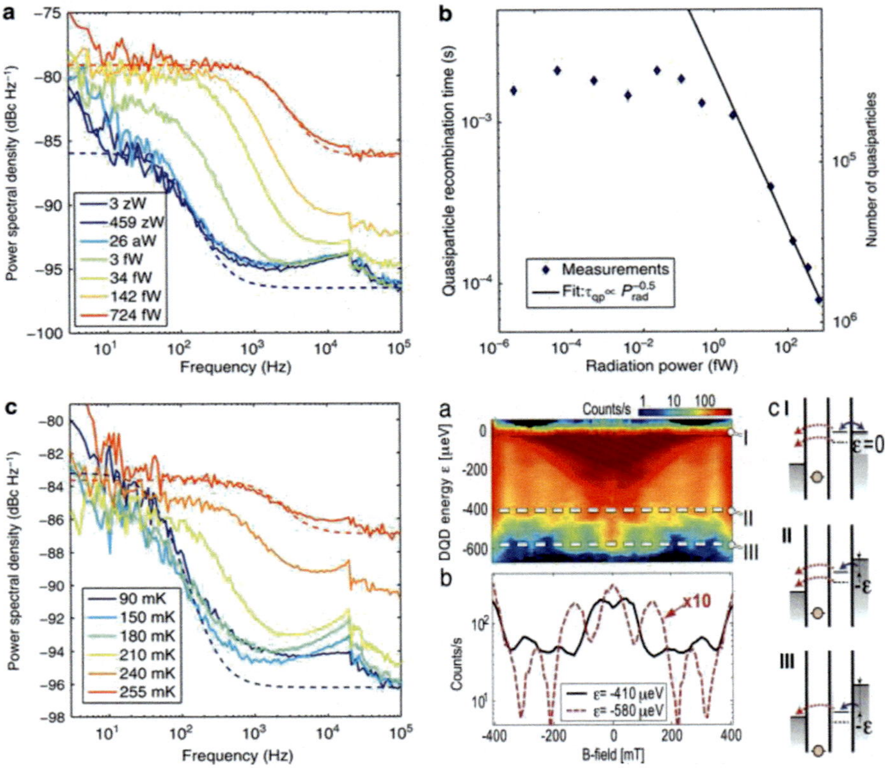

Fig. 5.9 Shows the (**a**) thermal photon power spectral density of the resonator amplitude as a function of frequencies. (**b**) The quasiparticle recombination time as a function of thermal photon power obtained from the roll-off frequency in the measured spectra. (**c**) The thermal photon power spectral density of the resonator amplitude as a function of frequency for different bath temperatures. (**d**) The average heating photon energy measurement (DQD) of *a*, *b*, *c* respectively counting photon excitation into the Higgs boson quantum *B*-field

References

1. A.K. Agger, A.H. Sørensen, Atomic and molecular structure and dynamics. Phys. Rev. A **55**, 402 (1997)
2. P. Arnold, Photon emission from ultrarelativistic plasmas. J. High Energy Phys. (2001). https://doi.org/10.1088/1126-6708/2001/11/057
3. N. Artemyev, U.D. Jentschura, V.G. Serbo, A. Surzhykov, Strong electromagnetic field effects in ultra-relativistic heavy-ion collisions. Eur. Phys. J. C **72**, 1935 (2012)
4. G. Baur, K. Hencken, D. Trautmann, S. Sadovsky, Y. Kharlov, Dense laser-driven electron sheets as relativistic mirrors for coherent production of brilliant X-ray and γ-ray beams. Phys. Rep. **364**, 359 (2002)
5. G. Baur, K. Hencken, D. Trautmann, Revisiting unitarity corrections for electromagnetic processes in collisions of relativistic nuclei. Phys. Rep. **453**, 1 (2007)

6. U. Becker, N. Grün, W. Scheid., K-shell ionisation in relativistic heavy-ion collisions. J. Phys. B: At. Mol. Phys. **20**, 2075 (1987)
7. H. Belkacem, B. Gould, R. Feinberg, W.E. Bossingham, Meyerhof., Semiclassical dynamics and relaxation. Phys. Rev. Lett. **71**, 1514 (1993)
8. N.D. Benavides, P.L. Chapman, Modeling the effect of voltage ripple on the power output of photovoltaic modules. IEEE Trans. Ind. Electron. **55**(7), 2638–2643 (2008)
9. B. Boukhezzar, H. Siguerdidjane, Nonlinear control with wind estimation of a DFIG variable speed wind turbine for power capture optimization. Energy Conver. Manage. **50**(4), 885–892 (2009)
10. A.N. Celik, N. Acikgoz, Modelling and experimental verification of the operating current of mono-crystalline photovoltaic modules using four- and five-parameter models. Appl. Energy **84** (1), 1–15 (2007)
11. W. De Soto, S.A. Klein, W.A. Beckman, Improvement and validation of a model for photovoltaic array performance. Sol. Energy **80**(1), 78–88 (2006)
12. J.S. Douglas, H. Habibian, C.-L. Hung, A.V. Gorshkov, H.J. Kimble, D.E. Chang, Quantum many-body models with cold atoms coupled to photonic crystals. Nat. Photon. **9**, 326–331 (2015)
13. J. Eichler, T. Stöhlker, Radiative electron capture in relativistic ion-atom collisions and the photoelectric effect in hydrogen-like high-Z systems. Phys. Rep. **439**, 1 (2007)
14. H. Faida, J. Saadi, Modelling, control strategy of DFIG in a wind energy system and feasibility study of a wind farm in Morocco. Int. Rev. Model. Simul. **3**(6), 1350–1362 (2010)
15. T. Ghennam, E.M. Berkouk, B. Francois, A vector hysteresis current control applied on three-level inverter. Application to the active and reactive power control of doubly fed induction generator based wind turbine. Int. Rev. Electr. Eng. **2**(2), 250–259 (2007)
16. C. Gopal, M. Mohanraj, P. Chandramohan, P. Chandrasekar, Renewable energy source water pumping systems—a literature review. Renew. Sustain. Energy Rev. **25**, 351–370 (2013). https://doi.org/10.1016/j.rser.2013.04.012
17. R.J. Gould, Pair production in photon-photon collisions. Phys. Rev. **155**, 1404 (1967)
18. M.C. Güclüa, J. Lib, A.S. Umarb, D.J. Ernstb, M.R. Strayer, Electromagnetic lepton pair production in relativistic heavy-ion collisions. Ann. Phys. **272**, 7 (1999)
19. N. Gupta, S.P. Singh, S.P. Dubey, D.K. Palwalia, Fuzzy logic controlled three-phase three-wired shunt active power filter for power quality improvement. Int. Rev. Electr. Eng. **6**(3), 1118–1129 (2011)
20. K. Hencken, Transverse momentum distribution of vector mesons produced in ultraperipheral relativistic heavy ion collisions. Phys. Rev. Lett. **96**, 012303 (2006)
21. F. Hossain, Solar energy integration into advanced building design for meeting energy demand and environment problem. J. Energy Res. ***17***, *49–55* (2016)
22. E. Kamal, M. Koutb, A.A. Sobaih, B. Abozalam, An intelligent maximum power extraction algorithm for hybrid wind-diesel-storage system. Int. J. Electr. Power Energy Syst. ***32***(*3*), 170–177 (2010)
23. S.A. Klein, Calculation of flat-plate collector loss coefficients. Sol. Energy **17**, 79–80 (1975)
24. L. Langer, S.V. Poltavtsev, I.A. Yugova, M. Salewski, D.R. Yakovlev, G. Karczewski, T. Wojtowicz, I.A. Akimov, M. Bayer, Access to long-term optical memories using photon echoes retrieved from semiconductor spins. Nat. Photon. **8**, 851–857 (2014)
25. Q. Li, D.Z. Xu, C.Y. Cai, C.P. Sun, Recoil effects of a motional scatterer on single-photon scattering in one dimension. Sci. Rep. **3**, 3144 (2013)
26. B. Najjari, A.B. Voitkiv, A. Artemyev, A. Surzhykov, Simultaneous electron capture and bound-free pair production in relativistic collisions of heavy nuclei with atoms. Phys. Rev. A **80**, 012701 (2009)
27. S. Pamuji, F. Agung, M. Ashari, Grid quality hybrid power system control of microhydro, wind turbine and fuel cell using fuzzy logic. Int. Rev. Model. Simul. **6**, 1271–1278 (2013)
28. J. Park, H.-g. Kim, Y. Cho, C. Shin, Simple modeling and simulation of photovoltaic panels using Matlab/Simulink. Adv. Sci. Technol. Lett. **73**, 147–155 (2014)

29. T. Pregnolato, E.H. Lee, J.D. Song, S. Stobbe, P. Lodahl, Single-photon non-linear optics with a quantum dot in a waveguide. Nat. Commun. **6**, 8655 (2015)
30. A. Reinhard, Strongly correlated photons on a chip. Nat. Photon. **6**, 93–96 (2011)
31. B. Robyns, B. Francois, P. Degobert, J.P. Hautier, *Vector Control of Induction Machines* (Springer-Verlag, London, 2012)
32. K.G. Sharma, A. Bhargava, K. Gajrani, Stability analysis of DFIG based wind turbines connected to electric grid. Int. Rev. Model. Simul. **6**, 879–887 (2013)
33. G. Sivasankar, V.S. Kumar, Improving low voltage ride through of wind generators using STATCOM under symmetric and asymmetric fault conditions. Int. Rev. Model. Simul. **6**, 1212–1218 (2013)
34. J.J. Soon, K.-S. Low, Optimizing photovoltaic model parameters for simulation, in *2012 IEEE International Symposium on Industrial Electronics* (2012)
35. M.S. Tame, K.R. McEnery, Ş.K. Özdemir, J. Lee, S.A. Maier, M.S. Kim, Quantum plasmonics. Nat. Phys. **9**, 329–340 (2013)
36. Y.T. Tan, D.S. Kirschen, N. Jenkins, A model of PV generation suitable for stability analysis. IEEE Trans. Energy Convers. **19**(4), 748–755 (2004)
37. M.W.Y. Tu, W.M. Zhang, Non-Markovian decoherence theory for a double- dot charge qubit. Phys. Rev. B **78**, 235311 (2008)
38. S.R. Valluri, U. Becker, N. Grün, W. Scheid, Relativistic collisions of highly-charged ions. J. Phys. B: At. Mol. Phys. **17**, 4359 (1984)
39. W. Xiao, W.G. Dunford, A. Capal, A novel modeling method for photovoltaic cells, in *35th Annual IEEE Power Electronics Specialists Conference, Aachen, Germany*, (IEEE, Piscataway, 2004), pp. 1950–1956
40. Y.F. Xiao et al., Asymmetric Fano resonance analysis in indirectly coupled microresonators. Phys. Rev. A **82**, 065804 (2010)
41. W.-B. Yan, H. Fan, Single-photon quantum router with multiple output ports. Sci. Rep. **4**, 4820 (2014)
42. L. Yang, S. Wang, Q. Zeng, Z. Zhang, T. Pei, Y. Li, L.-M. Peng, Efficient photovoltage multiplication in carbon nanotubes. Nat. Photonics **5**, 672–676 (2011)
43. W.M. Zhang, P.Y. Lo, H.N. Xiong, M.W.Y. Tu, F. Nori, General non-Markovian dynamics of open quantum systems. Phys. Rev. Lett. **109**, 170402 (2012)
44. Y. Zhu, X. Hu, H. Yang, Q. Gong, On-chip plasmon-induced transparency based on plasmonic coupled nanocavities. Sci. Rep. **4**, 3752 (2014)

Chapter 6
Implementation of Bose–Einstein (B–E) Photon Energy Reformation for Cooling and Heating the Building Naturally

Abstract Conventional heating and cooling systems consume fossil fuels and release toxic gases into the environment. Therefore, alternative sustainable systems for heating and air conditioning of premises are urgently demanded. For this purpose, photon particles can be decoded by invoking the Bose–Einstein photon distribution mechanism in a helium-assisted glazing wall. A building can be cooled by locally inducing the photonic bandgap state, which naturally cools the photons. This cooling-state photon, denoted as the *Hossain Cooling Photon, HcP*$^-$, can be transformed into a thermal-state photon, denoted as the *Hossain Thermal Photon, HtP*$^-$, through bremsstrahlung radiation emitted by quantum Higgs bosons ($H \rightarrow \gamma\gamma^-$). The resulting electromagnetic field can be generated by two-diode thermal semiconductors. Because the $H \rightarrow \gamma\gamma^-$ quantum field is initiated by an extremely short-range weak force, the electrically charged *HcP*$^-$ will be transformed into an *HtP*$^-$. The *HcP*$^-$ formation and *HtP*$^-$ transformation are demonstrated in a series of mathematical tests, confirming the feasibility of photon decoding in glazing walls as a natural cooling and heating mechanism for the premises.

Keywords Decoding photon particle · Bose–Einstein photon distribution · Helium · Higgs boson BR ($H \rightarrow \gamma\gamma^-$) quantum field · Natural cooling and heating technology

Introduction

Conventional heating and cooling systems installed in premises are causing serious environmental and atmospheric problems. Traditional heating technology consumes fossil fuels and releases CO_2, which is a major contributor to climate change. Excessive CO_2 emissions threaten the environment and can potentially trigger catastrophic natural disasters. Conventional cooling technologies also release chlorofluorocarbons, which create holes in the ozone layer. The ozone layer, which lies between 9.3 and 18.6 miles (between 15 and 30 km) above the Earth's surface, is a protective blanket layer that blocks most of the sun's high-frequency ultraviolet rays that induce skin cancer in humans and serious reproductive problems in all mammals [20, 27, 40]. Although clean-energy technology, climate change, and conventional heating/cooling systems have been extensively researched [9, 22, 25, 29], natural

M. F. Hossain, *Sustainable Design for Global Equilibrium*,
https://doi.org/10.1007/978-3-030-94818-4_6

cooling and heating systems in the building sector are rarely reported. Therefore, in this study, I propose a natural cooling and heating technology based on the Bose–Einstein photon distribution mechanism and activation of Higgs bosons ($H \rightarrow \gamma\gamma^{-}$) quantum. The proposed system decodes photons (solar energy) into cooling and heating states. Cooling photon emission panels consisting of nano-point breaks and waveguides, which install helium in a portion of the exterior curtain wall, have been already proposed [15, 21]. Quantum electrodynamics (QED) waveguides naturally cool the photons emitted by the sun using photon band edges (PBEs) [8, 31]. Mediated by two-diode semiconductors, the cooling-state photon can then be transformed into a heating-state photon through the bremsstrahlung radiation (BR) emitted by quantum Higgs bosons ($H \rightarrow \gamma\gamma^{-}$), creating an electromagnetic field that naturally heats the building [7, 13, 36]. This cooling and heating transformation process is a novel approach that mitigates unsustainable energy usage, damage to the environment, and the depletion of the ozone layer.

Methods and Simulation

Cooling Mechanism

Activated photons can be decoded into the cooling state in photon emission networks of nano-point breaks, waveguides, and helium-assisted curtain walls, which create point defects in the photon emission panel [1, 4, 17]. In the same way, the arrays of photonic bandgap (PBG) waveguide defects can be incorporated into curtain walls [10, 16, 37]. Such point defects and PBG waveguides decode the quantum dynamics of photons under helium cooling conditions, providing a means of cooling solar photons. The present study calculates the formation of cooling state by the conversion of solar photons in MATLAB (v. 9.0) software. In these calculations, helium waveguides embedded in the curtain wall are treated as photon reservoirs. The electrodynamics of the cooling-state photons can be expressed by the following Hamiltonian [31, 34, 49]:

$$H = \sum \omega_{ci} a_i^{\dagger} a_i + \sum_K \omega_k b_k^{\dagger} b_k + \sum_{ik} \left(V_{ik} a^{\dagger} b_k + V_{ik}^{*} b_k^{\dagger} a_i \right), \tag{6.1}$$

where $a_i\left(a_i^{\dagger}\right)$ and $b_k\left(b_k^{\dagger}\right)$ represent the drivers of the nano-point break modes and the photodynamic modes of the photon nanostructure, respectively, and the coefficient, V_{ik}, represents the magnitudes of the photonic modes among the nano-breakpoints and photon nanostructures. Figure 6.1 displays the transmissivity contours and the spectra of photon plane waves and pulses.

The proposed photon module, which generates HcP^{-} to cool a premise, comprises helium-assisted point breaks, two diodes, and two resistors (see Fig. 6.2). The current in the module is converted from the photon energy.

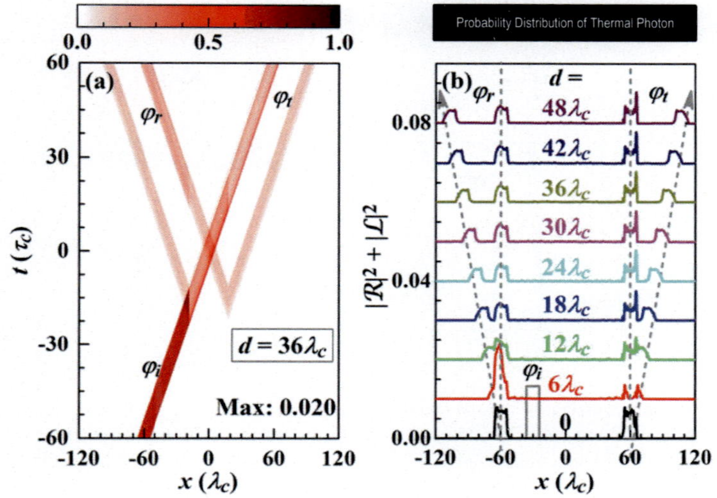

Fig. 6.1 (**a**) Contour map of the photon probability density (normalized to its maximum value 0.020) as functions of x and t for an incident square pulse [*gray solid line* in panel (**b**)]. (**b**) Probability distributions of the reflected and transmitted pulses of thermal photons

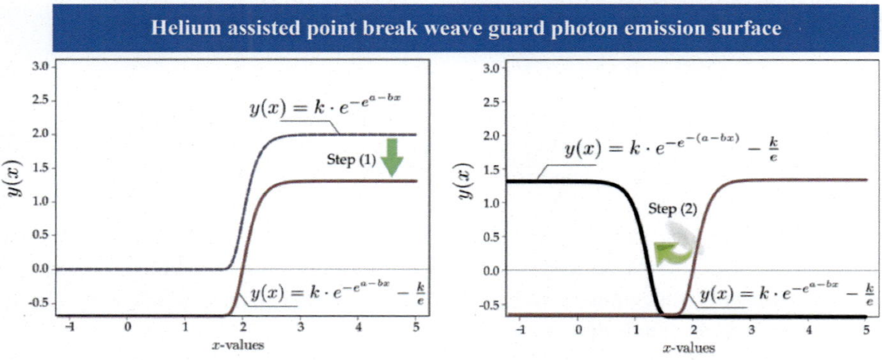

Fig. 6.2 Diagram of the two-diode model of the solar irradiance receptor. Electrons are cooled in the glazing wall skin by photon induction aided by helium-assisted point breaks considering (**a**) shifting the curve-step 1, (**b**) refection symmetry-step 2

The current–voltage (*I*–*V*) characteristics of the photon cells in the single-diode mode are given by

$$I = I_{\mathrm{L}} - I_{\mathrm{o}} \left\{ \exp^{\left[\frac{q(V + I_{\mathrm{RS}})}{AkT_{\mathrm{c}}}\right]} - 1 \right\} - \frac{(V + I_{\mathrm{RS}})}{R_{\mathrm{s}}}, \tag{6.2}$$

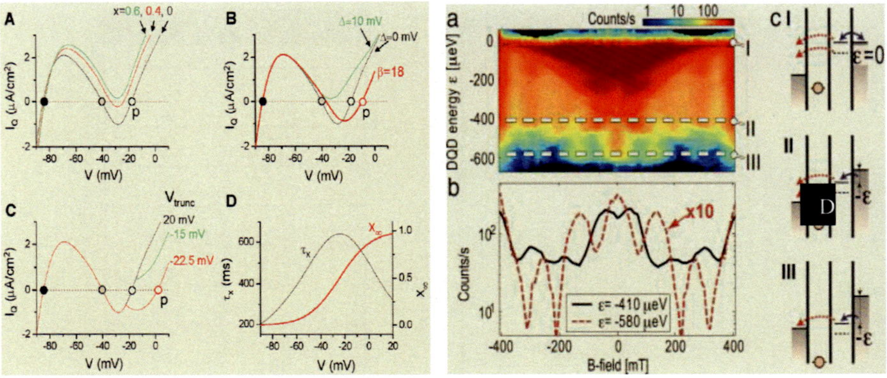

Fig. 6.3 *I–V* characteristics of the two-diode cooling mechanism: (**a**) glazing wall skin cell in the normal state, (**b**) glazing wall skin cell under the normalized condition, (**c**) glazing wall skin cell module in the normal state, and (**d**) glazing wall skin cell module under the normalized condition

where I_L represents the photon generating current, I_o represents the saturated current in the diode, and R_s is the series resistance. A represents the passive function of the diode, k ($= 1.38 \times 10^{-23}$ W/m^2 K) is the Boltzmann's constant, q ($= 1.6 \times 10^{-19}$ C) is the magnitude of the charge of an electron, and T_c represents the functional cell temperature. Subsequently, the *I–q* relationship in the photon cells varies with the diode and/or saturation current, which is given by [18, 32, 33]

$$I_o = I_{RS}\left(\frac{T_c}{T_{ref}}\right)^3 \exp\left[\frac{qEG\left(\frac{1}{T_{ref}} - \frac{1}{T_c}\right)}{KA}\right]. \tag{6.3}$$

In Eq. (6.3), I_{RS} represents the saturation current, which depends on the functional temperature and solar irradiance speed, and qEG represents the bandgap energy of the electrons per unit area of the photon cell. The *I–V* characteristics of various two-diode models are shown in Fig. 6.3.

In the photon module, the *I–V* equation integrates the *I–V* curves of all cells in the photon emission panel. The *V–R* relationship in the module is given by

$$V = -IR_s + K\log\left[\frac{I_L - I + I_o}{I_o}\right], \tag{6.4}$$

where K is a constant $\left(= \frac{AkT}{q}\right)$ and I_{mo} and V_{mo} are the current and voltage in the PV panel, respectively. Therefore, the relationship between I_{mo} and V_{mo} is the same as the *I–V* relationship in the PV cell:

$$V_{\mathrm{mo}} = -I_{\mathrm{mo}}R_{\mathrm{Smo}} + K_{\mathrm{mo}} \log \left(\frac{I_{\mathrm{Lmo}} - I_{\mathrm{mo}} + I_{\mathrm{Omo}}}{I_{\mathrm{Omo}}} \right), \tag{6.5}$$

where I_{Lmo} represents the photon-generated current, I_{Omo} represents the saturated current in the diode, R_{Smo} is the series resistance, and K_{mo} is a constant. When the resistances of all non-series (NS) cells are connected in series, the total resistance is $R_{\mathrm{Smo}} = N_{\mathrm{s}} \times R_{\mathrm{s}}$, and the constant is $K_{\mathrm{mo}} = N_{\mathrm{s}} \times K$. The current flowing into the series of connected cells is the same in each component, i.e., $I_{\mathrm{Omo}} = I_{\mathrm{o}}$ and $I_{\mathrm{Lmo}} = I_{\mathrm{L}}$. Thus, the I_{mo}–V_{mo} relationship in the N_{s} connected cells is given by

$$V_{\mathrm{mo}} = -I_{\mathrm{mo}}N_{\mathrm{s}}R_{\mathrm{s}} + N_{\mathrm{s}}K \log \left(\frac{I_{\mathrm{L}} - I_{\mathrm{mo}} + I_{\mathrm{o}}}{I_{\mathrm{o}}} \right). \tag{6.6}$$

Similarly, when all N_{p} cells are connected in parallel, the I_{mo}–V_{mo} relationship is given by [5, 24]

$$V_{\mathrm{mo}} = -I_{\mathrm{mo}}\frac{R_{\mathrm{s}}}{N_{\mathrm{p}}} + K \log \left(\frac{N_{\mathrm{sh}}I_{\mathrm{L}} - I_{\mathrm{mo}} + N_{\mathrm{p}}I_{\mathrm{o}}}{N_{\mathrm{p}}I_{\mathrm{o}}} \right). \tag{6.7}$$

Because the photon-generated current depends primarily on the solar irradiance and relativistic temperature conditions of the photon emission panel, the current can be calculated as follows:

$$I_{\mathrm{L}} = G[I_{\mathrm{sc}} + K_{\mathrm{I}}\,(T_{\mathrm{cool}})]V_{\mathrm{mo}} \tag{6.8}$$

$$T_{\mathrm{cool}} = \left(\frac{I_{\mathrm{L}}}{(G * V_{\mathrm{mo}}) \times (I_{\mathrm{sc}} + K_{\mathrm{I}})} \right),$$

where I_{sc} is the photon current per unit area at 25 °C, K_{I} denotes the relativistic photon panel coefficient, T_{cool} represents the cooling temperature of the photon cell, and G represents the solar energy per unit area [14, 41].

Heating Mechanism

To convert cooling photons to heating photons, researchers have exploited the Higgs boson electromagnetic field and determined the accuracy and parameters of the photon-heating relationship [2, 11, 51]. Thus, to create a local Higgs quantum field in the curtain wall skin, I simulated Abelian local symmetries in MATLAB 9.0 software. The penetration of solar irradiance breaks the gauge field symmetry, and the Goldstone scalar particle becomes the longitudinal mode of the vector boson [5, 26]. In the Abelian case, the local symmetry of each spontaneously broken particle T^{α} is the corresponding gauge field of $A_{\mu}^{\alpha}(x)$. The Higgs quantum field begins to operate in local $U(1)$ phase symmetry [3, 24, 38]. Thus, the model can comprise a complex scalar field $\Phi(x)$ of electric charge q coupled to the EM field $A^{\mu}(x)$, which can be expressed by the following Lagrangian function:

$$\mathcal{L} = -\frac{1}{4} F_{\mu\nu}F^{\mu\nu} + D_{\mu}\Phi^{*} D^{\mu}\Phi - V(\Phi^{*}\Phi), \tag{6.9}$$

where

$$D_{\mu}\Phi(x) = \partial_{\mu}\Phi(x) + iqA_{\mu}(x)\Phi(x)$$
$$D_{\mu}\Phi^{*}(x) = \partial_{\mu}\Phi^{*}(x) - iqA_{\mu}(x)\Phi^{*}(x), \tag{6.10}$$

and

$$V(\Phi^{*}\Phi) = \frac{\lambda}{2} (\Phi^{*}\Phi)^{2} + m^{2}(\Phi^{*}\Phi). \tag{6.11}$$

Suppose that $\lambda > 0$ but $m^2 < 0$ so that $\Phi = 0$ is a local maximum of the scalar potential, and the minima form a degenerate circle $\Phi = \frac{v}{\sqrt{2}} * e^{i\theta}$ with

$$v = \sqrt{\frac{-2m^{2}}{\lambda}} \quad \text{for any real } \theta. \tag{6.12}$$

Consequently, the scalar field Φ develops a non-zero vacuum expectation value $\langle\Phi\rangle \neq 0$, which spontaneously creates the $U(1)$ symmetry of the magnetic field. The breakdown of this symmetry creates a massless Goldstone scalar from the phase of the complex field $\Phi(x)$. However, in local $U(1)$ symmetry, the phase of $\Phi(x)$ is the x-dependent phase of the dynamic $\Phi(x)$ field rather than the phase of the expectation value $\langle\Phi\rangle$.

To confirm this mechanism, I express the scalar field space in polar coordinates:

$$\Phi(x) = \frac{1}{\sqrt{2}} \Phi_{\mathrm{r}}(x) * e^{i\Theta(x)}, \quad \text{real } \Phi_{\mathrm{r}}(x) > 0, \text{real } \Phi(x). \tag{6.13}$$

As the field in this formulation is singular at $\Phi(x) = 0$, it is inapplicable to theories with $\langle\Phi\rangle \neq 0$ but is adequate for spontaneously broken theories, where $\Phi\langle x\rangle \neq 0$ is expected almost everywhere. In terms of the real fields $\phi_{\mathrm{r}}(x)$ and $\Theta(x)$, the scalar potential depends only on the radial field ϕ_{r},

$$V(\phi) = \frac{\lambda}{8} \left(\phi_{\mathrm{r}}^{2} - v^{2}\right)^{2} + \text{const.} \tag{6.14}$$

If the radial field shifted by a variable scalar, $\Phi_{\mathrm{r}}(x) = v + \sigma(x)$, we have

$$\phi_{\mathrm{r}}^2 - v^2 = (v+\sigma)^2 - v^2 = 2v\sigma + \sigma^2 \tag{6.15}$$

$$V = \frac{\lambda}{8}\left(2v\sigma - \sigma^2\right)^2 = \frac{\lambda v^2}{2} * \sigma^2 + \frac{\lambda v}{2} * \sigma^3 + \frac{\lambda}{8} * \sigma^4. \tag{6.16}$$

Meanwhile, the covariant derivative $D_\mu\phi$ becomes

$$\begin{aligned} D_\mu\phi &= \frac{1}{\sqrt{2}}\left(\partial_\mu\left(\phi_{\mathrm{r}}e^{i\Theta}\right) + iqA_\mu * \phi_{\mathrm{r}}e^{i\Theta}\right) \\ &= \frac{e^{i\Theta}}{\sqrt{2}}\left(\partial_\mu\phi_{\mathrm{r}} + \phi_{\mathrm{r}} * i\partial_\mu\Theta + \phi_{\mathrm{r}} * iqA_\mu\right). \end{aligned} \tag{6.17}$$

$$\begin{aligned} \left|D_\mu\phi\right|^2 &= \frac{1}{2}\left|\partial_\mu\phi_{\mathrm{r}} + \phi_{\mathrm{r}} * i\partial_\mu\Theta + \phi_{\mathrm{r}} * iqA_\mu\right|^2 \\ &= \frac{1}{2}\left(\partial_\mu\phi_{\mathrm{r}}\right) + \frac{\phi_{\mathrm{r}}^2}{2} * \left(\partial_\mu\Theta qA_\mu\right)^2 \\ &= \frac{1}{2}\left(\partial_\mu\,\sigma\right)^2 + \frac{(v+\sigma)^2}{2} * \left(\partial_\mu\Theta + qA_\mu\right)^2. \end{aligned} \tag{6.18}$$

The Lagrangian is then given by

$$\mathcal{L} = \frac{1}{2}\left(\partial_\mu\,\sigma\right)^2 - v\,(\sigma) - \frac{1}{4}\,F_{\mu\nu}F^{\mu\nu} + \frac{(v+\sigma)^2}{2} * \left(\partial_\mu\Theta + qA_\mu\right)^2. \tag{6.19}$$

To incorporate the heating $(\mathcal{L}_{\mathrm{heat}})$ into the magnetic field properties of this Lagrangian, I expand $\mathcal{L}_{\mathrm{heat}}$ as a power series of the fields (and their derivatives) and extract the quadratic part describing the free particles:

$$\mathcal{L}_{\mathrm{heat}} = \frac{1}{2}\left(\partial_\mu\,\sigma\right)^2 - \frac{\lambda v^2}{2} * \sigma^2 - \frac{1}{4}\,F_{\mu\nu}F^{\mu\nu} + \frac{v^2}{2} * \left(qA_\mu + \partial_\mu\Theta\right)^2. \tag{6.20}$$

Obviously, to initiate high heating within the quantum field of the curtain wall, the free particles (with Lagrangian $\mathcal{L}_{\mathrm{free}}$) must be real scalar particles with positive $m^2 = \lambda v^2$ (where m denotes the particle mass; see Fig. 6.4).

Results and Discussion

Cooling Mechanism

To mathematically demonstrate the formation of cooling photons by the helium-assisted curtain wall skin, I determined the dynamic photon proliferation by integrating Eqs. (6.15) and (6.16). Owing to the cooling unit areal condition *J(ω)* and the

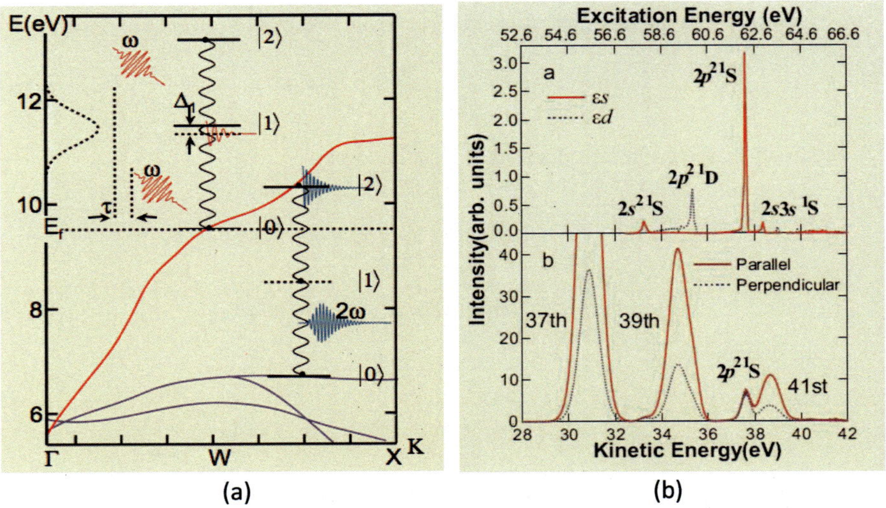

Fig. 6.4 Transformation mechanism of a photon from energy level ω to 2ω (eV) in the two-diode feed semiconductors (**a**) and quantum field intensity spectra of the electrons (**b**). The photon excitation energy (eV) is transformed into kinetic energy (eV) by the conversion of the heating state of photons

persistent weak-coupling limit, the curtain wall skin is expected to proliferate photons [24, 31]. Here, $J(\omega)$ is the quantum field area defining the density of states (DOS) field produced in the PV cell by the fine cooling photonic magnitude $V(\omega)$ within the photonic band (PB) and the PV cell [5, 31, 33]. Moreover, photon production should follow the Weisskopf–Winger approximation and/or the Markovian master equation. Consequently, all of the proliferated *HcP*s will possess a dynamic state mode (1D, 2D, and 3D) in the curtain wall skin, as described in Table 6.1 [24, 31].

In the 3D curtain wall skin, Ω_C is a fine frequency cutoff that avoids the bifurcation of the DOS. Similarly, the 1D and 2D glazing wall skins require a sharp frequency cutoff at Ω_d to avoid negative DOSs (Fig. 6.5). Hence, $\mathrm{Li}_2(x)$ and $\mathrm{erfc}(x)$ are di-logarithmic and additive variables, respectively. Subsequently, the DOS of various curtain wall skins, denoted as $\varrho_{\mathrm{PC}}(\omega)$, is determined by calculating the photon eigenfrequencies and eigenfunctions of Maxwell's rules in the nanostructures [10, 24, 32]. In a 1D glazing wall skin, the DOS is given by $\varrho_{\mathrm{PC}}(\omega) \propto \frac{1}{\sqrt{\omega-\omega_{\mathrm{e}}}}\Theta(\omega-\omega_{\mathrm{e}})$, where $\Theta(\omega-\omega_{\mathrm{e}})$ is the Heaviside step function, and ω_e represents the frequency of the PBE at the given DOS.

The DOS is required for accurately predicting the qualitative state of the non-Weisskopf–Winger mode and the photon-cooling state of the photon cell in a 3D isotropic analysis in the curtain wall skin. The DOS and projected DOS (PDOS) are displayed in Fig. 6.6. In a 3D glazing wall skin, the DOS close to the PBE is anisotropic and given by $\varrho_{\mathrm{PC}}(\omega) \propto \frac{1}{\sqrt{\omega-\omega_{\mathrm{e}}}}\Theta(\omega-\omega_{\mathrm{e}})$. This DOS is then clarified

Table 6.1 Photonic structures of the density of states (DOS) in different dimensional modes of the curtain wall skin. The unit area $J(\omega)$ and self-energy induction in the reservoir $\Sigma(\omega)$, determined by the photon dynamics into the extreme relativistic curtain wall skin, differ among the structures. The variables C, η, and χ function as coupled forces between the point break and PV in the curtain wall skin in 1, 2, and 3 dimensions [24, 31]

Photon	Unit area $J(\omega)$ for different DOS*	Reservoir-induced self-energy correction $\Sigma(\omega)$*
1D	$\frac{C}{\pi}\frac{1}{\sqrt{\omega-\omega_e}}\Theta(\omega-\omega_e)$	$-\frac{C}{\sqrt{\omega_e-\omega}}$
2D	$-\eta\left[\ln\left\lvert\frac{\omega-\omega_0}{\omega_0}\right\rvert-1\right]\Theta(\omega-\omega_e)\Theta(\Omega_d-\omega)$	$\eta\left[\mathrm{Li}_2\left(\frac{\Omega_d-\omega_0}{\omega-\omega_0}\right)-\mathrm{Li}_2\left(\frac{\omega_0-\omega_e}{\omega_0-\omega}\right)-\ln\frac{\omega_0-\omega_e}{\Omega_d-\omega_0}\ln\frac{\omega_e-\omega}{\omega_0-\omega}\right]$
3D	$\chi\sqrt{\frac{\omega-\omega_e}{\Omega_C}}\exp\left(-\frac{\omega-\omega_e}{\Omega_C}\right)\Theta(\omega-\omega_e)$	$\chi\left[\pi\sqrt{\frac{\omega_e-\omega}{\Omega_C}}\exp\left(-\frac{\omega-\omega_e}{\Omega_C}\right)\mathrm{erfc}\sqrt{\frac{\omega_e-\omega}{\Omega_C}}-\sqrt{\pi}\right]$

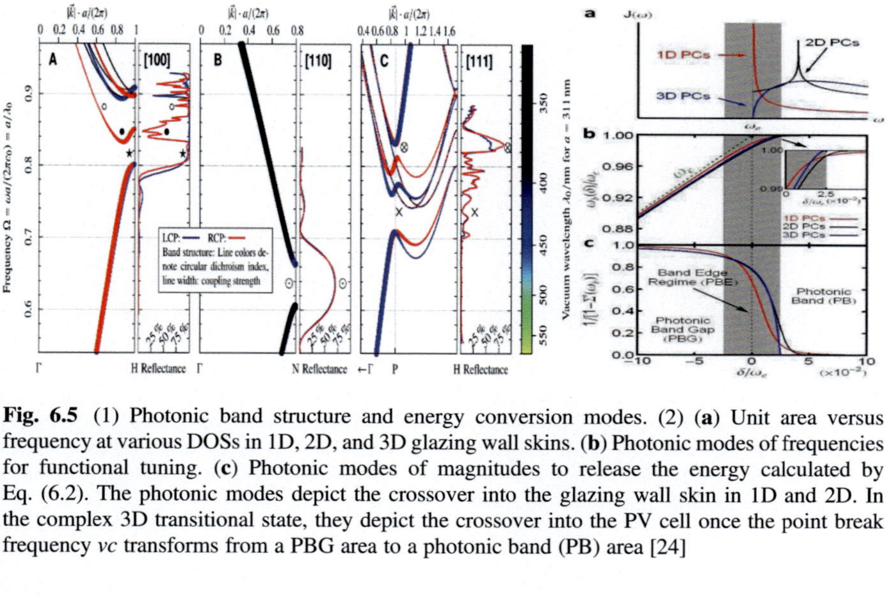

Fig. 6.5 (1) Photonic band structure and energy conversion modes. (2) (**a**) Unit area versus frequency at various DOSs in 1D, 2D, and 3D glazing wall skins. (**b**) Photonic modes of frequencies for functional tuning. (**c**) Photonic modes of magnitudes to release the energy calculated by Eq. (6.2). The photonic modes depict the crossover into the glazing wall skin in 1D and 2D. In the complex 3D transitional state, they depict the crossover into the PV cell once the point break frequency *vc* transforms from a PBG area to a photonic band (PB) area [24]

with respect to the electromagnetic field vector [24, 27, 31, 38]. In 2D and 1D curtain wall skins, the cooling photon DOS exhibits a pure logarithmic divergence close to the PBE, which is approximated as $\varrho_{PC}(\omega) \propto -[\ln|(\omega-\omega_0)/\omega_0|-1]\Theta(\omega-\omega_e)$, where ω_e represents the central point of the peak in the DOS distribution.

As mentioned above, $J(\omega)$ defines the DOS field produced in the PV cell by the fine cooling photonic magnitude $V(\omega)$ within the PB and the PV cell [5, 31, 33]:

$$J(\omega) = \varrho(\omega)|V(\omega)|^2. \tag{6.21}$$

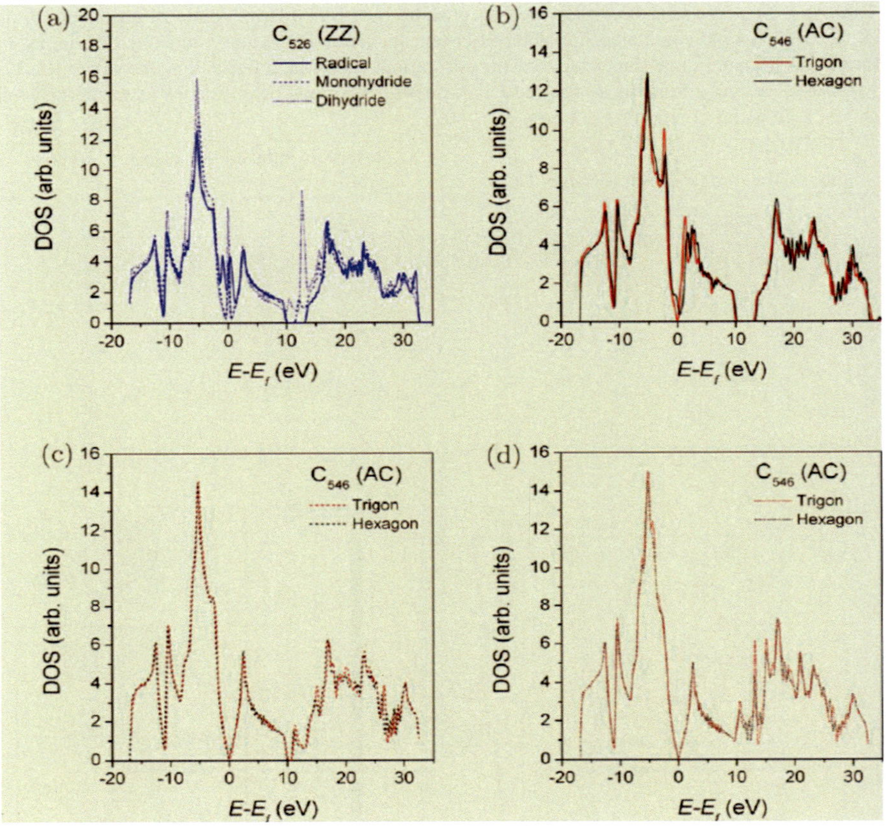

Fig. 6.6 Total density of states (DOS) and the projected density of states (PDOS) of decoded photons for transformation into the cooling state: Panel (**a**) (1) total DOS (T) and DOS projected onto the *s*, *p*, and *d* orbitals, (2) PDOS of *d* orbitals on the fourth level of Mo atoms, and (3) PDOS of *d* orbitals on the third level of Mo atoms; Panel (**b**) as in Panel (**a**) but for projected DOS of Mo atoms; Panel (**c**) (1–3) as in Panel (**a**) and (4) PDOS of *p* orbitals of O atoms. Panel (**d**) (1–3) as in Panel (**b**) and (4) PDOS of *p* orbitals of external S atoms

Hereafter, I consider the PB frequency ω_c and the proliferative photon dynamics $\langle a(t)\rangle = u(t, t_0)\langle a(t_0)\rangle$, where the function $u(t, t_0)$ describes the photon structure. $u(t, t_0)$ is calculated using the dissipative integral–differential equation given in Eq. (6.18):

$$u(t, t_0) = \frac{1}{1 - \Sigma'(\omega_b)} e^{-i\omega(t-t_0)} + \int_{\omega_e}^{\infty} d\omega \frac{J(\omega) e^{-i\omega(t-t_0)}}{[\omega - \omega_c - \Delta(\omega)]^2 + \pi^2 J^2(\omega)}, \quad (6.22)$$

where $\Sigma'(\omega_b) = [\partial\Sigma(\omega)/\partial\omega]_{\omega=\omega_b}$ and $\Sigma(\omega)$ represents the PB photon self-energy correction induced in the reservoir,

$$\Sigma(\omega) = \int_{\omega_e}^{\infty} d\omega' \frac{J(\omega')}{\omega - \omega'}. \tag{6.23}$$

Here, the frequency ω_b in Eq. (6.2) represents the cooling photonic frequency mode in the PBG ($0 < \omega_b < \omega_e$), calculated under the pole condition $\omega_b - \omega_c - \Delta(\omega_b) = 0$, where $\lesssim\Delta(\omega) = \mathcal{P}\left[\int d\omega' \frac{J(\omega')}{\omega-\omega'}\right]$ is a principal-value integral.

Figure 6.6 plots the cooling photonic dynamics of the proliferation magnitude $|u(t, t_0)|$, calculated in 1D, 2D, and 3D photon cells for various values of the detuning parameter δ and integrated from the PBG area to the PB area [23, 47, 52]. The cooling photonic dynamic rates $\kappa(t)$ are plotted in Fig. 6.6. The results indicate that, once ω_c has crossed from the PBG to the PB area, dynamic photons are produced at a high rate. Because the $u(t, t_0)$ range is $1 \geq |u(t, t_0)| \geq 0$, I have defined the crossover area to satisfy $0.9 \gtrsim |u(t \to \infty, t_0)| \geq 0$. This corresponds to $-0.025\omega_e \lesssim \delta \lesssim 0.025\omega_e$, with a cooling photon induction rate $\kappa(t)$ within the PBG ($\delta < -0.025\omega_e$) and near the PBE ($-0.025\omega_e \lesssim \delta \lesssim 0.025\omega_e$).

More specifically, I first consider the PB as the Fock cooling determination n_0, i.e., $\rho(t_0) = |n_0\rangle\langle n_0|$, which is obtained theoretically through real-time quantum feedback control [24, 35, 40], and then, by solving Eq. (6.1), considering the state of cooling photon induction at time t:

$$\rho(t) = \sum_{n=0}^{\infty} \mathcal{P}_n^{(n_0)}(t)|n_0\rangle\langle n_0| \tag{6.24}$$

$$\mathcal{P}_n^{(n_0)}(t) = \frac{[v(t,t)]^n}{[1 + v(t,t)]^{n+1}}[1 - \Omega(t)]^{n_0} \times \sum_{k=0}^{\min\{n_0, n\}} \binom{n_0}{k} \times \binom{n}{k}\left[\frac{1}{v(t,t)}\frac{\Omega(t)}{1 - \Omega(t)}\right]^k, \tag{6.25}$$

where $\Omega(t) = \frac{|u(t,t_0)|^2}{1+v(t,t)}$. This result suggests that a Fock state cooling photon is induced into dynamic states $\mathcal{P}_n^{(n_0)}(t)$ of $|n_0\rangle$. In fact, Fig. 6.7 plots the proliferation of photon dissipation $\mathcal{P}_n^{(n_0)}(t)$ in the primary state $|n_0 = 5\rangle$ and in the steady-state limit $\mathcal{P}_n^{(n_0)}(t \to \infty)$. Therefore, the proliferation of the produced cooling photons will ultimately reach a non-equilibrium cooling state that cools the building.

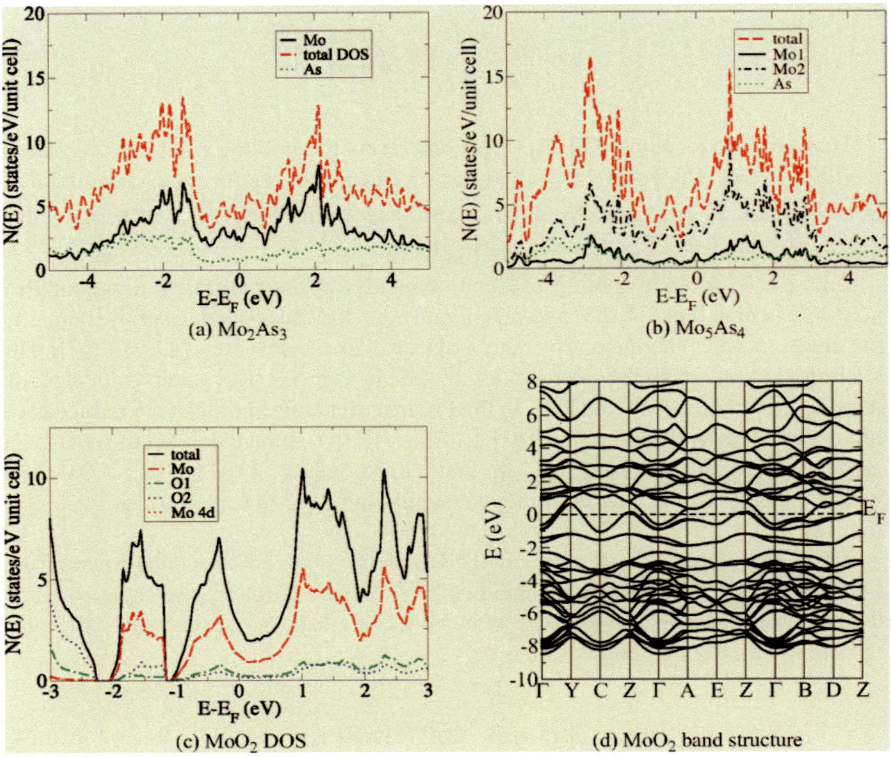

Fig. 6.7 Schematic of the main mechanisms responsible for magnetic-field-induced photon production. (**a**) Photon heating is simultaneously coupled into the fundamental mode and a higher-order mode of the quantum field. (**b**) Coincidence rate of the fundamental mode output. (**c**) Coincidence rate of the higher-order mode output with respect to the detuning parameter. (**d**) Classical coincidence rates of heating photons with respect to the detuning parameter into the band structure [28, 31, 42]

Heating Mechanism

In the proposed system, two semiconductors utilize the electromagnetic field created by the Higgs boson quantum field. Therefore, local $U(1)$ gauge-invariant QED allows an additional mass term for the gauge particle under $\varnothing' \to e^{i\alpha(x)}\,\varnothing$; that is, the cooling photons can be transformed into heating photons. This mechanism can be explained by a covariant derivative with a special transformation rule for the scalar field, given by [44, 48, 50]

$$\partial_\mu \rightarrow D_\mu = \partial_\mu = ieA_\mu \text{ [covariant derivatives]}$$

$$A'_\mu = A_\mu + \frac{1}{e}\,\partial_\mu\alpha \quad [A_\mu \text{ derivatives}], \tag{6.26}$$

where the local $U(1)$ gauge-invariant Lagrangian for a complex scalar field is given by

$$\mathcal{L} = (D^\mu)^\dagger\,(D_\mu\varnothing) - \frac{1}{4}\,F_{\mu\nu}F^{\mu\nu} - V(\varnothing). \tag{6.27}$$

The term $\frac{1}{4}\,F_{\mu\nu}F^{\mu\nu}$ is the kinetic term in the gauge field (heating photons) and $V(\varnothing)$ is an additional term expressed as $V(\varnothing^*\varnothing) = \mu^2(\varnothing^*\varnothing) + \lambda\,(\varnothing^*\varnothing)^2$.

According to the Lagrangian $\mathcal{L}$, perturbations in the quantum field initiate the production of massive scalar particles ϕ_1 and ϕ_2 and a mass μ. In this situation, $\mu^2 < 0$ admits an infinite number of quanta, each satisfying $\phi_1^2 + \phi_2^2 = -\mu^2/\lambda = v^2$. In terms of the shifted fields η and ξ, the quantum field is defined as $\phi_0 = \frac{1}{\sqrt{2}}\,[(v+\eta) + i\xi]$, and the covariant derivatives of the Lagrangian become

$$\begin{aligned}\text{Kinetic term: } \mathcal{L}_{\text{kin}}(\eta,\xi) &= (D^\mu\phi)^\dagger(D^\mu\phi)\\ &= (\partial^\mu + ieA^\mu)\phi^*\,(\partial_\mu - ieA_\mu)\,\phi\end{aligned} \tag{6.28}$$

Potential term (to second order): $V(\eta,\xi) = \lambda\,v^2\eta^2$. Thus, the full Lagrangian can be written as

$$\begin{aligned}\mathcal{L}_{kin}(\eta,\xi) &= \frac{1}{2}\,(\partial_\mu\eta)^2 - \lambda v^2\eta^2 + \frac{1}{2}\,(\partial_\mu\xi)^2 - \frac{1}{4}\,F_{\mu\nu}F^{\mu\nu} + \frac{1}{2}\,e^2v^2A_\mu^2\\ &\quad - evA_\mu(\partial^\mu\xi) + \text{int.terms}.\end{aligned} \tag{6.29}$$

Here, η is massive, ξ is massless (as before), μ is the mass term for the quantum, and A_μ is fixed up to a term $\partial_\mu\alpha$, as is evident in Eq. (6.27). In general, A_μ and ϕ change simultaneously, so Eq. (6.28) can be redefined to accommodate the heating photon particle spectrum within the quantum field:

$$\mathcal{L}_{\text{scalar}} = (D^\mu\phi)^\dagger(D^\mu\phi) - V(\phi^\dagger\phi)$$

$$= (\partial^\mu + ieA^\mu)\frac{1}{\sqrt{2}}\,(v+h)\,(\partial_\mu - ieA_\mu)\frac{1}{\sqrt{2}}\,(v+h) - V\,(\phi^\dagger\phi) \tag{6.30}$$

$$= \frac{1}{2}\,(\partial_\mu h)^2 + \frac{1}{2}\,e^2A_\mu^2\,(v+h)^2 - \lambda v^2h^2 - \lambda vh^3 - \frac{1}{4}\,\lambda h^4 + \frac{1}{4}\lambda h^4. \tag{6.31}$$

The expanded term in the Lagrangian of the scalar field suggests that the Higgs boson quantum field can initiate heating photons.

To confirm this heating photon transformation, I calculate the isotropic distribution of movement on the differential cone with respect to the angle θ from the vertical axis. The differential between θ and $\theta + d\theta$ is $\frac{1}{2}\sin\theta d\theta$. The differential photon density at energy $\in$ and angle θ is then given by

$$dn = \frac{1}{2} n(\in) \sin\theta d \in d\theta \tag{6.32}$$

Consequently, the functional speeds of the high-energy photons were calculated as $c(1 - \cos\theta)$, and the absorption per unit path length is

$$\frac{d\tau_{\text{abs}}}{dx} = \int\int \frac{1}{2} \sigma n\ (\in)(1 - \cos\theta) \sin\theta d \in d\theta. \tag{6.33}$$

Re-expressing these functions as integrals over s instead of θ, by (6.31) and (6.33), I obtain

$$\frac{d\tau_{\text{abs}}}{dx} = \pi r_0^2 \left(\frac{m^2c^4}{E}\right)^2 \int\limits_{\frac{m^2c^4}{E}}^{\infty} \in^{-2} n(\in)\ \overline{\varphi}[s_0(\in)] de, \tag{6.34}$$

where

$$\overline{\varphi}[s_0(\in)] = \int\limits_{1}^{s_0(\in)} s\overline{\sigma}\ (s) ds, \quad \overline{\sigma}(s) = \frac{2\sigma(s)}{\pi r_0^2}. \tag{6.35}$$

This result defines the dimensional variable $\overline{\varphi}$ and dimensionless cross section $\overline{\sigma}$. The variable $\overline{\varphi}[s_0]$ is calculated based on a detailed graphical frame for $1 < s_0 < 10$. I calculated $\overline{\varphi}$ by a functional asymptotic calculation

$$\overline{\varphi}[s_0] = \frac{1+\beta_0^2}{1-\beta_0^2} \ln\omega_0 - \beta_0^2 \ln\omega_0 - \ln^2\omega_0 - \frac{4\beta_0}{1-\beta_0^2} + 2\beta_0 + 4\ \ln\omega_0 \ln\ (\omega_0 + 1)$$
$$- L(\omega_0),$$

where

$$s_0 - 1 \ll 1 \text{ or } s_0 \gg 1,$$

$$\beta_0^2 = \frac{1-1}{s_0}, \quad \omega_0 = \frac{(1+\beta_0)}{(1-\beta_0)}, \quad \text{and } L(\omega_0) = \int\limits_{1}^{\omega_0} \omega^{-1} \ln(\omega + 1) d\omega. \tag{6.36}$$

The last integral can be written as

$$(\omega + 1) = \omega\left(\frac{1+1}{\omega}\right), \quad L(\omega_0) = \frac{1}{2}\ln^2\omega_0 + L'(\omega_0),$$

where

$$L'(\omega_0) = \int_1^{\omega_0} \omega^{-1}\ln\left(1 + \frac{1}{\omega}\right)d\omega, = \frac{\pi^2}{12} - \sum_{n=1}^{\infty}(-1)n^{-1}n^{-2}\omega_0^{-n} \tag{6.37}$$

This accurate representation of the heating photons readily allows the calculation of $\overline{\varphi}[s_0]$ to the desired accuracy for the expected value of s_0. Thus, the corrective functional asymptotic formulas are expressed as follows:

$$\overline{\varphi}[s_0] = 2s_0(\ln 4s_0 - 2) + \ln 4s_0(\ln 4s_0 - 2) - \frac{(\pi^2 - 9)}{3} + s_0^{-1}\left(\ln 4s_0 + \frac{9}{8}\right) + \cdots \quad (s_0 \gg 1), \tag{6.38}$$

$$\overline{\varphi}[s_0] = \left(\frac{2}{3}\right)(s_0 - 1)^{\frac{3}{2}} + \left(\frac{5}{3}\right)(s_0 - 1)^{\frac{5}{2}} - \left(\frac{1507}{420}\right)(s_0 - 1)^{\frac{7}{2}} + \cdots \quad (s_0 - 1 \ll 1). \tag{6.39}$$

The function $\frac{\overline{\varphi}[s_0]}{(s_0-1)}$ is shown in Fig. 6.5 for $1 < s_0 < 10$; at larger s_0, it becomes a natural logarithmic function of s_0. The power-law spectrum of the heating photons is expressed in the form $n(\in) \propto \in_m$ for two systems in a pristine state and for a system with BN in a glazing sheet.

Thus, the light absorption spectrum should feature a high-energy cutoff with $m > 0$. We now derive the heating photon spectrum with a high-energy cutoff.

Consider a spectrum of the form

$$n(\in) = D\in_\beta, \ \in < \in_m, \quad \beta \le 0 \tag{6.40}$$

$$= 0, \ \in > \in_m \tag{6.41}$$

For this spectrum, I obtained

$$\frac{d\tau_{abs}}{dx} = \pi r_0^2 D\left(\frac{m^2c^4}{E}\right)^{1+\beta} \times \begin{cases} 0, & E < E_m, \\ F_\beta(\sigma_m), & E > E_m, \end{cases} \tag{6.42}$$

where

$$\sigma_m = \frac{E}{E_m} = \frac{\in_m E}{m^2c^4}, \tag{6.43}$$

$$F_\beta(\sigma_m) = \int_1^{\sigma_m} s_0^{\beta-2}\overline{\varphi}[s_0]ds_0. \tag{6.44}$$

Again, by Eqs. (6.40) and (6.41), we can obtain the asymptotic forms

$$\beta = 0: \quad F_\beta\,(\sigma_m) \rightarrow A_\beta + \ln^2\sigma_m - 4\,\ln\sigma_m + \cdots,$$

$$\beta \neq 0: \quad F_\beta\,(\sigma_m) \rightarrow A_\beta + 2\beta^{-1}\sigma_m^\beta\,\left(\ln 4\sigma_m - \beta^{-1} - 2\right) + \cdots, \quad \sigma_m > 10 \tag{6.45}$$

$$\text{All}\,\beta: \quad F_\beta\,(\sigma_m) \rightarrow \left(\frac{4}{15}\right)(\sigma_m - 1)^{\frac{5}{2}} + \left[\frac{2(2\beta+1)}{21}\right](\sigma_m - 1)^{\frac{7}{2}} + \cdots, \quad \sigma_m - 1 \ll 1. \tag{6.46}$$

Figure 6.6 plots $\sigma_m^{-\beta}F_\beta(\sigma_m)$ for $\beta = 0$–$3.0\,A_\beta$ in 0.5–A_β intervals, which contribute to the integral in the region [1, 15]. The values were calculated as $A_\beta = 8.111$ ($\beta = 0$), 13.53 ($\beta = 0.5$), 9.489 ($\beta = 1.0$), 15.675 ($\beta = 1.5$), 34.54 ($\beta = 2.0$), 85.29 ($\beta = 2.5$), and 222.9 ($\beta = 3.0$). Subsequently, I calculated the heating photon terms corresponding to the spectra for both negative and positive indexes:

$$\begin{aligned} n(\in) &= 0, \quad \in < \in_0 \\ &= C_\in^{-\alpha} \text{ or } D_\in^\beta, \quad \in_0 < \in < \in_m \\ &= 0, \ \in > \in_m. \end{aligned} \tag{6.47}$$

I then obtain

$$\left(\frac{d\tau_{\text{abs}}}{dx}\right)_\alpha = \pi r_0^2 C\left(\frac{m^2c^4}{E}\right)^{1-\alpha} \times \begin{cases} 0, & E < E_m, \\ [F_\alpha(1) - F_\alpha(\sigma_m)], & E_m < E < E_0, \\ [F_\alpha(\sigma_0) - F_\alpha(\sigma_m)], & E > E_0, \end{cases} \tag{6.48}$$

$$\left(\frac{d\tau_{\text{abs}}}{dx}\right)_\beta = \pi r_0^2 D\left(\frac{m^2c^4}{E}\right)^{1+\beta} \times \begin{cases} 0, & E < E_m, \\ [F_\beta(\sigma_m)], & E_m < E < E_0, \\ [F_\beta(\sigma_m) - F_\beta(\sigma_0)], & E > E_0, \end{cases} \tag{6.49}$$

In this case, the heating photon spectrum can be adequately described by asymptotic formulas. The term Γ_{γ}^{LPM} denotes the photon's contribution to the emitted irradiance per unit volume due to bremsstrahlung processes [12, 43, 46, 53]:

$$\Gamma_{\gamma} \equiv \frac{dn_{\gamma}}{dVdt}. \tag{6.50}$$

After summing the contributions $\Gamma_{\gamma}^{\mathrm{LPM}}$, the emission rate was confirmed as $O(\alpha_{\mathrm{EM}}\alpha_{\mathrm{s}})$. Here, I have assumed the following expression for the polarized emission rate $\Gamma_{\gamma}^{\mathrm{LPM}}$ at the thermodynamically controlled equilibrium of the plasma surface at temperature T and photophysical reaction μ:

$$\frac{d\Gamma_{\gamma}^{\mathrm{LPM}}}{d^3k} = \frac{d_{\mathrm{F}} q_{\mathrm{s}}^2 \alpha_{\mathrm{EM}}}{4\pi^2 k} \int_{-\infty}^{\infty} \frac{dp_{\parallel}}{2\pi} \int \frac{d^2 p_{\perp}}{(2\pi)^2} A\left(p_{\parallel}, k\right) \, \mathrm{Re}\left\{2P_{\perp} \cdot f\left(p_{\perp}; p_{\parallel}, k\right)\right\}, \tag{6.51}$$

where d_{F} represents the functional strategy of a quark particle [N_{c} in $\mathrm{SU}(N_{\mathrm{c}})$], q_{s} represents the Abelian charge of a quark, $k \equiv |k|$, and the kinetic function $A(p_{\parallel}, k)$ of the emitted particle is given by

$$A\left(p_{\parallel}, k\right) \equiv \begin{cases} \dfrac{n_{\mathrm{b}}\left(k + p_{\parallel}\right)\left[1 + n_{\mathrm{b}}\left(p_{\parallel}\right)\right]}{2p_{\parallel}\left(p_{\parallel} + k\right)}, & \text{scalars,} \\[2ex] \dfrac{n_{\mathrm{f}}\left(k + p_{\parallel}\right)\left[1 - n_{\mathrm{f}}\left(p_{\parallel}\right)\right]}{2\left[p_{\parallel}\left(p_{\parallel} + k\right)\right]^2}\left[{p_{\parallel}}^2 + \left(p_{\parallel} + k\right)^2\right], & \text{fermions,} \end{cases} \tag{6.52}$$

with

$$n_{\mathrm{b}}(p) \equiv \frac{1}{\exp\left[\beta(p - \mu)\right] - 1}, \quad n_{\mathrm{f}}(p) \equiv \frac{1}{\exp\left[\beta(p - \mu)\right] + 1}. \tag{6.53}$$

The function $f(p_{\perp}; p_{\parallel}, k)$ in Eq. (6.51) was introduced to solve the following linear integral equation, which confirms multiple photon production by three-diode scattering [6, 19, 45]:

$$2p_{\perp} = i\delta E f\left(p_{\perp};p_{\parallel},k\right) + \frac{\pi}{2} C_{\mathrm{F}} g_{\mathrm{s}}^{2} m_{\mathrm{D}}^{2} \int \frac{d^{2}q_{\perp}}{(2\pi)^{2}} \frac{dq_{\parallel}}{2\pi} \frac{dq^{0}}{2\pi} 2\pi\delta\left(q^{0} - q_{\parallel}\right)$$
$$\times \frac{T}{|q|} \left[\frac{2}{|q^{2} - \Pi_{\mathrm{L}}(Q)|^{2}} + \frac{\left[1 - \left(q^{0}/\left|q_{\parallel}\right|\right)^{2}\right]^{2}}{\left|(q^{0})^{2} - q^{2} - \Pi_{\mathrm{T}}(Q)\right|^{2}} \right]$$
$$\times \left[f\left(p_{\perp};p_{\parallel},k\right) - f\left(q + p_{\perp};p_{\parallel},k\right) \right]. \tag{6.54}$$

In Eq. (6.54), C_{F} is a quadratic quark [$C_{\mathrm{F}} = (N_{\mathrm{c}}^{2} - 1)/2N_{\mathrm{c}} = 4/3$ in QCD], m_{D} is the leading-order Debye mass, and δE is the energy difference between quasi-particles, which considers the photon emission and the state of the thermodynamic temperature equilibrium:

$$\delta E \equiv k^{0} + E_{p}\mathrm{sign}\left(p_{\parallel}\right) - E_{p+k}\mathrm{sign}\left(p_{\parallel} + k\right). \tag{6.55}$$

For an SU(N) gauge theory with N_{s} complex scalars and N_{f} Dirac fermions, the Debye mass in the fundamental representation is given by [47]

$$m_{\mathrm{D}}^{2} = \frac{1}{6}(2N + N_{\mathrm{s}} + N_{\mathrm{f}})g^{2}T^{2} + \frac{N_{\mathrm{f}}}{2\pi^{2}} g^{2}\mu^{2}. \tag{6.56}$$

To accurately determine the photon energy emission rate in the region $p_{\parallel} > 0$, I calculated the distribution of $n(k + p_{\parallel})[1 \pm n(p_{\parallel})]$ in the integral containing $A(p_{\parallel}, k)$ in Eq. (6.51), which determines the distribution of pair annihilations, using the following equation:

$$n_{\mathrm{b}}(-p) = -[1 + \overline{n}_{\mathrm{b}}(p)], \quad n_{\mathrm{f}}(-p) = [1 - \overline{n}_{\mathrm{f}}(p)], \tag{6.57}$$

where $n(p) \equiv 1/[e\beta(p + \mu) \mp 1]$ is the appropriate anti-particle distribution function. Therefore, the factor $A(p_{\parallel}, k)$ in this interval may be rewritten as

$$A\left(p_{\parallel},k\right) \equiv \begin{cases} \dfrac{n_{\mathrm{b}}\left(k - \left|p_{\parallel}\right|\right)\overline{n}_{\mathrm{b}}\left(\left|p_{\parallel}\right|\right)}{2\left|p_{\parallel}\right|\left(k - \left|p_{\parallel}\right|\right)}, & \text{scalars,} \\[2ex] \dfrac{n_{\mathrm{f}}\left(k - \left|p_{\parallel}\right|\right)\overline{n}_{\mathrm{f}}\left(\left|p_{\parallel}\right|\right)}{2\left[\left|p_{\parallel}\right|\left(k - \left|p_{\parallel}\right|\right)\right]^{2}} \left[{p_{\parallel}}^{2} + \left(k - \left|p_{\parallel}\right|\right)^{2}\right], & \text{fermions.} \end{cases} \tag{6.58}$$

Thus, the energy E_{p} of a hard quark with momentum $|p|$ is explicitly given by

$$E_p = \sqrt{p^2 + m_\infty^2} \simeq |p| + \frac{m_\infty^2}{2|p|} \simeq \left|p_\parallel\right| + \frac{p_\perp^2 + m_\infty^2}{2\left|p_\parallel\right|}, \tag{6.59}$$

where the asymptotic thermal "mass" is

$$m_\infty^2 = \frac{C_\mathrm{f} g^2 T^2}{4}. \tag{6.60}$$

Substituting the explicit form of E_p into definition (6.60), we get

$$\delta E = \left[\frac{p_\perp^2 + m_\infty^2}{2}\right]\left[\frac{k}{p_\parallel\left(k + p_\parallel\right)}\right]. \tag{6.61}$$

Above, I have derived explicit forms of Eqs. (6.52) and (6.55). Figure 6.8 plots the leading-order heating photoemission rates that maximize the power given the electron time-of-flight of the photons into the glazing wall plane.

Conclusions

Traditional cooling and heating systems adopted by the building sector are problematic as they contribute to climate change and threaten the ozone layer. To mitigate these potentially catastrophic effects, the present paper has proposed transforming solar irradiation into cooled photons by implementing the Bose–Einstein (B–E) photon distribution mechanism on helium-assisted glazing walls, which can be installed on the external surfaces of buildings. The cooled photons are then available for cooling the premises. The feasibility of the approach is demonstrated in mathematical tests implemented in MATLAB software. Moreover, the cooling photons can be converted into the heating state by the Higgs boson [BR ($H \rightarrow \gamma\gamma^{-}$)] quantum field, which can be created by two thermal semiconductor diodes installed in the helium-assisted glazing wall, providing a natural heat source for buildings. Simply, it has been concluded that cooling-state photons (HcP^{-}) can be extracted from the solar irradiance. Then it can be transformed into heating-state photons (HtP^{-}), providing a natural cooling and heating system for the premises to reduce global energy consumption, environmental damage, and ozone layer depletion dramatically.

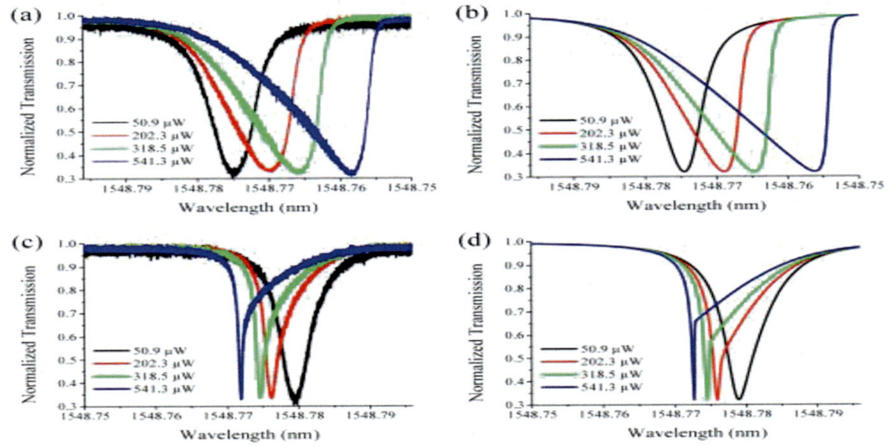

(9A)

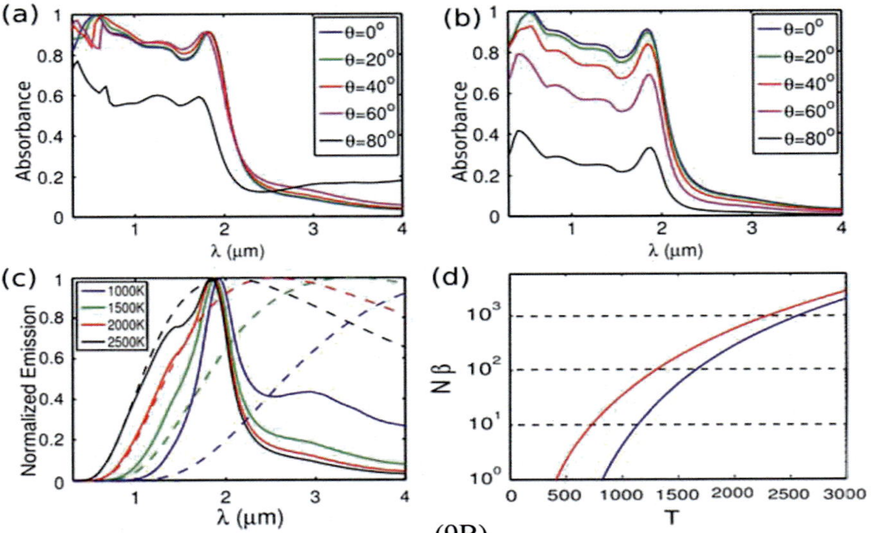

(9B)

Fig. 6.8 (**A**) (a) Thermal photon transmission vs wavelength at various normalized condition power spectral density of the resonator amplitude as a function of frequency. (b) Quasi-particle recombination time as a function of thermal roll-off frequency. (c) Resonator amplitude as a function of frequency for different bath temperatures. (d) Energy measurements (DQD) incorporating photon excitation into the Higgs boson quantum field. (**B**) Numerical demonstration of the wide-angle absorbance thermal spectra into the glazing film for (a) p polarized and (b) s polarized incident radiation for various incidence angles. (c) Normalized thermal emission spectra (per unit frequency) for dashed lines and glazing (*solid lines*) surfaces. The different colors correspond to different emitter temperatures. (d) Solar concentration versus equilibrium temperature for the glazing surface

Acknowledgments This research was supported by Green Globe Technology under grant RD-02017-05 for building a better environment. Any findings, predictions, and conclusions described in this article are solely performed by the authors and it is confirmed that there is no conflict of interest for publishing this research paper in a suitable journal.

References

1. A.K. Agger, A.H. Sørensen, Atomic and molecular structure and dynamics. Phys. Rev. A **55**, 402–413 (1997)
2. P. Arnold, G.D. Moore, L.G. Yaffe, Photon emission from ultrarelativistic plasmas. J. High Energy Phys. **11**, 057 (2001)
3. N. Artemyev, U.D. Jentschura, V.G. Serbo, A. Surzhykov, Strong electromagnetic field EFFECTS in ultra-relativistic heavy-ion collisions. Eur. Phys. J. **72**, 1935 (2012)
4. G. Baur, K. Hencken, D. Trautmann, S. Sadovsky, Y. Kharlov, Dense laser-driven electron sheets as relativistic mirrors for coherent production of brilliant X-ray and γ-ray beams. Phys. Rep. **364**, 359–450 (2002)
5. G. Baur, K. Hencken, D. Trautmann, Revisiting unitarity corrections for electromagnetic processes in collisions of relativistic nuclei. Phys. Rep. **453**, 1–27 (2007)
6. U. Becker, N. Grün, W. Scheid, K-shell ionisation in relativistic heavy-ion collisions. J. Phys. B: At. Mol. Phys. **20**, 2075 (1987)
7. A. Belkacem, H. Gould, B. Feinberg, R. Bossingham, W.E. Meyerhof, Semiclassical dynamics and relaxation. Phys. Rev. Lett. **71**, 1514–1517 (1993)
8. N.D. Benavides, P.L. Chapman, Modeling the effect of voltage ripple on the power output of photovoltaic modules. IEEE Trans. Ind. Electron. **55**, 2638–2643 (2008)
9. B. Boukhezzar, H. Siguerdidjane, Nonlinear control with wind estimation of a DFIG variable speed wind turbine for power capture optimization. Energy Convers. Manag. **50**, 885–892 (2009)
10. A.N. Celik, N. Acikgoz, Modelling and experimental verification of the operating current of mono-crystalline photovoltaic modules using four- and five-parameter models. Appl. Energy **84**, 1–15 (2007)
11. W. Chihhui, Metamaterial-based integrated plasmonic absorber/emitter for solar thermophotovoltaic systems. J. Opt. **14**, 024005 (2012)
12. W. De Soto, S.A. Klein, W.A. Beckman, Improvement and validation of a model for photovoltaic array performance. Sol. Energy **80**, 78–88 (2006)
13. J.S. Douglas, H. Habibian, C.-L. Hung, A.V. Gorshkov, H.J. Kimble, D.E. Chang, Quantum many-body models with cold atoms coupled to photonic crystals. Nat. Photonics **9**, 326–331 (2015)
14. J. Eichler, T. Stöhlker, Radiative electron capture in relativistic ion-atom collisions and the photoelectric effect in hydrogen-like high-Z systems. Phys. Rep. **439**, 1–99 (2007)
15. H. Faida, J. Saadi, Modelling, control strategy of DFIG in a wind energy system and feasibility study of a wind farm in Morocco. IREMOS **3**, 1350–1362 (2010)
16. M. Faruque Hossain, Theory of global cooling. Energy Sustain. Soc. **6**, 1–5, 24 (2016)
17. M. Faruque Hossain, Design and construction of ultra-relativistic collision PV panel and its application into building sector to mitigate total energy demand. J. Build. Eng. **9**, 147–154 (2017a)
18. M. Faruque Hossain, Green science: Advanced building design technology to mitigate energy and environment. Renew. Sustain. Energy Rev. **81**, 3051–3060 (2017b)
19. M. Faruque Hossain, Green science: Independent building technology to mitigate energy, environment, and climate change. Renew. Sustain. Energy Rev. **73**, 695–705 (2017c)

20. M. Faruque Hossain, Photonic thermal control to naturally cool and heat the building. Appl. Therm. Eng. **131**, 576–586 (2018)
21. T. Ghennam, E.M. Berkouk, B. Francois, A vector hysteresis current control applied on three-level inverter. Application to the active and reactive power control of doubly fed induction generator based wind turbine. Int. Rev. Electr. Eng. **2**, 250–259 (2007)
22. C. Gopal, M. Mohanraj, P. Chandramohan, P. Chandrasekar, Renewable energy source water pumping systems—A literature review. Renew. Sustainable Energy Rev. **25**, 351–370 (2013). https://doi.org/10.1016/j.rser.2013.04.012
23. R.J. Gould, G.P. Schréder, Pair production in photon-photon collisions. Phys. Rev. **155**, 1404–1407 (1967)
24. M.C. Güçlü, J. Li, A.S. Umar, D.J. Ernst, M.R. Strayer, Electromagnetic lepton pair production in relativistic heavy-ion collisions. Ann. Phys. **272**, 7–48 (1999)
25. N. Gupta, S.P. Singh, S.P. Dubey, D.K. Palwalia, Fuzzy logic controlled three-phase three-wired shunt active power filter for power quality improvement. Int. Rev. Electr. Eng. **6**, 1118–1129 (2011)
26. K. Hencken, G. Baur, D. Trautmann, Transverse momentum distribution of vector mesons produced in ultraperipheral relativistic heavy ion collisions. Phy. Rev. Lett. **96**, 012303 (2006)
27. F. Hossain, Solar energy integration into advanced building design for meeting energy demand and environment problem. Int. J. Energy Res. **17**, 49–55 (2016)
28. S.C. Huot, Photon and dilepton production in supersymmetric Yang-Mills plasma. J. High Energy Phys. (2006). https://doi.org/10.1088/1126-6708/2006/12/015
29. E. Kamal, M. Koutb, A.A. Sobaih, B. Abozalam, An intelligent maximum power extraction algorithm for hybrid wind-diesel-storage system. Int. J. Electr. Power Energy Syst. **32**, 170–177 (2010)
30. S.A. Klein, Calculation of flat-plate collector loss coefficients. Sol. Energy **17**, 79–80 (1975)
31. L. Langer, S.V. Poltavtsev, I.A. Yugova, M. Salewski, D.R. Yakovlev, G. Karczewski, T. Wojtowicz, I.A. Akimov, M. Bayer, Access to long-term optical memories using photon echoes retrieved from semiconductor spins. Nat. Photonics **8**, 851–857 (2014)
32. Q. Li, D.Z. Xu, C.Y. Cai, C.P. Sun, Recoil effects of a motional scatterer on single-photon scattering in one dimension. Sci. Rep. **3**, 3144 (2013)
33. B. Najjari, A.B. Voitkiv, A. Artemyev, A. Surzhykov, Simultaneous electron capture and bound-free pair production in relativistic collisions of heavy nuclei with atoms. Phys. Rev. A **80**, 012701 (2009)
34. J. Park, H. Kim, Y. Cho, C. Shin, Simple modeling and simulation of photovoltaic panels using Matlab/Simulink. Adv. Sci. Technol. Lett. **73**, 147–155 (2014)
35. T. Pregnolato, E.H. Lee, J.D. Song, S. Stobbe, P. Lodahl, Single-photon non-linear optics with a quantum dot in a waveguide. Nat. Commun. **6**, 8655 (2015)
36. A. Reinhard, T. Volz, M. Winger, A. Badolato, K.J. Hennessy, E.L. Hu, A. Imamoğlu, Strongly correlated photons on a chip. Nat. Photonics **6**, 93–96 (2012)
37. B. Robyns, B. Francois, P. Degobert, J.P. Hautier, *Vector Control of Induction Machines* (Springer-Verlag, London, 2012)
38. K.G. Sharma, A. Bhargava, K. Gajrani, Stability analysis of DFIG based wind turbines connected to electric grid. IREMOS **6**, 879–887 (2013)
39. G. Sivasankar, V.S. Kumar, Improving low voltage ride through of wind generators using STATCOM under symmetric and asymmetric fault conditions. IREMOS **6**, 1212–1218 (2013)
40. A. Soedibyo, F.A. Pamuji, M. Ashari, Grid quality hybrid power system control of microhydro, wind turbine and fuel cell using fuzzy logic. IREMOS **6**, 1271–1278 (2013)
41. J.J. Soon, K.S. Low, Optimizing photovoltaic model parameters for simulation, in *IEEE International Symposium on Industrial Electronics*, (IEEE, Piscataway, NJ, 2012), pp. 1813–1818
42. A.H. Sørensen, The pairproduction channel in atomic processes. Radiat. Phys. Chem. **75**, 656–695 (2006)

43. R. Szafron, A. Czarnecki, High-energy electrons from the muon decay in orbit: Radiative corrections. Phys. Lett. B **753**, 61–64 (2016)
44. M.S. Tame, K.R. McEnery, Ş.K. Özdemir, J. Lee, S.A. Maier, M.S. Kim, Quantum plasmonics. Nat. Phys **9**, 329–340 (2013)
45. Y.T. Tan, D.S. Kirschen, N. Jenkins, A model of PV generation suitable for stability analysis. IEEE Trans. Energy Convers. **19**, 748–755 (2004)
46. M.W.Y. Tu, W.M. Zhang, Non-Markovian decoherence theory for a double-dot charge qubit. Phys. Rev. B **78**, 235311 (2008)
47. S.R. Valluri, U. Becker, N. Grün, W. Scheid, Relativistic collisions of highly-charged ions. J. Phys. B: At. Mol. Phys. **17**, 4359–4370 (1984)
48. W. Xiao, W.G. Dunford, A. Capal, A novel modeling method for photovoltaic cells, in *35th Annula IEEE Power Electronics Specialists Conference, Aachen, Germany*, 2004, pp. 1950–1956
49. Y.F. Xiao, M. Li, Y.C. Liu, Y. Li, X. Sun, Q. Gong, Asymmetric Fano resonance analysis in indirectly coupled microresonators. Phys. Rev. A **82**, 065804 (2010)
50. W.B. Yan, H. Fan, Single-photon quantum router with multiple output ports. Sci. Rep. **4**, 4820 (2014)
51. L. Yang, S. Wang, Q. Zeng, Z. Zhang, T. Pei, Y. Li, L.M. Peng, Efficient photovoltage multiplication in carbon nanotubes. Nat. Photonics **5**, 672–676 (2011)
52. W.M. Zhang, P.Y. Lo, H.N. Xiong, M.W.Y. Tu, F. Nori, General non-Markovian dynamics of open quantum systems. Phys. Rev. Lett. **109**, 170402 (2012)
53. Y. Zhu, X. Hu, H. Yang, Q. Gong, On-chip plasmon-induced transparency based on plasmonic coupled nanocavities. Sci. Rep. **4**, 3752 (2014)

Chapter 7
Photon Application in the Design of Sustainable Buildings to Console Global Energy and Environment

Abstract Photon energy has been implemented to design the sustainable building where at least 25% of its exterior curtain skin wall could be used as the acting photovoltaic (PV) panel to trap the solar energy to transform into electricity to satisfy energy demand for a building itself without any outsource connection. Given the current rate of conventional fuel consumption, atmospheric greenhouse gas emission (GHGs) increasing rapidly where building sector along responsible for 40% GHGs emission. These GHGs ultimately cause environmental vulnerabilities such as climate change, stratospheric ozone depletion, acid rain, flooding, and air toxicity which threaten the survival of all living beings on Earth. Therefore, the mechanism of photophysical transformation by the acting PV panel of the building exterior skin in response to solar radiation shall indeed be a cutting-edge technology to console the global energy demand and mitigate the climate change perplexity dramatically.

Keywords Global environmental vulnerability · Solar radiation · PV panel acting build skin · Clean energy production · Climate change mitigation

Highlights

- Building curtain wall has been utilized as PV panel to capture solar energy.
- Photon has been clarified by PV panel to convert into electricity energy.
- Energy production was calculated to meet the energy need for a building.

Introduction

Massive development of conventional urbanization throughout the world is consuming fossil energy tremendously. Consequently, it is causing severe environmental perplexity such as acid deposition, stratospheric ozone depletion, and climate change severely, where traditional building development alone is responsible for 40% of this environmental and climate change disaster [3, 5, 60]. Besides, conventional energy deposition is getting finite level. At present, the total amount of fossil fuel reserved in

M. F. Hossain, *Sustainable Design for Global Equilibrium*,
https://doi.org/10.1007/978-3-030-94818-4_7

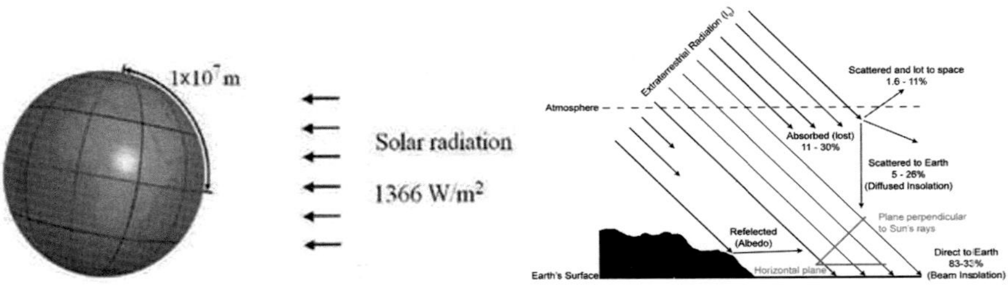

Fig. 7.1 (**a**) Shows the yearly solar irradiance arrives on surface of Earth. The mean solar energy on the Earth is 1366 W/m^2, respectively. The length of the meridian of Earth is 10,000,000 m. The total solar irradiance that arrives at the surface of Earth per year is 5,460,000 EJ calculatively. (**b**) The effect of the atmosphere on the solar radiation reaching the Earth's surface

the whole world is 36,600 EJ (crude oil 1.65×10^{11} t or unit energy 4.2×10^{10} J/t is equivalent to 6930 EJ; natural gas 1.81×10^{14} t or unit energy 3.6×10^{7} J/m^3 is equivalent to 6500 EJ; high-quality coal 4.90×10^{11} t or unit energy 3.1×10^{10} J/t equivalent to 15,000 EJ; low-quality coal 4.3×10^{11} t or 1.9×10^{10} J/t unit energy equivalent to 8200 EJ) [6, 7, 9]. The annual energy consumption worldwide was 283 EJ in 1980, 347 EJ in 1990, 400 EJ in 2000, 511 EJ in 1994, and 590 EJ in 2025. This rate is expected to be increased at 607 EJ in the year 2020, 702 EJ in 2030, 855 EJ in 2040, and 988 EJ in 2050 [2, 8, 17]. In the year 2017 the total fossil energy consumption was 560 EJ where building sector alone consumed 224 EJ. This means if the current level of fossil fuel consumption continues, the total fossil fuel energy source will be run out in 65 years. Since the fossil fuel is causing severe environmental disaster and is also getting finite level rapidly, clean and renewable energy source is an urgent demand for the sustainability of the Earth. Here, the solar radiation has been defined as a sustainable energy to implement in the building exterior skin by the process of photophysical reaction to act as the PV panel to produce the clean energy to satisfy its total energy demand without any outsource connection. The average solar radiation on the Earth's surface is 1366 W/m^2, commonly known as solar constant. The radius of earth is $(2/\pi) \times 10^{7}$ m and thus the total solar radiation reaching the Earth is solar $= 1366 \times (4/\pi) \times 10^{14} \cong 1.73 \times 10^{17}$ W [9, 36]. There are 86,400 s a day, with an average of 365.2422 days a year. Total annual solar radiation energy per year $= 1.73 \times 10^{17} \times 86{,}400 \times 365.2422 = 5.46 \times 10^{24}$ J, Equivalent to 5,460,000 EJ/year (Fig. 7.1). The world energy consumption in 2017 was 5.60×10^{20} J $= 560$ EJ, of which nearly 40% was spent by building sector. Only 0.01% of the solar energy reaching the earth can meet the global energy demand where building sector can play vital role to harvest solar energy by its exterior skin to fulfill the mission of clean energy technology for the sustainability of the mother Earth.

In the past several decades, huge research has been performed on solar energy and its supply technology to the national grid for commercial application as the source of alternative energy technology [10, 11, 15, 16]. Shi et al. showed that solar energy can be harvested by installing massive solar panel in particular places in tropical or sub-tropical area and then supplying to the national grid as a source of sustainable energy technology [54]. Gleyzes et al. suggested an advanced mechanism of solar panel by the application of graphene silicon surface that can have the ability of maximum 30% efficiency to capture the solar energy to convert into clean energy [14]. Reinhard showed that the breakdown of photon energy and its application by quantum machines into a PV panel can produce tremendous amount of clean energy [48]. All these research findings are indeed interesting, but these technologies required additional place and technology and supply mechanism to utilize the solar energy, and none of these investigations revealed that this solar energy can be utilized directly by the building itself by using its exterior curtain wall to act as the PV panel to produce energy. In this research, therefore, an innovative technology has been proposed to design all buildings to have at least 25% of the exterior curtain walls to be used as the photovoltaic (PV) panel to capture the solar radiation and then convert it into clean energy to meet the total energy demand for a building.

Methods and Materials

The building is proposed to be designed in such a way that 25% of the exterior curtain walls are to be built with solar panels. Prior to that this solar panel acting curtain wall skin must be determined the factor involved angle, latitude, longitude, and coordinate transformation in a Cartesian coordinate system to ensure that the maximum solar thermal radiation can be captured by the panel (Fig. 7.2). Considering angle, the acting solar panel needs to be designed for the effect of latitude and module tilt on the solar radiation received throughout the year and the module is to be facing south in the northern hemisphere and north in the southern hemisphere [12, 20, 21]. Cartesian coordinates for the horizon system need to be used accordingly where south should be x, west should be y, and the zenith should be z [13, 14, 61]. These positions of the celestial body are to be determined by two angles, height h and azimuth angle A; Cartesian coordinates for the equatorial system of z'-axis points to the North Pole, east-west axis y'-axis and the x'-axis are to be perpendicular to the both directions. Then the position of the celestial body is to be determined by declination δ and hour angle ω; refer to figure in the preceding texts.

Since the light is an electromagnetic wave which is produced when an electric charge vibrates due to a hot object, acting PV panel installation must follow the longitude and latitude, polar coordinates, and three-dimension axis (x, y, z) to get maximum sunlight obtainable on the building the whole year considering the Stefan–Boltzmann laws. The clarification of Stefan–Boltzmann laws is that the electromagnetic waves follow the equal-partition radiation intensity once it is emitted on the plane of the PV panel [4, 35].

Once the angles and Cartesian coordinates of the solar panel are determined, the panel wiring diagram is focused on ensuring that the photovoltaic DC current can transmit into AC current at high efficiency [23, 37]. Thus, two of the Kyocera plates

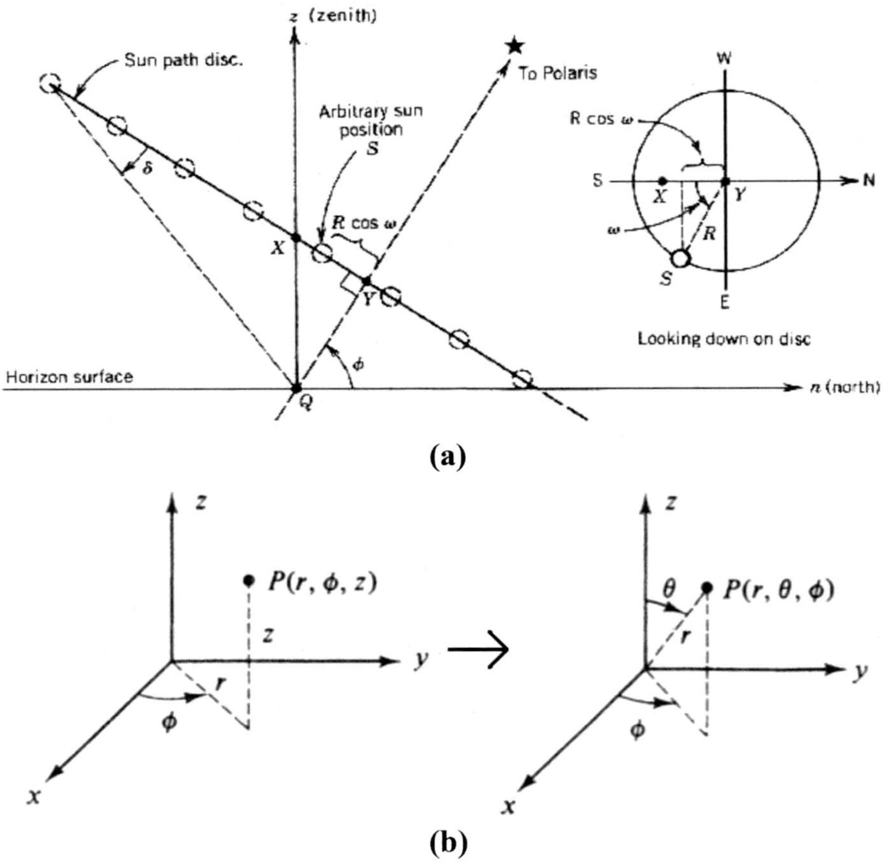

Fig. 7.2 (**a**) Photovoltaic (PV) panel angle has been calculated in consideration of the impact of latitude and mode tilt on the solar energy received within the year. (**b**) The Cartesian coordinate has been utilized for the horizon clarification considering use the conventions where south is x, west is y, and z the position of a celestial body and it is determined by two angles of the equatorial systems

are mounted on aluminum bars to carry light from photovoltaic modules DC and then multiplies into the short-circuit current by 125% and uses this value for all 80% efficiency and then get AC current 125% of the continuous flow of the current by using three-parallel circuit.

Thereafter, number of parameters of the current–voltage (*I*–*V*) characteristic are necessarily explained considering two/single-diode equivalent circuit model of the PV cell [35, 38, 52]. Subsequently, the photovoltaic array at various parameters (*voltage proliferation, transformation rate, an PVVI curves*) and least control strategy are being used to confirm the active electricity production (I_{v+}) from solar energy to the utilization of it by the building itself.

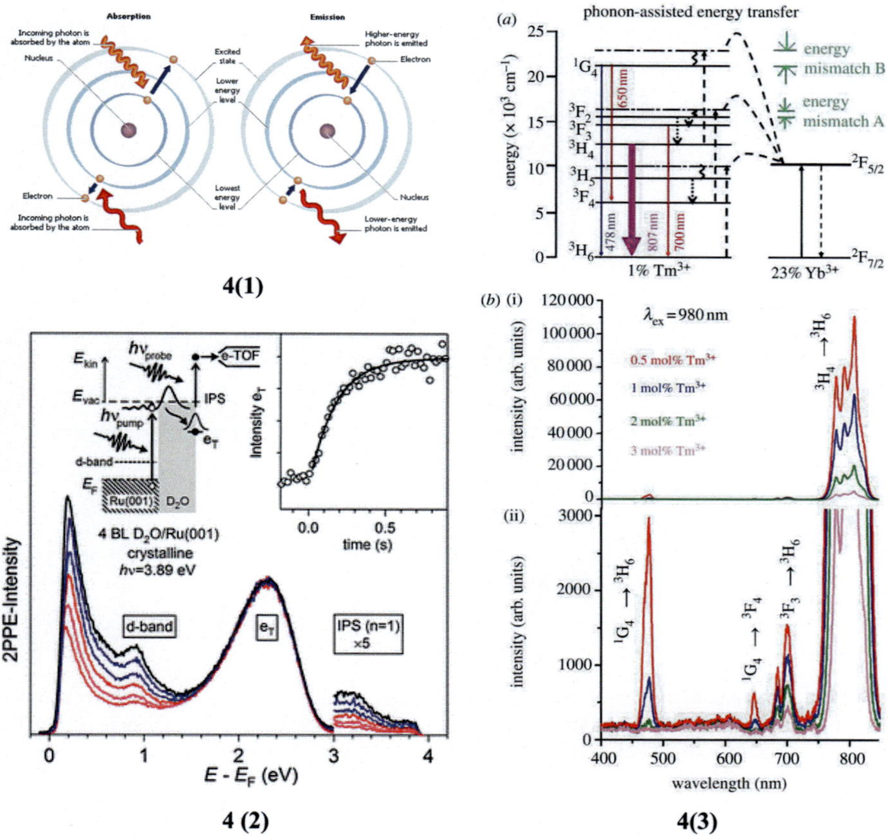

Fig. 7.3 Diagram of PV system model. (1) The module photon absorption and emission reaction mechanism once solar irradiance are on the photovoltaic mode, (2) solar intensity rate, (3) (a) the irradiation energy at 980 nm; *solid arrows* represent photon absorption or emission, *dotted arrows* represent multi-phonon relaxation, *dashed arrows* represent phonon assisted energy transfer, and *wavy arrows* represent energy mismatches. (b) (i) The up-conversion spectrum of UCNPs excited at 980 nm (90 mW), (ii) vertical axis is magnified to show the details of the visible emission spectra

The next step is to determine the photovoltaic current production by I_{pv} calculation from the model of one diode (Fig. 7.4a), considering *I*–*V*–*R* relationship (Fig. 7.4b), and using the illumination received by the photovoltaic array to convert from DC to AC and then use for domestic energy and low voltage current demand (Fig. 7.4c).

Since the solar thermal energy is taken by the photophysical reaction of photovoltaic panel (PV) which is a continuum flow of photons into the PV panel, it is necessary to determine the excellent photophysical reaction into the curtain wall acting PV panel. For this, it is essential to clarify solar thermal conductivity and solar cell anti-reflective coatings by analyzing quantum electrodynamics, the most

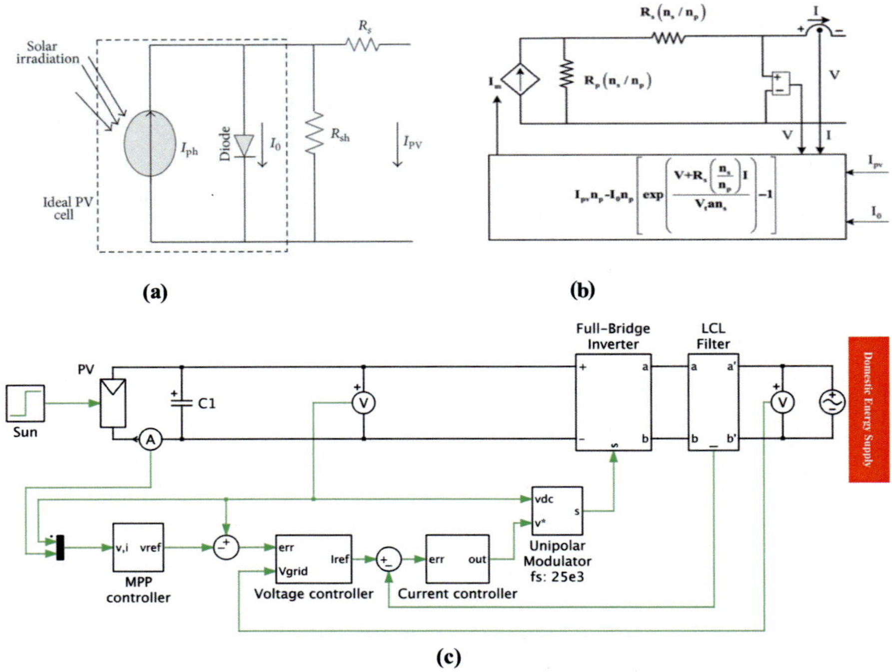

Fig. 7.4 Single-diode circuit of a photovoltaic (PV) cell modeled by MATLAB simulation, (**a**) the photovoltaic current production, (**b**) the model with a diode considering *I–V–R* relationship (**c**), the conversion process of DC to AC for the use of domestic energy and low voltage current for the building

effective fields in modern physics to capture much more solar energy [18, 22, 24]. Subsequently, classical statistical physics has been used to determine the radiation energy density of inner surface on its curtain wall considering the electromagnetic wave and Maxwell–Boltzmann constant statistics (Fig. 7.4). Therefore, a mathematical model of photovoltaic dynamic, in this research, has been developed to capture maximum solar energy by the acting PV panel of building exterior skin curtain wall where the following equation calculates the energy output of a photovoltaic (PV) cell:

$$P_{\mathrm{pv}} = \eta_{\mathrm{pvg}} A_{\mathrm{pvg}} G_{\mathrm{t}} \tag{7.1}$$

In this equation, η_{pvg} refers to the PV-generation efficiency, A_{pvg} refers to the PV generator area (m^2), and G_{t} refers to the solar radiation in a titled module plane (W/m^2). η_{pvg} can be further defined as:

$$\eta_{\text{pvg}} = \eta_{\text{r}}\eta_{\text{pc}}[1 - \beta(T_{\text{c}} - T_{\text{cref}})] \tag{7.2}$$

η_{pc} refers to the power conditioning efficiency; when MPPT is applied, it is equal to 1; β refers to temperature coefficient (0.004–0.006 per °C); η_r refers to the reference module efficiency; and T_{cref} refers to the reference cell temperature in °C. The reference cell temperature (T_{cref}) can be obtained from the relation below:

$$T_{\text{c}} = T_{\text{a}} + \left(\frac{\text{NOCT} - 20}{800}\right)G_{\text{t}} \tag{7.3}$$

T_a refers to the ambient temperature in °C, G_t refers to the solar irradiance in a tilted module plane (W/m^2), and NOCT refers to the standard operating cell temperature in Celsius (°C) degree. The total irradiance in the solar cell, considering both standard and diffuse solar irradiance, can be estimated by the following equation:

$$I_{\text{t}} = I_{\text{b}}R_{\text{b}} + I_{\text{d}}R_{\text{d}} + (I_{\text{b}} + I_{\text{d}})R_{\text{r}} \tag{7.4}$$

The solar cell is essentially a P-N junction semiconductor able to produce electricity via the PV effect; solar cells are interconnected in a series-parallel configuration to form a photovoltaic (PV) cell [61, 62, 64]. Besides, to improve the efficiency of the resulting photovoltaic (PV), graphene is integrated into the PV module [53, 54, 63].

Using a standard single diode, as depicted in Fig. 7.2, for a cell with N_s series-connected arrays and N_p parallel-connected arrays, the cell current must be related to the cell voltage as

$$I = N_{\text{p}}\left[I_{\text{ph}} - I_{\text{rs}}\left[\exp\left(\frac{q(V + IR_{\text{s}})}{AKTN_{\text{s}}} - 1\right)\right]\right] \tag{7.5}$$

where

$$I_{\text{rs}} = I_{\text{rr}}\left(\frac{T}{T_{\text{r}}}\right)^3 \exp\left[\frac{E_{\text{G}}}{AK}\left(\frac{1}{T_{\text{r}}} - \frac{1}{T}\right)\right] \tag{7.6}$$

In Eqs. (7.5) and (7.6), q refers to the electron charge (1.6×10^{-9} C), K refers to Boltzmann's constant, A refers to the diode idealist factor, and T refers to the cell temperature (K). I_{rs} refers to the cell reverse saturation current at T, T_r refers to the cell referred temperature, I_{rr} refers to the reverse saturation current at T_r, and E_G refers to the bandgap energy of the semiconductor used in the cell. The photocurrent I_{ph} varies with the cell's temperature and radiation as follows:

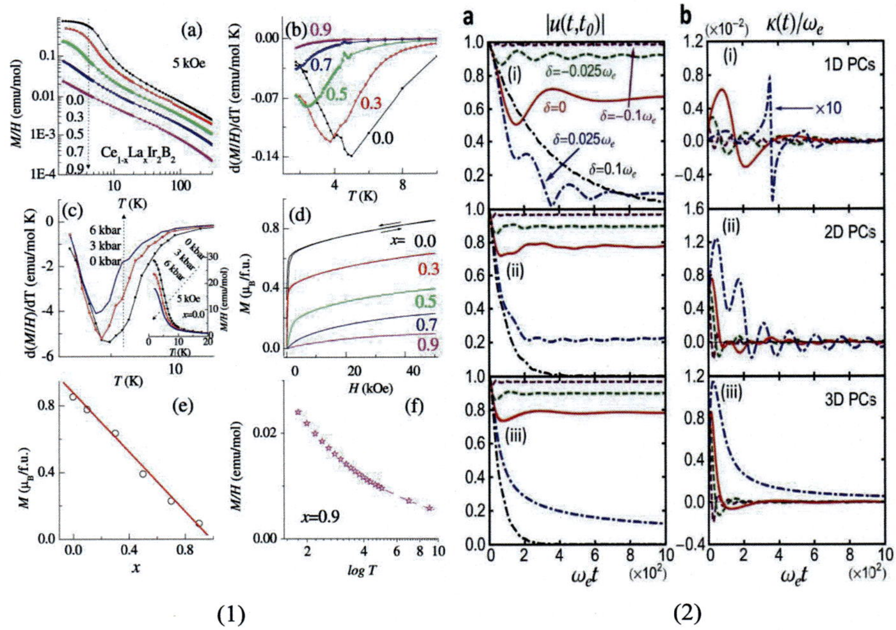

Fig. 7.5 (1) Photonic thermal activation mechanism in various PB parameters, (2) dynamic photons emission in PV cells. (a) Shows the charge generation in the PB area $\langle a(t)\rangle = 5u(t, t_0)\langle a(t_0)\rangle$, (b) demonstrates the dynamic photon emission rate $k(t)$, plotted for (i) 1D, (ii) 2D, and (iii) 3D areal field into the PV cells [25, 27]

$$I_{\mathrm{ph}} = \left[I_{\mathrm{SCR}} + k_{\mathrm{i}}(T - T_{\mathrm{r}})\frac{S}{100}\right] \tag{7.7}$$

I_{SCR} refers to the cell short-circuit current at the reference temperature and irradiance, k_{i} refers to the short-circuit current temperature coefficient, and S refers to the solar irradiance (mW/cm^2). The I–V characteristics of the photovoltaic (PV) cell can be derived using a single-diode model which includes an additional shunt resistance concurrent with the optimal shunt diode model as follows:

$$I = I_{\mathrm{ph}} - I_{\mathrm{D}} \tag{7.8}$$

$$I = I_{\mathrm{ph}} - I_0\left[\exp\left(\frac{q(V + R_{\mathrm{s}}I)}{AKT} - 1\right) - \frac{V + R_{\mathrm{s}}I}{R_{\mathrm{sh}}}\right] \tag{7.9}$$

I_{ph} refers to the photocurrent (A), I_{D} refers to the diode current (A), I_0 refers to the inverse saturation current (A), A refers to the diode constant, q refers to the charge of the electron (1.6×10^{-9} C), K refers to Boltzmann's constant, T refers to the cell temperature (°C), R_{s} refers to the series resistance (ohm), R_{sh} refers to the shunt resistance (Ohm), I refers to the cell current (A), and V refers to the cell voltage (V). The output current of the PV cell using the diode model can be described as follows:

$$I = I_{pv} - I_{D1} - I_{D2} - \left(\frac{V + IR_s}{R_{sh}}\right) \tag{7.10}$$

where

$$I_{D1} = I_{01}\left[\exp\left(\frac{V + IR_s}{a_1 V_{T1}}\right) - 1\right] \tag{7.11}$$

$$I_{D2} = I_{02}\left[\exp\left(\frac{V + IR_s}{a_2 V_{T2}}\right) - 1\right] \tag{7.12}$$

I_{01} and I_{02} are the reverse saturation currents of diode 1 and diode 2, respectively, and V_{T1} and V_{T2} are the thermal voltages of the respective diodes. The diode idealist constants are represented by a_1 and a_2. The simplified model of the photovoltaic (PV) system model is presented below:

$$v_{oc} = \frac{V_{oc}}{cKT/q} \tag{7.13}$$

$$P_{max} = \frac{\frac{V_{oc}}{cK\,T/q} - \ln\left(\frac{V_{oc}}{cK\,T/q} + 0.72\right)}{\left(1 + \frac{V_{oc}}{nK\,T/q}\right)}\left(1 - \frac{V_{oc}}{\frac{V_{oc}}{I_{sc}}}\right) \times \left(\frac{V_{oc0}}{1 + \beta \ln \frac{G_0}{G}}\right)\left(\frac{T_0}{T}\right)^{y} I_{sc0}\left(\frac{G}{G_0}\right)^{a} \tag{7.14}$$

where ν_{oc} refers to the normalized value of the open-circuit voltage V_{oc} related to the thermal voltage $V_t = nkT/q$, K refers to Boltzmann's constant, n refers to the idealist factor ($1 < n < 2$), T refers to the temperature of the photovoltaic (PV) module in Kelvin, α refers to the factor responsible for all the nonlinear effects on which the photocurrent depends, q refers to the electron charge, γ refers to the factor representing all the nonlinear temperature–voltage effects, while β refers to a photovoltaic (PV) module technology-specific dimensionless coefficient. Equation (7.14) only represents the maximum energy output of a single photovoltaic (PV) module while a real system consists of several photovoltaic (PV) modules connected in series and in parallel. Therefore, the equation of total power output for an array with N_s cells connected in series and N_p cells connected in parallel with power P_M for each module would be for PV panel

$$P_{array} = N_s N_p P_M \tag{7.15}$$

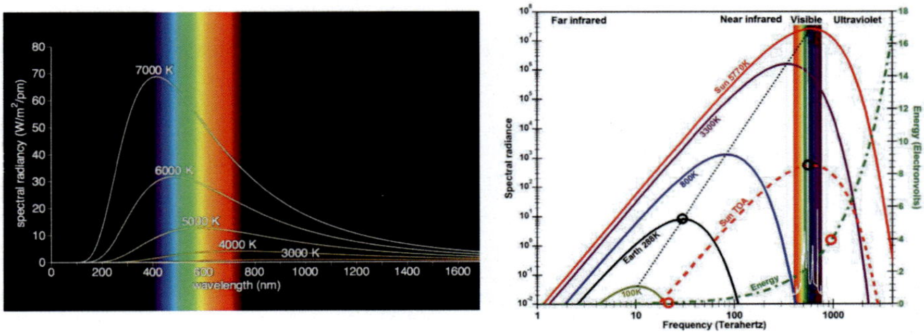

(a) (b)

Fig. 7.6 The thermal energy density of the solar radiation frequencies shown by the classical statistical physics and the figure depicts the solar radiation in various temperatures. (**a**) The spectral irradiance of the light in different wavelengths. (**b**) The radiation in difference frequencies at different temperatures where the maximum irradiance by sun nearly at 5800 K (actual 5770 K) power is equivalent to 6.31×10^7 (W/m^2); peak E is 1.410 (eV); peak λ is 0.88 (μm); peak μ is 2.81×10^7 (W/m^2 eV)

Naturally the movement of photon flux applied to the solar panel will be activated by the photophysical reactions to deliver energy level charges [26, 33]. Since the energy density of the solar radiation considering the photon wave frequency has been modeled by using the classical statistical physics in Fig. 7.6a, the maximum solar energy formation considering a single photon excitation at the rate of 1.4 eV with an energy value of 27.77 MW/m^2 eV has been determined in Fig. 7.6b.

Results and Discussion

The result of the PV model is determined by the *I–V* equation of PV cells in the single-diode mode. The *I–V* relationship equation in the PV panel can be expressed as

$$I = I_{\mathrm{L}} - I_{\mathrm{o}}\left\{\exp\left[\frac{q(V+I_{\mathrm{RS}})}{AkT_{\mathrm{c}}}\right] - 1\right\} - \frac{(V+I_{\mathrm{RS}})}{R_{\mathrm{Sh}}} \tag{7.16}$$

I_{L} represents the photon generating current, I_{o} represents the saturated current in the diode, R_{s} represents resistance in a series, A represents the diode passive function, k ($= 1.38 \times 10^{-23}$ W/m^2 K) represents Boltzmann's constant, q ($= 1.6 \times 10^{-19}$C) represents the charge amplitude of an electron, and T_{c} represents the functional cell temperature. Subsequently, the *I–q* relationship in the PV cells varies owing to the diode current and/or saturation current, which can be expressed as [28, 34]

$$I_o = I_{RS}\left(\frac{T_c}{T_{ref}}\right)^3 \exp\left[\frac{qEG\left(\frac{1}{T_{ref}} - \frac{1}{T_c}\right)}{KA}\right] \tag{7.17}$$

where I_{RS} represents the saturated current considering the functional temperature and solar irradiance and qEG represents the bandgap energy into the silicon and graphene PV cell considering the normal, normalized, and perfect modes.

Considering a PV module, the I–V equation, apart from the I–V curve, is a conjunction of I–V curves among all cells of the PV panel. Therefore, the equation can be rewritten as follows to determine the V–R relationship:

$$V = -IR_s + K\ \log\left[\frac{I_L - I + I_o}{I_o}\right] \tag{7.18}$$

Here, K is as a constant $\left(= \frac{AkT}{q}\right)$ and I_{mo} and V_{mo} are the current and voltage in the PV panel. Therefore, the relationship between I_{mo} and V_{mo} shall be same as the PV cell I–V relationship:

$$V_{mo} = -I_{mo}R_{Smo} + K_{mo}\log\left(\frac{I_{Lmo} - I_{mo} + I_{Omo}}{I_{Omo}}\right) \tag{7.19}$$

where I_{Lmo} represents the photon-generated current, I_{Omo} represents the saturated current into the diode, R_{Smo} represent the resistance in series, and K_{mo} represents the factorial constant. Once all non-series (NS) cells are interconnected in series, then the series resistance shall be counted as the summation of each cell series resistance $R_{Smo} = N_s \times R_{s,}$ and the constant factor can express as $K_{mo} = N_s \times K$. There is a certain amount of current flow into the series-connected cells; thus, the current flow in Eq. (7.5) remains the same in each component, i.e., $I_{Omo} = I_o$ and $I_{Lmo} = I_L$. Thus, the module $I_{mo} - V_{mo}$ equation for the N_s series of connected cells will be written as

$$V_{mo} = -I_{mo}N_sR_s + N_sK\log\left(\frac{I_L - I_{mo} + I_o}{I_o}\right) \tag{7.20}$$

Similarly, the current–voltage calculation can be rewritten for the parallel connection once all N_p cells are connected in parallel mode and can be expressed as follows [50, 56, 65]:

$$V_{mo} = -I_{mo}\frac{R_s}{N_p} + K\log\left(\frac{N_{sh}I_L - I_{mo} + N_pI_o}{N_pI_o}\right) \tag{7.21}$$

Because the photon-generated current primarily will depend on the solar irradiance and relativistic temperature conditions of the PV panel, the current can be calculated using the following equation:

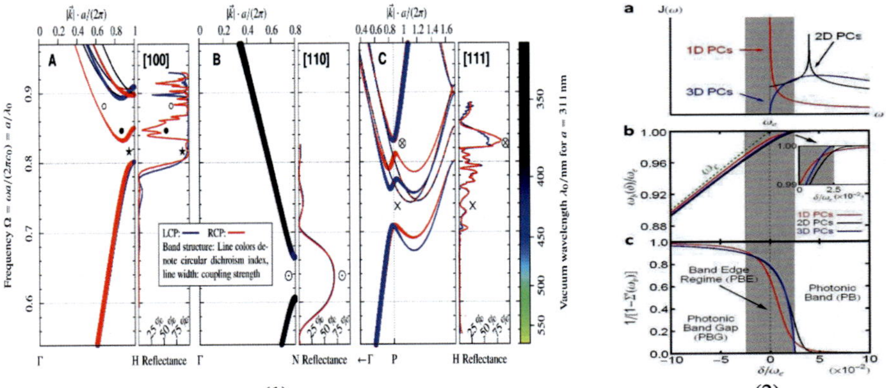

(1) (2)

Fig. 7.7 (1) Photonic thermal energy conversion modes. (2) (a) Unit area vs frequency at various DOSs in 1D, 2D, and 3D glazing wall skins. (b) Photonic modes of frequencies for functional tuning. (c) Photonic modes of magnitudes to release the energy calculated by Eq. (7.2). The photonic modes depict the crossover into the glazing wall skin in 1D and 2D. In the complex 3D transitional state, they depict the crossover into the PV cell once the point break frequency *vc* transforms from a PBG area to a photonic band (PB) area [47, 66]

$$I_{\mathrm{L}} = G[I_{\mathrm{sc}} + K_{\mathrm{I}}\,(T_{\mathrm{c}} - T_{\mathrm{ref}})] * V_{\mathrm{mo}} \tag{7.22}$$

where I_{sc} represents PV current at 25 °C and kW/m^2, K_I represents the relativistic PV panel coefficient factor, T_{ref} represents the PV panel's functional temperature, and G represent the solar energy in kW/m^2.

Conversion of Electricity

To convert solar thermal energy into electricity, a single-diode circuit has been used consisting of a small disk of semiconductor attached by wire to a circuit consisting of a positive and a negative film of silicon placed under a thin slice of glass and attached to graphene, using a building's exterior curtain wall skin. Necessarily, the PV panel shall have an *open circuit point*, where the current and voltage is at the maximum, open circuit voltage V_{oc}, and the *maximum power point* can be clarified using the maximum current and voltage calculation immediately upon capturing the non-equilibrium photons [29, 51, 55]. The power delivered by a PV panel will thus have the capability to attain a maximum value at the points (I_{mp},V_{mp}) [19, 30, 49]. It is confirmed by the PV panel's capability of such current–voltage flow to get net energy production by the PV panel by analyzing circuit models of the PV module (a) normal, (b) normalized, and (c) perfect modes where perfect modes are the best option.

To establish a connection between the number of light-quanta solar energy by steady state, the intensity of solar irradiance is considered to convert it into electricity energy by PV panel [41, 42, 46]. The number of stationary states of light-quanta is a certain type of polarization whose frequency is in the range of ν_r to $\nu_r + d\nu_r$ [44, 48]. From there maximum solar radiation it can be achieved at 1.4 eV with an energy value of 27.77 mW/m^2 eV based on an average of 5 h solar irradiance harvesting in a day peak levels, which is the equivalent of 27,770 kW/year or 7.6 kW/day energy [31, 67]. Due to physical principles, there are losses in the conversion of solar energy into DC power and converting direct current into alternating current (AC). This ratio of AC to DC is called "derating factor," which is typically 0.8 [1, 32, 45]. Thus, the surface texture of selective solar metal is excellent in energy conversation [35, 37, 56], since the current net conversion by solar panels is 125% higher level with an efficiency of 80% [10, 15, 16] of solar panels, which means that (27,770 × 1.25 × 0.8) = 27,770 kW/year or 7.6 kW/day. Energy remains equal to the solar initially what was before the introduction to the solar panel. Necessarily, the maximum solar irradiance is depicted 1.4 eV with an energy value of 27.77 mW/m^2 eV in Fig. 7.8 per year in an average of 5 h a day maximum levels for 365 days referents by panel Solar and black body [39, 50, 59]. A standard residential house requires average 6 kW/day [58, 62, 65]. Since the produced energy is equivalent to 27,770 kW/year or 7.6 kW/day, which in fact will meet the energy demand for a residential house required, 6 kW/day, by using only one solar panel 1 m^2. The average energy consumption rate monthly of commercial office or buildings is about 10,000 kW/day for a foot printing 32 m × 31 m with 30 m (10 floors), respectively [40, 43, 57]. In calculation of a building is an average of 32 m × 31 m footprint with a height of 30 m, total installed 1 m^2 PV panels requires 1195 units (945 + 250) with the capacity of 7.6 kW/unit energy production can provide total energy × 1195 = 9082 kW/day to meet the daily energy demand of about10,000 kW/day for a commercial office or building.

Savings on Energy Cost

On the other hand, the net cost for 30 years energy purchase from traditional utility sources for a standard industry (100 people capacity) at 0.12/kWh of 4000 kWh per month is (30 × 12 × 4000 × 0.12) $172,800. This difference between traditional energy use and curtain wall-assisted PV panel energy production clearly indicates the cost saving of $68,400 once curtain wall-assisted PV panel is used as the energy source.

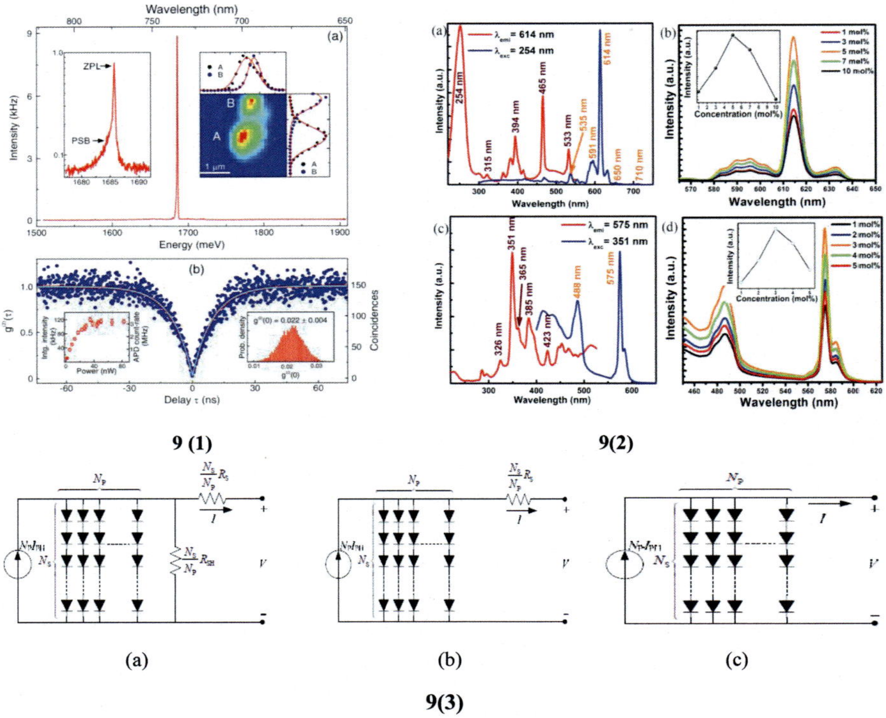

Fig. 7.8 Solar cell current–voltage characteristic features for conceptual function of (1) net optimal solar energy production intensity with respect to the delay of coincidental factor and (2) intensity of solar energy production in various wavelength, (3) the current–voltage module of the current source near the short circuit point and as a voltage source in the vicinity of the open circuit point at the condition of the PV module are (a) normal, (b) normalized, and (c) perfect modes

Table 7.1 The estimates are (1 m^2 each solar panel and total 1195 panels) prepared by confirming recent (June 2017) cost of material for selective manufacturer and labor rate added in accordance with international of union specified trade workers considering US location. The equipment rental cost is calculated as current rental market in conjunction with the standard practice of construction of the production rate

Design and construction cost of the curtain wall acting solar panel for energy production of a ten stories building					
List of component	Materials cost	Labor cost	Equipment cost	GC and OH cost	Total cost
Acting solar panel	\$10,000	\$5000	\$2500	\$3500	\$21,000
Instrumentation	\$2000	\$1000	\$2000	\$1000	\$6000
Electrical, and mechanical control	\$2500	\$1000	\$1000	\$900	\$5400
Supply for 30 years cost at \$0.05/kWh for monthly 4000 kWh for 100 people					\$72,000
				Total cost	\$104,400

Conclusions

In recent decades, concern about the deadly risks of greenhouse gases (GHGs) has been growing due to their level of accumulation in the atmosphere and the adverse impact on Earth. Given that conventional energy consumption is the final factor that leads to environmental perplexity and climate change, the intelligent deployment of solar energy, in effect, for a better methodological application to mitigate global environmental perplexity and climate changes is an urgent demand. In addition, the limited nature of the fossil fuel reserve poses a serious challenge for future energy supply as the consumption of fossil fuels will end in the next 65 years. In this study, it is therefore proposed to ensure an economical, reliable sustainable energy technology to meet the future energy demand which is also climate-friendly. Simply capturing solar thermal energy using exterior curtain wall demonstrated in this research shall indeed be the innovative technology to meet the total energy demand for the building sector and contribute with the quota to mitigate the global energy needs. Therefore, application of this sustainable energy technology is very much computational that need to be encouraged globally for the maximum utilization of solar energy harvesting by building sector to secure a sustainable and environmentally friendly earth.

Acknowledgments This research was supported by Green Globe Technology, Inc. under grant RD-02018-03 for building a better environment. Any findings, predictions, and conclusions described in this article are solely those of the authors, who confirm that the article has no conflicts of interest for publication in a suitable journal.

References

1. R. Andreas, K. Norbert, R. Gerhard, R. Stephan, A quantum gate between a flying optical photon and a single trapped atom (RESEARCH: LETTER) (Report). Nature (10 Apr 2014)
2. D.K. Armani, T.J. Kippenberg, S.M. Spillane, K.J. Vahala, Ultra-high-Q toroid microcavity on a chip. Nature **421**, 925 (2003)
3. N. Artemyev, U.D. Jentschura, V.G. Serbo, A. Surzhykov, Strong electromagnetic field effects in ultra-relativistic heavy-ion collisions. Eur. Phys. J. C **72**, 1935 (2012)
4. H.F. Beyer, T. Gassner, M. Trassinelli, R. Heß, U. Spillmann, D. Banaś, K.-H. Blumenhagen, F. Bosch, C. Brandau, W. Chen, E. Chr Dimopoulou, R.E. Förster, A.G. Grisenti, S. Hagmann, P.-M. Hillenbrand, P. Indelicato, P. Jagodzinski, T. Kämpfer, M. Chr Kozhuharov, D.L. Lestinsky, Y.A. Litvinov, R. Loetzsch, B. Manil, R. Märtin, F. Nolden, N. Petridis, M.S. Sanjari, K.S. Schulze, M. Schwemlein, A. Simionovici, M. Steck, C.I. Th Stöhlker, S.T. Szabo, I. Uschmann, G. Weber, O. Wehrhan, N. Winckler, D.F.A. Winters, N. Winters, E. Ziegler, Crystal optics for precision X-ray spectroscopy on highly charged ions—Conception and proof. J. Phys. B At. Mol. Opt. Phys. **48**, 144010 (2015)
5. K.M. Birnbaum et al., Photon blockade in an optical cavity with one trapped atom. Nature **436**, 87–90 (2005)
6. K. Busch, G. von Freymann, S. Linden, S.F. Mingaleev, L. Tkeshelashvili, M. Wegener, Periodic nanostructures for photonics. Phys. Rep. **444**, 101 (2007)

7. D.E. Chang, A.S. Sørensen, E.A. Demler, M.D. Lukin, A single-photon transistor using nanoscale surface plasmons. Nat. Phys. **3**, 807–812 (2007)
8. J. Chen, C. Wang, R. Zhang, J. Xiao, Multiple plasmon-induced transparencies in coupled-resonator systems. Opt. Lett. **37**, 5133–5135 (2012)
9. M. Cheng, Y. Song, Fano resonance analysis in a pair of semiconductor quantum dots coupling to a metal nanowire. Opt. Lett. **37**, 978–980 (2012)
10. B. Dayan et al., A photon turnstile dynamically regulated by one atom. Science **319**, 1062–1065 (2008)
11. J.S. Douglas, H. Habibian, C. Hung, A. Gorshkov, H. Kimble, D. Chang, Quantum many-body models with cold atoms coupled to photonic crystals. Nat. Photonics **9**, 326–331 (2015)
12. J. Eichler, T. Stöhlker, Radiative electron capture in relativistic ion-atom collisions and the photoelectric effect in hydrogen-like high-Z systems. Phys. Rep. **439**, 1 (2007)
13. D. Englund et al., Resonant excitation of a quantum dot strongly coupled to a photonic crystal nanocavity. Phys. Rev. Lett. **104**, 073904 (2010)
14. S. Gleyzes et al., Quantum jumps of light recording the birth and death of a photon in a cavity. Nature **446**, 297 (2007)
15. R.J. Gould, Pair production in photon-photon collisions. Phys. Rev. **155**, 1404 (1967)
16. C. Guerlin et al., Progressive field-state collapse and quantum non-demolition photon counting. Nature **448**, 889 (2007)
17. N. Gupta, S.P. Singh, S.P. Dubey, D.K. Palwalia, Fuzzy logic controlled three-phase three-wired shunt active power filter for power quality improvement. Int. Rev. Electr. Eng. **6**(3), 1118–1129 (2011)
18. Z. Han, S.I. Bozhevolnyi, Plasmon-induced transparency with detuned ultracompact Fabry-Pérot resonators in integrated plasmonic devices. Opt. Exp. **19**, 3251–3257 (2011)
19. K. Hencken, Transverse momentum distribution of vector mesons produced in ultraperipheral relativistic heavy ion collisions. Phys. Rev. Lett. **96**, 012303 (2006)
20. M.F. Hossain, Solar energy integration into advanced building design for meeting energy demand. Int. J. Energy Res. **40**, 1293–1300 (2016)
21. M.F. Hossain, Design and construction of ultra-relativistic collision PV panel and its application into building sector to mitigate total energy demand. J Build. Eng. (2017a). https://doi.org/10.1016/j.jobe.2016.12.005
22. M.F. Hossain, Green science: Independent building technology to mitigate energy, environment, and climate change. Renew. Sustain. Energy Rev. (2017b). https://doi.org/10.1016/j.rser.2017.01.136
23. M.F. Hossain, Photonic thermal energy control to naturally cool and heat the building. Adv. Ther. Eng. **131**, 576–586 (2018a)
24. M.F. Hossain, Green science: Advanced building design technology to mitigate energy and environment. Renew. Sustain. Energy Rev. **81**(2), 3051–3060 (2018b)
25. M.F. Hossain, Transforming dark photon into sustainable energy. Int. J. Energy Environ. Eng. (2018c). https://doi.org/10.1007/s40095-017-0257-1
26. Y. Huang, C. Min, G. Veronis, Subwavelength slow-light waveguides based on a plasmonic analogue of electromagnetically induced transparency. Appl. Phys. Lett. **99**, 143117 (2011)
27. J.F. Huang, T. Shi, C.P. Sun, F. Nori, Controlling single-photon transport in waveguides with finite cross section. Phys. Rev. A **88**, 013836 (2013)
28. U. Jentschura, K. Hencken, V. Serbo, Revisiting unitarity corrections for electromagnetic processes in collisions of relativistic nuclei. Eur. Phys. J. C **58**(2), 281–289 (2008)
29. J.D. Joannopoulos, P.R. Villeneuve, S. Fan, Photonic crystals: Putting a new twist on light. Nature **386**, 143 (1997)
30. S.A. Klein, Calculation of flat-plate collector loss coefficients. Sol. Energy **17**, 79–80 (1975)
31. A.G. Kofman, G. Kurizki, B. Sherman, Spontaneous and induced atomic decay in photonic band structures. J. Mod. Opt. **41**, 353 (1994)
32. P. Kolchin, R.F. Oulton, X. Zhang, Nonlinear quantum optics in a waveguide: Distinct single photons strongly interacting at the single atom level. Phys. Rev. Lett. **106**, 113601 (2011)

33. C. Lang et al., Observation of resonant photon blockade at microwave frequencies using correlation function measurements. Phys. Rev. Lett. **106**, 243601 (2011)
34. C.U. Lei, W.M. Zhang, A quantum photonic dissipative transport theory. Ann. Phys. **327**, 1408 (2012)
35. Q. Li, D.Z. Xu, C.Y. Cai, C.P. Sun, Recoil effects of a motional scatterer on single-photon scattering in one dimension. Sci. Rep. **3**, 3144 (2013)
36. J.Q. Liao, C.K. Law, Correlated two-photon transport in a one-dimensional waveguide side-coupled to a nonlinear cavity. Phys. Rev. A **82**, 053836 (2010)
37. J.Q. Liao, C.K. Law, Correlated two-photon scattering in cavity optomechanics. Phys. Rev. A **87**, 043809 (2013)
38. P. Lo, H. Xiong, W. Zhang, Breakdown of Bose-Einstein distribution in photonic crystals. Sci. Rep **5**, 9423 (2015)
39. P. Longo, P. Schmitteckert, K. Busch, Few-photon transport in low-dimensional systems. Phys. Rev. A **83**, 063828 (2011)
40. X. Lü, W. Zhang, S. Ashhab, Y. Wu, F. Nori, Quantum-criticality-induced strong Kerr nonlinearities in optomechanical systems. Sci. Rep **3**, 2943 (2013)
41. M.T. Manzoni, D.E. Chang, J.S. Douglas, Simulating quantum light propagation through atomic ensembles using matrix product states. Nat. Commun. **8**, 1743 (2017)
42. H. Matteo Mariantoni, R.C. Wang, M.L. Bialczak, et al., Photon shell game in three-resonator circuit quantum electrodynamics. Nat. Phys. **7**, 287–293 (2011)
43. B. Najjari, A. Voitkiv, A. Artemyev, A. Surzhykov, Simultaneous electron capture and bound-free pair production in relativistic collisions of heavy nuclei with atoms. Phys. Rev. A **80**, 012701 (2009)
44. D. O'Shea, C. Junge, J. Volz, A. Rauschenbeutel, Fiber-optical switch controlled by a single atom. Phys. Rev. Lett. **111**, 193601 (2013)
45. H. Okamoto, K. Yamaguchi, M. Haraguchi, T. Okamoto, Development of plasmonic racetrack resonators with a trench structure, in *Plasmonics Metallic Nanostructures and Their Optical Properties X*, (SPIE, San Diego, CA, 2012)
46. A.V. Poshakinskiy, A.N. Poddubn, Biexciton-mediated superradiant photon blockade. Phys. Rev. A **93**, 033856 (2016)
47. H. Rauh, Optical transmittance of photonic structures with linearly graded dielectric constituents. N. J. Phys. **12**, 073033 (2010)
48. A. Reinhard, Strongly correlated photons on a chip. Nat. Photonics **6**, 93–96 (2012)
49. D. Roy, Two-photon scattering of a tightly focused weak light beam from a small atomic ensemble: An optical probe to detect atomic level structures. Phys. Rev. A **87**, 063819 (2013)
50. E. Saloux, A. Teyssedou, M. Sorin, Explicit model of photovoltaic panels to determine voltages and currents at the maximum power point. Sol. Energy **85**, 713–722 (2011)
51. C. Sánchez Muñoz, F. Laussy, E. Valle, C. Tejedor, A. González-Tudela, Filtering multiphoton emission from state-of-the-art cavity quantum electrodynamics. Optica **5**(1), 14–26 (2018)
52. C. Sayrin et al., Real-time quantum feedback prepares and stabilizes photon number states. Nature **477**, 73 (2011)
53. J.T. Shen, S. Fan, Strongly correlated two-photon transport in a one-dimensional waveguide coupled to a two-level system. Phys. Rev. Lett. **98**, 153003 (2007)
54. T. Shi, S. Fan, C.P. Sun, Two-photon transport in a waveguide coupled to a cavity in a two-level system. Phys. Rev. A **84**, 063803 (2011)
55. M.S. Tame, K.R. McEnery, Ş.K. Özdemir, J. Lee, S.A. Maier, M.S. Kim, Quantum plasmonics. Nat. Phys. **9**, 329–340 (2013)
56. J. Tang, W. Geng, X. Xiulai, Quantum interference induced photon blockade in a coupled single quantum dot-cavity system. Sci. Rep. **5**, 9252 (2015)
57. M.W.Y. Tu, W.M. Zhang, Non-Markovian decoherence theory for a double-dot charge qubit. Phys. Rev. B **78**, 235311 (2008)
58. S.R. Valluri, U. Becker, N. Grün, W. Scheid, Relativistic collisions of highly-charged ions. J. Phys. B At. Mol. Phys. **17**, 4359 (1984)

59. Y. Wang, Y. Zhang, Q. Zhang, B. Zou, U. Schwingenschlogl, Dynamics of single photon transport in a one-dimensional waveguide twopoint coupled with a Jaynes-Cummings system. Sci. Rep. **6**, 33867 (2016)
60. Y.F. Xiao et al., Asymmetric Fano resonance analysis in indirectly coupled microresonators. Phys. Rev. A **82**, 065804 (2010)
61. W. Yan, H. Fan, Single-photon quantum router with multiple output ports. Sci. Rep. **4**, 4820 (2014)
62. W. Yan, J. Huang, H. Fan, Tunable single-photon frequency conversion in a Sagnac interferometer. Sci. Rep. **3**, 3555 (2013)
63. L. Yang, S. Wang, Q. Zeng, Z. Zhang, T. Pei, Y. Li, L. Peng, Efficient photovoltage multiplication in carbon nanotubes. Nat. Photonics **5**, 672–676 (2011)
64. G. Yu, A novel two-mode MPPT control algorithm based on comparative study of existing algorithms. Sol. Energy **76**(4), 455–463 (2004)
65. Z. Yu, X. Hu, H. Yang, Q. Gong, On-chip plasmon-induced transparency based on plasmonic coupled nanocavities. Sci. Rep. **4**, 3752 (2014)
66. W.M. Zhang, P.Y. Lo, H.N. Xiong, M.W.Y. Tu, F. Nori, General Non-Markovian dynamics of open quantum systems. Phys. Rev. Lett. **109**, 170402 (2012)
67. W. Zhou, A novel model for photovoltaic array performance prediction. Appl. Energy **84**(12), 1187–1198 (2007)

Chapter 8
Rerouting the Transpiration Vapor of Trees to Mitigate Global Water Supply in Rural Area Naturally

Abstract The transpiration mechanism has been proposed for rerouting as it is the main cause of groundwater loss in rural area which is also causing significant climate change by releasing water vapor into the air. Since electrostatic force has the tendency to tug down the water, a static electricity force-creating plastic tank has been proposed to be installed at the bottom of plants of each home of rural area to capture the transpiration water vapor and treat it in site by applying UV technology to meet the daily water demand throughout the world.

Keywords Transpiration · Water vapor · Static electricity force · Capturing water vapor · UV technology · Potable water · Climate change mitigation

Introduction

Plants give O_2 and take CO_2 by the process of photosynthesis to keep the global environment in balance. Plants are simply the hero for the environment; unfortunately, hero plants are also the villain for the environment who plays the significant role in causing global warming. The body of plants needs water for the reaction of biochemical metabolism for its growth [5]. This water is taken up by the cohesion-tension mechanism of the soil (groundwater) through the roots, transported by osmosis through the xylem to the leaves of the plants [6, 15]. Interestingly only a mere 0.5–3% of water is used by plants for their metabolism and the rest of water is released into the air through stomatal cells by transpiration process [11, 12]. This process of transpiration is not only causing the largest loss of groundwater but also causing global warming, and this water vapor is a notable cause for global warming. Recent studies on transpiration and groundwater relationship has been discussed respectively terrestrial water fluxes especially in rural area where their water models revealed that streamflow getting lower due to the plants transpiration [6, 12]. These are very interesting findings, but no mechanism has been studied yet to trap this transpiration water for meeting global water demand. In this research, therefore, a technology has been proposed to eliminate this water loss by diverting this transpiration mechanism by collecting this water vapor instead of allowing it to enter the air and transform it into water potable and clean energy. Simply static electricity creator

M. F. Hossain, *Sustainable Design for Global Equilibrium*,
https://doi.org/10.1007/978-3-030-94818-4_8

plastic tank near the plants has been proposed to be installed of each home of rural area to trap all the water vapor as the water vapor is attracted by the force of static electricity. Just because water vapor has positive and negative charges and the electrons that ended up on static electrical force has a positive charge, while water molecules have a negative charge on one side, the positive charge of static electric force and negative charges of water vapor pull each other closer together, the positive side tugs the direction and forces the water get collected in a tank and be treated in site to meet the daily water demand.

Calculation revealed that only four standard oak trees can meet the total water demand for a small family throughout the year. Since the groundwater strata are getting to lower fast to finite level, and global water and global warming getting dangerous seriously to putting earth on vulnerable condition, thus these two vital needs must be resolved immediately. Interestingly, this new finding has the total solution to solve the global water and environmental crisis for the survival of this planet which will indeed open a new door in science.

Materials, Methods, and Simulations

Static Electric Force Generation

To capture the water vapor from air which is released by stomatal cells of the plants during the daytime, a model has been proposed to create *Hossain Static Electric Force* (*HSEF* = $\mathfrak{h}$) by implementing the friction of insulator into the plastic tank to pull down the water vapor into the plastic tank [1, 14]. To create *HSEF* into the plastic tank, I have implemented abelian local symmetries calculation by using MATLAB software considering gauge field symmetry and the Goldstone scalar with respect to longitudinal mode of the vector [2, 8]. Thus, for each spontaneously broken particle T^{α} of the local symmetry will be the corresponding gauge field of $A^{\alpha}_{\mu}(x)$, where *HSEF* will start to work at a local $U(1)$ phase symmetry [7, 10]. Therefore, the model will be comprised as a complex scalar field $\Phi(x)$ of static electric charge q coupled to the EM field $A^{\mu}(x)$ which is expressed by $\mathfrak{h}$:

$$\mathfrak{h} = -\frac{1}{4} F_{\mu\nu}F^{\mu\nu} + D_{\mu}\Phi^{*} D^{\mu}\Phi - V(\Phi^{*}\Phi) \tag{8.1}$$

where

$$\begin{aligned} D_{\mu}\Phi(x) &= \partial_{\mu}\Phi(x) + iqA_{\mu}(x)\Phi(x) \\ D_{\mu}\Phi^{*}(x) &= \partial_{\mu}\Phi^{*}(x) - iqA_{\mu}(x)\Phi^{*}(x) \end{aligned} \tag{8.2}$$

and

$$V(\Phi^*\Phi) = \frac{\lambda}{2}(\Phi^*\Phi)^2 + m^2(\Phi^*\Phi) \tag{8.3}$$

Suppose $\lambda > 0$ but $m^2 < 0$, so that $\Phi = 0$ is a local maximum of the scalar potential, while the minima form a degenerate circle $\Phi = \frac{v}{\sqrt{2}} * e^{i\theta}$,

$$v = \sqrt{\frac{-2m^2}{\lambda}}, \quad \text{any real } \theta \tag{8.4}$$

Consequently, the scalar field Φ develops a non-zero vacuum expectation value $\langle\Phi\rangle \neq 0$, which spontaneously creates the $U(1)$ symmetry of the static electric field. The breakdown would lead to a massless Goldstone scalar stemming from the phase of the complex field $\Phi(x)$. But for the local $U(1)$ symmetry, the phase of $\Phi(x)$—not just the phase of the expectation value $\langle\Phi\rangle$ but the x-dependent phase of the dynamical $\Phi(x)$ field. To analyze this static electricity force mechanism, I have used polar coordinates in the scalar field space, thus

$$\Phi(x) = \frac{1}{\sqrt{2}}\Phi_r(x) * e^{i\Theta(x)}, \quad \text{real } \Phi_r(x) > 0, \text{real } \Phi(x) \tag{8.5}$$

This field redefinition is singular when $\Phi(x) = 0$, so I never used it for theories with $\langle\Phi\rangle \neq 0$, but it's alright for spontaneously broken theories where I can expect $\Phi\langle x\rangle \neq 0$ almost everywhere. In terms of the real fields $\phi_r(x)$ and $\Theta(x)$, the scalar potential depends only on the radial field ϕ_r,

$$V(\phi) = \frac{\lambda}{8}\left(\phi_r^2 - v^2\right)^2 + \text{const}, \tag{8.6}$$

or in terms of the radial field shifted by its VEV, $\Phi_r(x) = v + \sigma(x)$,

$$\phi_r^2 - v^2 = (v+\sigma)^2 - v^2 = 2v\sigma + \sigma^2 \tag{8.7}$$

$$V = \frac{\lambda}{8}\left(2v\sigma - \sigma^2\right)^2 = \frac{\lambda v^2}{2} * \sigma^2 + \frac{\lambda v}{2} * \sigma^3 + \frac{\lambda}{8} * \sigma^4 \tag{8.8}$$

At the same time, the covariant derivative $D_\mu\phi$ becomes

$$D_\mu\phi = \frac{1}{\sqrt{2}}\left(\partial_\mu\left(\phi_r e^{i\Theta}\right) + iqA_\mu * \phi_r e^{i\Theta}\right)$$
$$= \frac{e^{i\Theta}}{\sqrt{2}}\left(\partial_\mu\phi_r + \phi_r * i\partial_\mu\Theta + \phi_r * iqA_\mu\right) \tag{8.9}$$

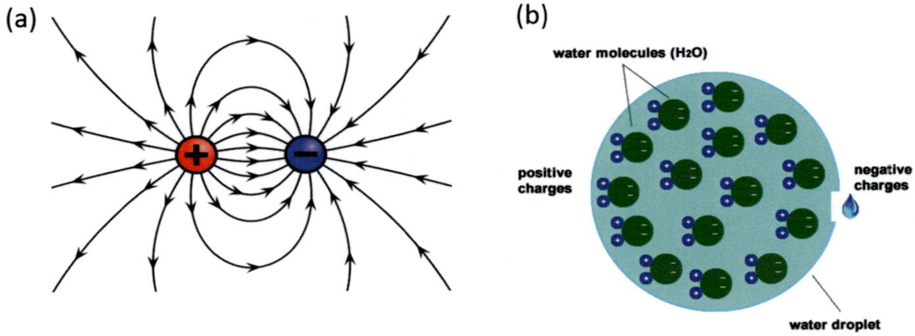

Fig. 8.1 (**a**) The creating of static electricity force, and (**b**) its mechanism of conversion of static energy into an electromotive force of positive and negative charges that mobilize the "static" electricity to tug down the water molecules

$$\begin{aligned} |D_\mu \phi|^2 &= \frac{1}{2} \left|\partial_\mu \phi_r + \phi_r * i\partial_\mu \Theta + \phi_r * iqA_\mu\right|^2 \\ &= \frac{1}{2} \left(\partial_\mu \phi_r\right) + \frac{\phi_r^2}{2} * \left(\partial_\mu \Theta q A_\mu\right)^2 \\ &= \frac{1}{2} \left(\partial_\mu \sigma\right)^2 + \frac{(v+\sigma)^2}{2} * \left(\partial_\mu \Theta + qA_\mu\right)^2 \end{aligned} \tag{8.10}$$

Altogether,

$$\mathfrak{h} = \frac{1}{2} \left(\partial_\mu \sigma\right)^2 - v(\sigma) - \frac{1}{4} F_{\mu\nu} F^{\mu\nu} + \frac{(v+\sigma)^2}{2} * \left(\partial_\mu \Theta + qA_\mu\right)^2 \tag{8.11}$$

To confirm the creating of this static electric force ($\mathfrak{h}_{\text{sef}}$) into the static electric field properties of this *HSEF*, it has been expanded in powers of the fields (and their derivatives) and focus on the quadratic part describing the free particles,

$$\mathfrak{h}_{\text{sef}} = \frac{1}{2} \left(\partial_\mu \sigma\right)^2 - \frac{\lambda v^2}{2} * \sigma^2 - \frac{1}{4} F_{\mu\nu} F^{\mu\nu} + \frac{v^2}{2} * \left(qA_\mu + \partial_\mu \Theta\right)^2 \tag{8.12}$$

Here this *HSEF* ($\mathfrak{h}_{\text{free}}$) function obviously will suggest a real scalar particle of positive mass^2 $= \lambda v^2$ involving the $A_\mu(x)$ and the $\Theta(x)$ fields to initiate to create tremendous static electricity force within the electric field of the plastic tank (Fig. 8.1).

In Site Water Treatment

Since the collected water in the plastic tank is just nothing but the liquid form of vapor, it will not require any sedimentation, coagulation, and chlorination to clean

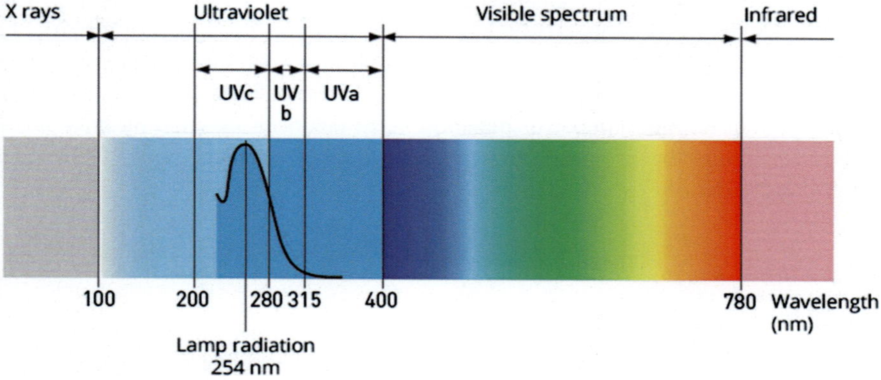

Fig. 8.2 The photophysics radiation application for the purification of water which shows that once UV radiation of 320 nm is applied into the water, it starts to disinfect all microorganism immediately once temperature reaches 50 °C

the water. Only mixing physics (UV application) and filtration will be required to treat the water to meet the US National Primary Drinking Water Standard code [3]. It is the simplest way to treat water by using *SODIS* system (*SOlar DISinfection*), where a transparent container is filled with water and exposed to full sunlight for several hours. As soon as the water temperature reaches 50 °C with a UV radiation of 320 nm, the inactivation process will be accelerated in order to lead to complete microbiological disinfection immediately and the treated water shall be used to meet the total domestic water demand (Fig. 8.2).

Results and Discussion

Electrostatic Force Analysis

To mathematically determine the electric static force proliferation around the plastic tank to confirm to tug down the water, I have initially solved the dynamic photon proliferation by integrating *HSEF* electric field creation; thus, the local $U(1)$ gauge invariant did allow to add a mass-term for the gauge particle under $\varnothing' \rightarrow e^{i\alpha(x)}\varnothing$. In detail, it can be explained by a covariant derivative with a special transformation rule for the scalar field expressed by Refs. [9, 17]:

$$\partial_\mu \rightarrow D_\mu = \partial_\mu = ieA_\mu \quad [\text{covariant derivatives}] A'_\mu = A_\mu + \frac{1}{e}\,\partial_\mu \alpha \quad [A_\mu \text{ derivatives}] \tag{8.13}$$

where the local $U(1)$ gauge invariant *HSEF* for a complex scalar field is given by:

$$\mathfrak{H} = (D^\mu)^\dagger\,(D_\mu \varnothing) - \frac{1}{4}\,F_{\mu\nu}F^{\mu\nu} - V(\varnothing) \tag{8.14}$$

The term $\frac{1}{4}\ F_{\mu\nu}F^{\mu\nu}$ is the kinetic term for the gauge field (heating photon) and $V(\varnothing)$ is the extra term in the *HSEF* that be: $V(\varnothing^{*}\varnothing) = \mu^2(\varnothing^{*}\varnothing) + \lambda\ (\varnothing^{*}\varnothing)^2$.

Therefore, the *HSEF*($\mathfrak{H}$) under perturbations into the quantum field have initiated with the massive scalar particles ϕ_1 and ϕ_2 along with a mass μ. In this situation, $\mu^2 < 0$ had an infinite number of quantum, each has been satisfied by $\phi_1^2 + \phi_2^2 = -\mu^2/\lambda = v^2$ and the $\mathfrak{H}$ through the covariant derivatives using again the shifted fields η and ξ defined the quantum field as $\phi_0 = \frac{1}{\sqrt{2}}\ [(v + \eta) + i\xi]$.

$$\begin{aligned}\text{Kinetic term} : \mathfrak{H}(\eta,\xi) &= (D^{\mu}\phi)^{\dagger}(D^{\mu}\phi)\\ &= (\partial^{\mu} + ieA^{\mu})\phi^{*}\ (\partial_{\mu} - ieA_{\mu})\ \phi\end{aligned} \tag{8.15}$$

Thus, this expanding term in the $\mathfrak{H}$ associated to the scalar field is suggesting that *HSEF* electric field is prepared to initiate the proliferation of static electricity force into its quantum field to tug down the water [3, 4, 16].

To confirm this tug down of water by static electricity force, hereby, I have readily implemented the calculation of $\overline{\varphi}\ [s_0]$ for the confirmation of the expected value of s_0 for capturing water vapor [13, 18]. Thus, the corrective functional asymptotic formulas are being used as follows:

$$\begin{aligned}\overline{\varphi}[s_0] &= 2s_0(\ln 4s_0 - 2) + \ln 4s_0(\ln 4s_0 - 2) - \frac{(\pi^2 - 9)}{3}\\ &\quad + s_0^{-1}\left(\ln 4s_0 + \frac{9}{8}\right) + \cdots \quad (s_0 \gg 1);\end{aligned} \tag{8.16}$$

$$\begin{aligned}\overline{\varphi}[s_0] &= \frac{2}{3}(s_0 - 1)^{\frac{3}{2}} + \left(\frac{5}{3}\right)(s_0 - 1)^{\frac{5}{2}} - \left(\frac{1507}{420}\right)(s_0 - 1)^{\frac{7}{2}}\\ &\quad + \cdots \quad (s_0 - 1 \ll 1)\end{aligned} \tag{8.17}$$

The function $\frac{\overline{\varphi}[s_0]}{(s_0 - 1)}$ is thus described as $1 < s_0 < 0$ for larger s_0, and it contains natural logarithmic which is s_0 to confirm the tug down of 100% water vapor by the *HSEF* into the plastic tank.

On average, 100 gallons of water are required per day per person in a standard daily life [4, 15]. Thus, it will require total (100_{gallons}/Day/Person $\times\ 4_{\text{persons}} \times 365_{\text{days}}$) of 146,000 gallons of water per year for a small family of four person. Since a standard oak tree can transpire 40,000 gallons (151,000 l) per year, tug down of 100% water vapor by *HSEF* described above will require only four standard oak trees to satisfy the total water demand for a small family.

Conclusions

Water and environmental vulnerability are the top two problems on rural area where trees play a significant role in creating these problems by the process of transpiration. To mitigate these problems, transpiration mechanism has been proposed to transform and convert it into clean water to meet the rural water demand and reduce the global warming by the utilization of electrostatic force to capture this transpiration water vapor and treat in site by UV application would indeed be a novel, integrated, and innovative field in science to console the rural water and global warming crisis.

Acknowledgments This research was supported by Green Globe Technology under grant RD-02017-07 for building a better environment. Any findings, predictions, and conclusions described in this article are solely performed by the authors and it is confirmed that there is no conflict of interest for publishing in a suitable journal.

References

1. R. Andreas, Strongly correlated photons on a chip. Nat. Photonics **6**, 93–96 (2012)
2. S. Douglas, H. Habibian, et al., Quantum many-body models with cold atoms coupled to photonic crystals. Nat. Photonics **9**, 326–331 (2015)
3. M. Hossain, Solar energy integration into advanced building design for meeting energy demand and environment problem. Int. J. Energy Res. **40**, 1293–1300 (2016a)
4. F. Hossain, Theory of global cooling. Energy Sustain. Soc. **7**, 6–24 (2016b)
5. E. Jaivime, J.M.D. Scott, Global separation of plant transpiration from groundwater and streamflow. Nature **525**, 91–94 (2015)
6. M. Josette, R. Scott, The ERECTA gene regulates plant transpiration efficiency in Arabidopsis. Nature **436**, 866–870 (2005)
7. L. Langer, S. Poltavtsev, M. Bayer, Access to long-term optical memories using photon echoes retrieved from semiconductor spins. Nat. Photonics **8**, 851–857 (2014)
8. Y. Leijing, W. Sheng, Z. Qingsheng, Z. Zhiyong, P. Tian, L. Yan, Efficient photovoltage multiplication in carbon nanotubes. Nat. Photonics **8**, 672–676 (2011)
9. Q. Li, D. Xu, Recoil effects of a motional scatterer on single-photon scattering in one dimension. Sci. Rep. **8**, 3144 (2013)
10. T. Pregnolato, E. Lee, J. Song, D. Stobbe, P. Lodahl, Single-photon non-linear optics with a quantum dot in a waveguide. Nat. Commun. **6**, 8655 (2015)
11. M. Reed, L. Maxwell, Connections between groundwater flow and transpiration partitioning. Science **353**, 377–380 (2015)
12. J. Scott, D. Zachary, Terrestrial water fluxes dominated by transpiration. Nature **496**, 347–350 (2013)
13. W. Soto, S. Klein, et al., Improvement and validation of a model for photovoltaic array performance. Sol. Energy **80**, 78–88 (2006)
14. M. Tame, S. McEnery, et al., Quant. Plasmonics **9**, 329–340 (2013)

15. T.D. Wheeler, A.D. Stroock, The transpiration of water at negative pressures in a synthetic tree. Nature **455**, 208–212 (2008)
16. W. Yan, F. Heng, Single-photon quantum router with multiple output ports. Sci. Rep. **4**, 4820 (2014)
17. W. Yuwen, Z. Yongyou, Z. Qingyun, Z. Bingsuo, S. Udo, Dynamics of single photon transport in a one-dimensional waveguide two- point coupled with a Jaynes-Cummings system. Sci. Rep. **6**, 33867 (2016)
18. Y. Zhu, H. Xiaoyong, Y. Hong, G. Qihuang, On-chip plasmon-induced transparency based on plasmonic coupled nanocavities. Sci. Rep. **4**, 3752 (2014)

Chapter 9
Application of Hybrid Wind and Solar Energy in the Transportation Section

Abstract A model of hybrid wind and solar energy implementation technology has been developed to investigate the use of wind and solar power by transportation vehicles to naturally meet energy requirements. Mathematical modeling confirmed that utilization of hybrid wind and solar energy technology as natural power sources for transportation vehicles is very feasible and would be an interesting field of engineering technology to mitigate global energy crisis in transportation sector. Simply, this finding suggests that the large-scale adoption of hybrid wind and solar energy to run transportation vehicles would be an excellent and environmentally benign solution to meet the net energy demand of the transportation sector throughout the world.

Keywords Wind and solar power · Turbine modeling and solar panel implementation · Control design · Energy conversion · Transportation vehicle

Introduction

The current magnitude of wind turbine installations for producing electricity seriously impacts global habitats and ecosystems [1–3]. In addition, possible radar interference is also a concern for the construction of wind turbines to produce electricity [4, 5]. Prominent concerns about the production of solar energy by using solar panels installed over vast land areas include visibility, and the misuse of land, which affects the harmonium of the landscape [7, 8]. In addition, the technological application for the conversion of wind and solar power into electricity supply through the national grid, including cost concerns, and engineering challenges along with transmission and operational mechanisms are likely to restrict the global commercial use of wind and solar energy [1, 3, 6].

Several studies in the past conducted on wind energy and solar energy have been carried out to use renewable energy as an alternative source to mitigate global energy crisis. Boumassata et al. [5] suggested that grid power control based on a wind energy conversion system is quite applicable throughout the world. Another study by Elmansouri et al. [7] suggested that a wind energy conversion system using a doubly fed induction generator (DFIG) controlled by back-stepping and an RST

M. F. Hossain, *Sustainable Design for Global Equilibrium*,
https://doi.org/10.1007/978-3-030-94818-4_9

controller may be the most interesting technology for utilizing this energy in the urban transportation sector to reduce energy and environmental crises. Recently, Hossain and Fara [16] revealed that solar energy has a tremendous capability to power vehicles using such a natural system implementation on transportation vehicles, and another study by Hossain [14, 15] revealed that wind energy has tremendous potential to meet the net energy demand of running a vehicle for the development of sustainable transportation system [6, 9]. Although their research on both wind and solar energy hypothetically suggested analyzing the possibility of utilizing this renewable energy for many sectors, applications of these renewable hybrid energy conversion technologies have not yet been attempted in the transportation sector [6, 9, 10].

Simply, the knowledge gap and the novelty of their research for the application of hybrid wind and solar power has led to its being ignored despite the tremendous potential it has to meet the near-term and long-term energy mitigation needs in the transportation sector globally.

Thus, to utilize these renewable energies, in this study, a hybrid model of wind turbines and solar panels was installed on a vehicle to capture wind and solar energy and create naturally the required energy to power the transportation vehicle. Subsequently, the drive train model, energy conversion mechanism, and the process of electricity energy generation through the main subsystems were mathematically calculated using MATLAB Simulink [7, 11, 12].

Simply, this innovative mechanism of hybrid renewable energy of wind and solar power will enable transportation vehicles to utilize this abundant energy resource, which indeed would be a cutting-edge technology to reduce the cost of energy and environmental vulnerability worldwide in the transportation sector globally.

Material and Methods

Modeling of Wind Turbine

Wind energy is measured here in volume from the air as the kinetic force [8, 14–16] in order to form energy by the rotation of wind turbine to convert kinetic wind into electric energy [13–15]. The process is such that the wind turbines rotate in the clockwise direction, and the main shaft connected to the gearbox in the nacelle causes the spinning process and then transfers energy to the generator, which eventually converts the kinetic energy into electric energy to power the transportation vehicle [17, 18]. Thus, in this research, the model is used through the application of MATLAB Simulink to develop wind turbines and then installed on transport vehicles, which will be driven by airflow due to the motion of the vehicle [19–21]. Consequently, a series of mathematical analyses have been integrated to perform simulations in a doubly fed induction generator (DFIG) to convert energy into electricity [20–22]. As a result, the conversion of wind energy is also analyzed using DC and AC converter circuits to power vehicles [11, 23]. The active stator is thus applied here in regulating the whole mechanism of wind conversion, which is represented by the following equation:

$$P_{\mathrm{w}} = \frac{1}{2} C_{\mathrm{p}}(\lambda, \beta) \rho A V^3 \tag{9.1}$$

where ρ is the density of air in kg/m^3, C_p is the power coefficient, V = average wind speed in m/s, λ = tip speed ratio, and A = intercepting area of wind rotor blades in m^2. C_p is 0.593 at its peak [24, 25]. The average tip speed ratio (TSR) is given in terms of velocity ratios, as shown.

$$\lambda = \frac{R\omega}{V} \tag{9.2}$$

R = radius of turbine (m), ω = angular velocity (rad/s), v = mean wind velocity (m/s). The power in the turbine is given by:

$$Q_{\mathrm{w}} = P \times T \tag{9.3}$$

Some of the factors that may have been interfered here with wind flow and velocity that are used as the factored in the equation below and take into account the error proneness due to obstacles, and expressed as:

$$v(z) \ln\left(\frac{z_{\mathrm{r}}}{z_0}\right) = v(z_{\mathrm{r}}) \ln\left(\frac{z_{\mathrm{r}}}{z_0}\right) \tag{9.4}$$

where z_r = height of the vehicle (m), z_0 = surface roughness (0.1–0.25), z = wind speed, $v(z)$ = wind velocity at height z (m/s), and $v(z_r)$= wind velocity at height z_r (m/s) (Fig. 9.1).

Furthermore, the velocity of wind is calculated by integrating the turbine rotation that helps in establishing the attainment of equilibrium wind velocity in a steady way within the turbine positioned on the transport vehicle. The equation for calculation is as shown below.

$$P_{\mathrm{w}}(v) = \begin{cases} \dfrac{v^k - v_{\mathrm{C}}^k}{v_{\mathrm{R}}^k - v_{\mathrm{C}}^k} . P_{\mathrm{R}} & v_{\mathrm{C}} \le v \le v_{\mathrm{R}} \\ P_{\mathrm{R}} & v_{\mathrm{R}} \le v \le v_{\mathrm{F}} \\ 0 & v \le v_{\mathrm{C}} \text{ and } v \ge v_{\mathrm{F}} \end{cases} \tag{9.5}$$

where P_R = rated power, v_C = cut-in wind velocity, v_R = rated wind velocity, v_F = rated cut-out velocity, and k = Weibull shape cofactor.

The maximum power is obtained from the angular velocity of the generator. Therefore, maximum power point tracking (MPPT) is derived from the optimum rotor velocity using the equation below:

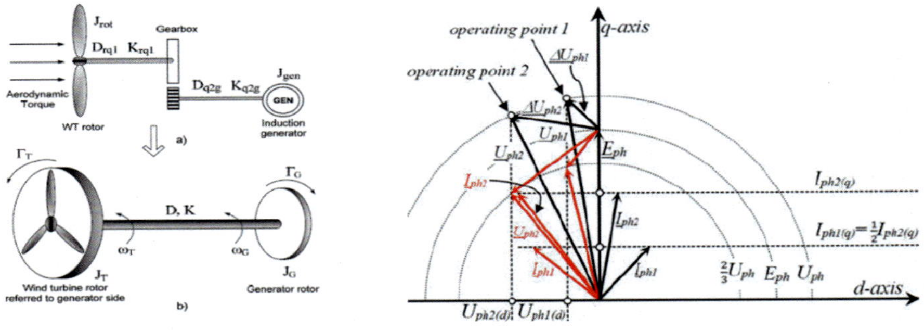

(a) (b)

Fig. 9.1. (**a**) Shows the representation of a schematic of a wind turbine installed on a transport vehicle while (**b**) shows the different orientation of *d*-axis (z) and *q*-axis (z_r) at maximum rotor speed ($z = Iph$) controlled by the stator that has the resistor and regulator to control the conversion of the wind energy process

$$\omega_{opt} = \frac{\lambda_{opt}}{R} V_{wn} \tag{9.6}$$

The equation is converted to:

$$V_{wn} = \frac{R\omega_{opt}}{\lambda_{opt}} \tag{9.7}$$

where ω_{opt} = maximum rotor angular velocity (rad/s), λ_{opt} = maximum tip velocity factor, R = radius of the turbine (m), and V_{wn}= wind velocity in m/s.

The entry of turbine force into the gearbox is being modeled by applying the torsional multibody dynamic, which is expressed by the following equation [26, 27]:

$$\begin{bmatrix} \dot{\omega}_l \\ \dot{\omega}_g \\ \dot{T}_{l_2} \end{bmatrix} = \begin{bmatrix} -\frac{K_l}{J_l} & 0 & -\frac{1}{J_l} \\ 0 & -\frac{K_g}{J_g} & \frac{1}{n_g J_g} \\ \left(B_{l_2} - \frac{K_{l_2} K_r}{J_r}\right) & \frac{1}{n_g}\left(\frac{K_{l_2} K_r}{J_g} - B_{l_2}\right) & -K_{l_2}\left(\frac{J_r + n_g^2 J_g}{n_g^2 J_g J_r}\right) \end{bmatrix} \begin{bmatrix} \omega_l \\ \omega_g \\ T_{l_2} \end{bmatrix} + \begin{bmatrix} \frac{1}{J_r} \\ 0 \\ \frac{K_{l_2}}{J_r} \end{bmatrix} T_m + \begin{bmatrix} 0 \\ -\frac{1}{J_g} \\ \frac{K_{l_2}}{n_g J_g} \end{bmatrix} T_g \tag{9.8}$$

The relationship is being analyzed here to obtain the moment of inertia, which is calculated using the following equation and represents a combined mass of the hub and blades.

$$J_t\dot{\omega}_t = T_a - K_t\omega_t - T_g \tag{9.9}$$

and

$$K_t = K_r + n_g^2 K_g \tag{9.10}$$

where J_t = turbine rotor moment of inertia (kg m^2), ω_t = angular velocity of the rotor (rad/s^2), K_t = turbine cofactor (Nm rad^{-1} s^{-1}), and K_g = generator factor (NM/rad/s).

The other factor that is modeling the drive train, which is a component of the torque at the rotor shaft and passes through the gearbox, has been analyzed to reduce the complexity associated with the turbine mass. Thus, the one-mass model derived from the torsional model is being considered using the following matrix [24, 29].

Consequently, a combination of mutual dumping from one mass and the weight of the blades is used in calculating the turbine inertia [30, 31]. The blades are designed to distribute weight equally and provide the best outcome in calculating the torque values as the sum of all the components. The one-mass model is used in the calculation by factoring the mass moment of inertia of the turbines, as shown below:

andwhere J_t = rotor moment of inertia for turbine (kg m^2), ω_t = angular speed of low shaft (rad/s^2), K_t = the damping coefficient of the turbine (Nm/rad/s), which is linked to the aerodynamic resistance, and K_g = the coefficient of the generator damping (Nm/rad/s), which shows the mechanical friction and windage.

The one-mass model is thus used in the analysis of wind speed based on the air mass flow, followed by the calculation of the mechanism used in converting energy and finally the analysis of the whole process of conversion [20, 21, 32].

Conversion of Wind Energy

Since the wind is transferred to the electrical power system using the DFIG, the conversion process is coupled by several stages that occur in the aerodynamic system and thereafter releasing electric energy (Fig. 9.2). The real log determination of velocity in the DFIG systems is applied in the analysis of the velocity systems and controlled through the aerodynamic system [17, 33]. The deterministic and stochastic elements are calculated using the equation below, and it is a component of wind speed V transferred in the rotor.

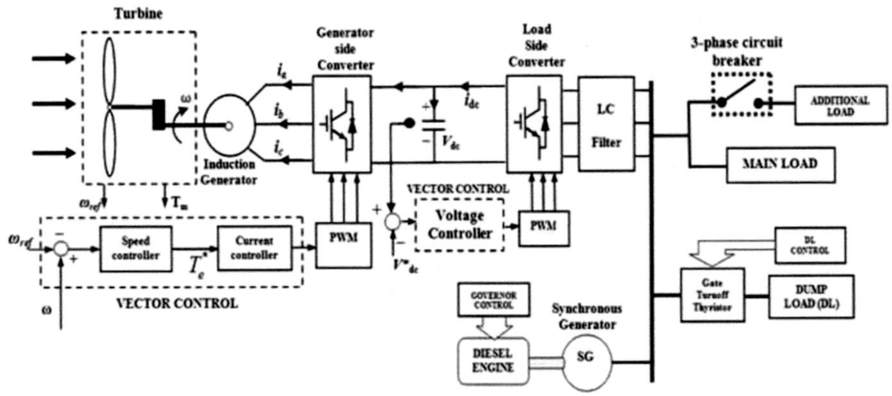

Fig. 9.2 Shows a block diagram of all the components. The modeling process of the control levels occurs in this stage, and the main components involved are the generator side convertor, turbine, as well as load side convertor for transforming the electric energy to the 3-phase circuit breaker

$$V(t) = V_0 + \sum_{i=1}^{n} A_i \sin(\omega_i t + \varphi_i) \tag{9.14}$$

In the above equation, V_0 = average component, A_i = magnitude, ω_i = pulsation, and ψ_i = initial phase for every turbulence.

Consequently, the conversion energy from kinetic to electrical is determined by taking into consideration the turbulence function, which is a component of the rotational blades. The kinetic energy is dependent on the speed of the rotor and power coefficient C_p, as shown in the equation below [9, 27].

$$P_{aer} = \frac{1}{2} C_p(\lambda, \beta) \rho R \pi^2 V^3, \tag{9.15}$$

where ρ = air density, R = blade length, and V = wind speed. The captured wind coefficient $C_p(\lambda)$ is calculated by considering the pitch angle and the wind speed (Ω_t), as shown in the equation below [26, 34].

$$\lambda = \frac{\Omega_T * R}{V}. \tag{9.16}$$

Modeling of Solar Energy

Since the sun produces electromagnetic waves that propagate in the bulk of photons, it is trapped by using solar panels installed on transportation vehicle [3, 19]. The

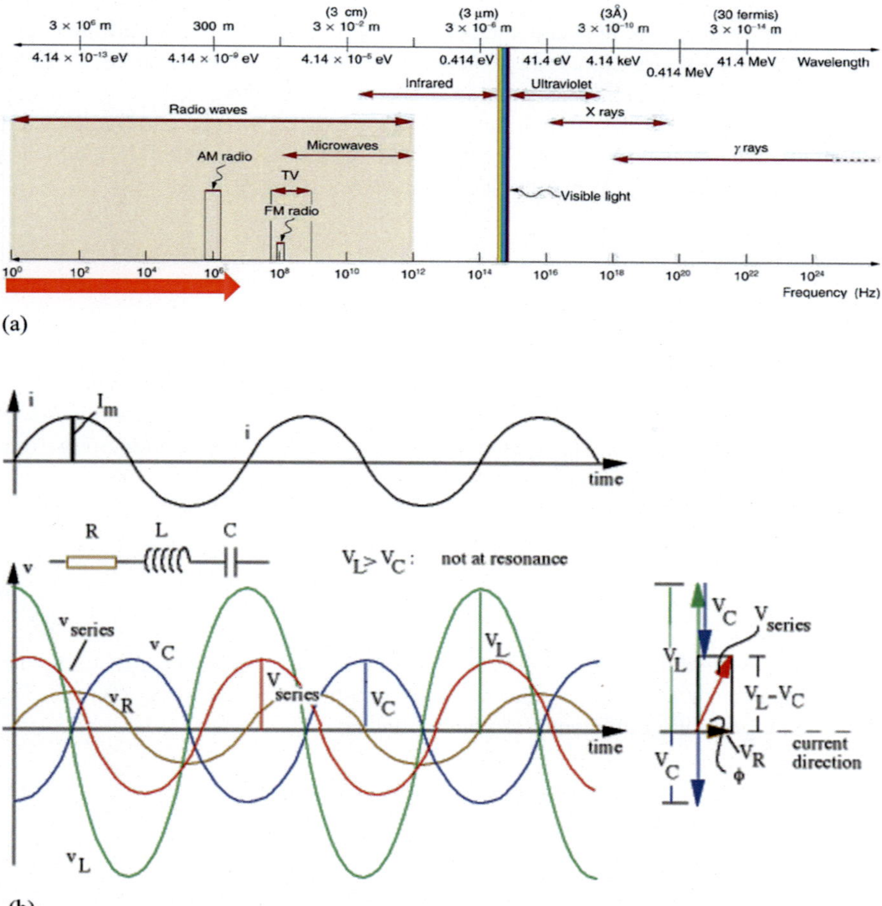

Fig. 9.3 Shows the representation of the PV system model. (**a**) The representation of the photovoltaic solar radiance in the photovoltaic mode and having temperature, thermocouple, and radiation as the variations. (**b**) Diagram of a Simulink block that constitutes the active solar volt parameter into the single-diode circuit as the current direction

modeling process of the transferred energy in this research has been achieved by employing the concept of classical statistical physics on the photon wave frequency (Fig. 9.3). Here, the components used in modeling the maximum solar energy are photon excitation (1.4 eV) and energy value (27.77 MW/m^2) in order to confirm the use of classical statistical physics to show the (Fig. 9.3a) solar radiation frequencies, and (Fig. 9.3b) shows the representation of maximum solar energy formation.

Subsequently, the maximum energy intake in the solar panels installed in the transportation vehicle has also been modeled using the photovoltaic module model and used to describe how photovoltaics are generated using the thermocouple, radiation shield and temperature component on the solar panel (Fig. 9.3). The maximum power output simplified the photovoltaic solar radiance model embedded in the photovoltaic module and attached to the solar radiation logarithmic and temperature module [6, 20, 21, 31]. Naturally, the configuration model in a single-diode circuit is applied in the calculation of the parameter characteristics of current-voltage (*I*–*V*) and considering their numbers [13, 17, 35]. The solar panel collects the solar radiance that is used in deriving the active solar volt parameter (I_{v+}), which is expressed as a single-diode circuit, as shown above in Fig. 9.3.

Solar Energy Conversion

Solar energy conversion is implemented by determining the current production by calculating the Ipv from a one-diode model, which is derived from the *I*–*V*–*R* relationship by employing the photovoltaic array collected and converting from DC to AC for use on the transport vehicle (Fig. 9.4).

The equation shown below is derived from calculating the energy output in the solar panel and linked to the solar radiation.

$$P_{\text{pv}} = \eta_{\text{pvg}} A_{\text{pvg}} G_{\text{t}} \tag{9.17}$$

where η_{pvg} = the efficiency of solar panel generation, A_{pvg} represents the solar panel generator area (m^2), and G_{t} = solar radiation in a tilted module plane (W/m^2). The solar panel generation efficiency can be written as:

$$\eta_{\text{pvg}} = \eta_{\text{r}} \eta_{\text{pc}} [1 - \beta(T_{\text{c}} - T_{\text{cref}})] \tag{9.18}$$

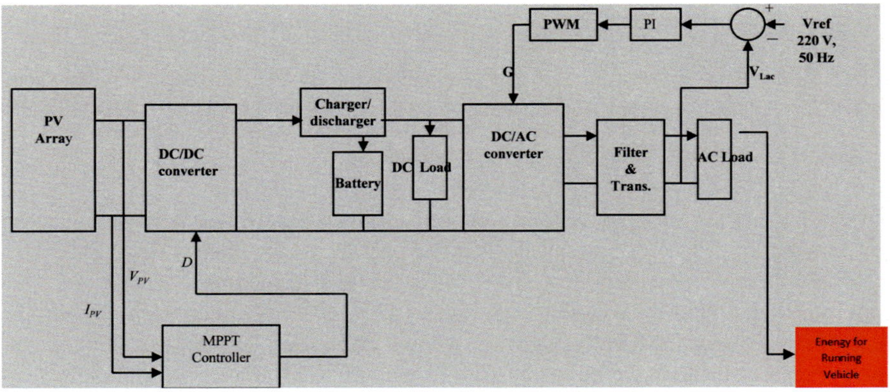

Fig. 9.4 The MATLAB is used for modeling the single diode for the photovoltaic cell (PV), resulting in the generation of current followed by the conversion of DC to AC by using the *I*–*V*–*R* relationships for various applications such as domestic use

where η_{pc} is the efficiency of power conditioning, and during the application of MPPT, it is equivalent to 1:β, which is also 0.004–0.006 per °C (temperature coefficient); η_r is the module efficiency; and T_{cref} is the reference cell temperature measured in °C. The quantity can be calculated using the equation below.

$$T_c = T_a + \left(\frac{\text{NOCT} - 20}{800}\right) G_t \tag{9.19}$$

The variables in the equations are defined as T_a = ambient temperature (°C), G_t = solar radiance in the tilted module plane (W/m^2), and finally NOCT—standard operating temperature (°C). The equation below is utilized to obtain the total radiance in the solar cell, which is a variable of the diffused solar irradiance.

$$I_t = I_b R_b + I_d R_d + (I_b + I_d) R_r \tag{9.20}$$

Here, the energy is generated in solar cells and because of the P–N junction semiconductor relationships that is interconnection of solar panel cells in series [2, 6, 33, 37].

Electrical Subsystem

After the analysis of the solar energy conversion system, using mathematical models, the result is then integrated into the hybrid energy system to introduce the driving force through the electrical subsystem, as shown below (Fig. 9.4). The analysis of the DFIG system is utilized in the analysis of the hybrid system that is used to establish the electric system that can generate electrical energy. The process is summarized using the equations below [16, 34, 36].

$$\phi_s = \phi_{ds} \Rightarrow \phi_{qs} = 0. \tag{9.21}$$

The above equation is used in calculating the stator constant flux ϕ_s and functional voltage V_s as shown below:

$$\begin{aligned} v_{ds} &= 0 \\ v_{qs} &= \omega_s \times \phi_s = V_s \end{aligned} \tag{9.22}$$

Therefore, the stator voltage vectors shown in Fig. 9.4 above are a function of the direction of quadrate advances and explained using the equations below to derive the rotor voltage.

$$\begin{cases} v_{dr} = \sigma L_r \dfrac{di_{dr}}{dt} + R_r i_{dr} + fem_d \\ v_{qr} = \sigma L_r \dfrac{di_{qr}}{dt} + R_r i_{qr} + fem_q \end{cases} \tag{9.23}$$

where fem_d represents the coupling term in the d-axis and fem_q represents the coupling term in the q-axis. The two can be represented using the following equation.

$$\begin{cases} fem_d = -\sigma L_r L_r i_{qr} \\ fem_q = \sigma L_r \omega_r i_{dr} + s\dfrac{M}{L_s} V_s \end{cases}. \tag{9.24}$$

Equations (9.10) are used to obtain the current fluxes, as shown below:

$$\begin{cases} \phi_{ds} = L_s i_{ds} + M I_{dr} \\ 0 = L_s i_{qs} + M I_{qr} \end{cases} \tag{9.25}$$

Equation (9.15) is used to calculate the currents, as shown below:

$$\begin{cases} i_{ds} = \dfrac{\phi_{ds} - M i_{dr}}{L_s} \\ i_{qs} = -\dfrac{M}{L_s} i_{qr} \end{cases}. \tag{9.26}$$

The stator is associated with rotor currents in the equations below:

$$\begin{cases} P_s = -V_s * \dfrac{M}{L_s} i_{qr} \\ Q_s = -V_s * \dfrac{M}{L_s} \left(i_{dr} - \dfrac{\phi_{ds}}{M} \right) \end{cases}, \tag{9.27}$$

where i_{qr} represents the stator active and i_{dr} shows the reactive powers. The integration of DFIGs on the transport vehicle together with the control variables from the power system is an important approach to creating hybrid renewable energy to power the transportation sector naturally.

Results and Discussions

The Wind Turbine Model

The modeled wind turbine system installed on the transport vehicle is used to enhance the operation of the whole system in both the MPPT control and stator flux to generate wind energy. MATLAB Simulink is subsequently applied in the mathematical calculation of all mechanisms involved in rotor converter control. Since the performance of the wind turbine operations at different values of wind

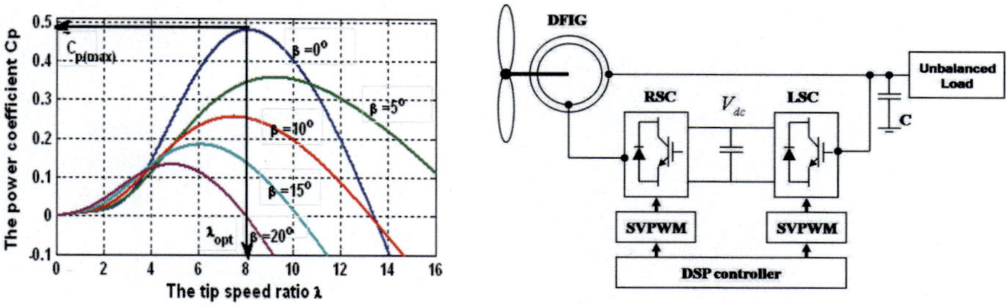

(a) (b)

Fig. 9.5. (**a**) Shows how the maximum values of C_p are attained from the diagram when $\beta = 2°$. The maximum value of $C_p = 0.5$ at $\lambda_{opt} = 0.91$. The λ_{opt} stands for the optimal speed ratio and has a rated speed of wind of 10 m/s and the mean value rated at 8 m/s. The standards conditions for testing are such that the stator voltage is estimated at 50% for 4–5 s, 25% for 6–6.5 s, and 50% for 8–8.5 s. The DFIG is a factor that was obtained from the relationships between the wind signals that allowed the use of DSP control and the wind turbine that resulted in V_{dc} energy in the turbine

speeds is related to the pitch angles, the degrees of angular speed are obtained by performing a robustness test by incorporating both the voltage dips and wind speed signals (Fig. 9.5).

The maximum angular speed acquired by the shaft in the generator is due to the increased speed of the wind, through the process of tracking the maximum power point speed through the functionalities of the generator, supported by stator active power. Thus, the load-side converter controls the power to derive the power factor that results in energy output [38, 39].

Then the determination of MPPT control was calculated through the electromagnetic torque and integrated to confirm the stator active power; therefore, the process was facilitated by the power-speed ratio that helped in tracking the maximum power point during the process of energy conversion [28, 40]. The maximum power cofactor is here achieved by the use of wind turbine shaft speed, which indicates that the optimum electromagnetic torque is obtained from MPPT control [1, 35]. The determination of rotor dynamics was from the active and reactive powers, and this tool was placed when the MPPT control deduced the wind turbine efficiency, which is expressed as:

$$\begin{cases} i_{qr}^{*} = -\dfrac{L_s}{MV_s} P_s^{*} \\ i_{dr}^{*} = -\dfrac{L_s}{MV_s}\left(Q_s^{*} - \dfrac{V_s^2}{\omega_s L_s}\right) \end{cases}. \tag{9.28}$$

Here, the assessment of this mechanism revealed an accurate speed airflow that was taken in by the wind turbine, and the tracking of wind velocity was made effective through the use of MPPT control, which prompted adjustment of the electromagnetic torque in the DFIG, producing the energy in a consistent way (Fig. 9.6).

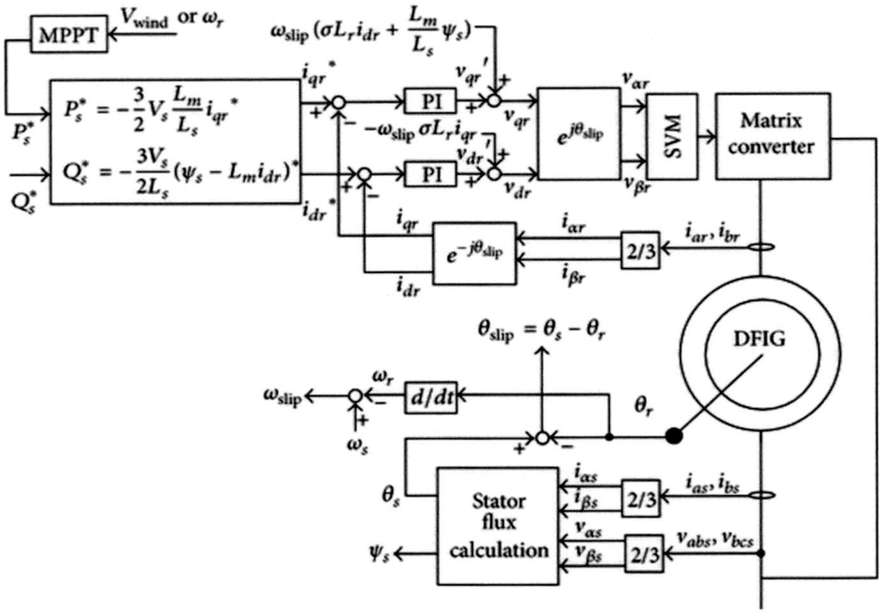

Fig. 9.6 Shows the block diagram of the MPPT (V_{wind}) that consists of the control system with enslaved DFIG velocity in the aerodynamic sub-system. The system performs important functions that include the conversion of wind energy to electrical as well as powering of the vehicles using the stator flux rotor

Hence, it is important to control the variable-speed wind turbine shaft speed using the gearbox to calculate the maximum power coefficient that is used in the output of maximum energy for the vehicle [7, 18]. Thus, the drive train model is formed following the synchronization of *d*–*q* comments and is given by the equation below:

$$V_q = -R_s i_q - L_q \frac{di_q}{dt} - \omega L_d i_d + \omega \lambda_m \tag{9.29}$$

The electronic torque is obtained using the equation below:

$$T_e = 1.5p\left[\lambda i_q + \left(L_d - L_q\right) i_d i_q\right] \tag{9.30}$$

where L_q is the resistance along the *q*-axis, L_d is the inductance resistance along the *d*-axis, i_q is the current along the *q*-axis, i_d is the current along the *d*-axis, V_q is the voltage along the *q*-axis, V_d is the voltage along the *d*-axis, ω_r represents the angular velocity of the rotor, λ is the amplitude of the induced influx, and p is the number of pairs of poles. The following equations of the *q*–*q* frame are applicable for the squirrel cage induction generator (SCIG).

$$\begin{bmatrix} V_{qs} \\ V_{ds} \\ V_{qr} \\ V_{dr} \end{bmatrix} = \begin{bmatrix} R_s + pL_s & 0 & pL_m & 0 \\ 0 & R_s + pL_s & 0 & pL_m \\ pL_m & -\omega_r L_m & R_r + pL_r & -\omega_r L_r \\ \omega_r L_m & pL_m & \omega_r L_r & R_r + pL_r \end{bmatrix} \begin{bmatrix} i_{qs} \\ i_{ds} \\ i_{qr} \\ i_{dr} \end{bmatrix} \tag{9.31}$$

The equation used for the stator side is as shown:

$$V_{qs} = R_s i_{qs} + \frac{d}{dt}\lambda_{qs} \tag{9.32}$$

The equation used for the rotor side is as shown below:

$$V_{qr} = R_r i_{qr} + \frac{d}{dt}\lambda_{qr} - \omega_r \lambda_{dr} \tag{9.33}$$

The equations for gap flux linkage are as shown:

$$\lambda_{qr} = L_m \left(i_{qr} + i_{qs} \right) \tag{9.34}$$

R_s represents the stator winding resistance, R_r is the motor winding resistance, L_m represents the magnetizing inductance, L_{ls} represents the stator leakage inductance, L_{ir} is the rotor leakage inductance, ω_r is the electrical rotor angular speed, i_d is the current, V_q and V_d represent the voltages, and λ_d and λ_q show the fluxes of the *d–q* model [17, 39].

Equation (9.34) can be substituted in Eq. (9.35) to generate the output power and torque of the turbine (T_t), as shown below:

$$P_w = \frac{1}{2}\rho A C_p(\lambda, \beta) \left(\frac{R\omega_{opt}}{\lambda_{opt}} \right)^3 \tag{9.35}$$

$$T_t = \frac{1}{2}\rho A C_p(\lambda, \beta) \left(\frac{R}{\lambda_{opt}} \right)^3 \omega_{opt} \tag{9.36}$$

The following equations are used to calculate the power coefficient C_p, which is a nonlinear function.

$$C_p(\lambda, \beta) = c_1 \left(c_2 \frac{1}{\lambda} - c_3 \beta - c_4 \right) e^{-c_5 \frac{1}{\lambda_i}} + c_6 \lambda \tag{9.37}$$

with

$$\frac{1}{\lambda_i} = \frac{1}{\lambda + 0.08\beta} - \frac{0.035}{\beta^3 + 1} \tag{9.38}$$

The constants $C1$–$C6$ are outlined in the subsequent sections.

Therefore, the energy conversion circuit diagram that was implemented through Simulink played an important role in the generation of energy that was collected through the wind.

Conversion of Wind Energy

The conversion of wind energy is then implemented through nonlinear procedures and characteristics that are integrated to facilitate the conversion from wind energy to electric energy, and this is done through the stator control circuit system [17]. The components of the control systems are the fuzzifications that determine the net energy output after the conversion process through the use of a fuzzy inference system (FIS) [8, 28, 32]. Subsequently, the FLC is used in calculating the wind turbine and the DFIG subsystems, which is helpful in solving several issues during the transformation process, as shown by the following equation:

$$\begin{cases} e_{\Omega_g(n)} = \Omega_g^*(n) - n_2(n-1) \\ \Delta e_{\Omega_g(n)} = \Omega_g^*(n) - \Omega_g(n-1) \end{cases} \tag{9.39}$$

Here, the fuzzy controller is used in the determination of the net realistic energy variants. The triangular, trapezoidal, and symmetrical components of the wind turbine generator were used to confirm the wind energy production. Variables such as the net current formation were used in calculating the FLC and the DFIG of the model using the equations below:

$$\begin{cases} e_{i_{dr}}(n) = i_{dr}^*(n) - i_{dr}(n) \\ e_{i_{dr}}(n) = i_{qr}^*(n) - i_{qr}(n) \end{cases} \tag{9.40}$$

Here, the quantification process of the fuzzy controller has been achieved successfully using both the input and output variables of the energy systems. After quantification, the wind energy transformation rate was determined by the behavior of the wind speed rotor of the wind turbine [8, 10, 11]. The determination of the accuracy and complexity of the wind energy conversion is achieved through the computation of the changes that take place in the output power. The process is also coupled with the simulation activities that help to produce the energy that is used to run the transportation vehicle.

Solar Energy Modeling

Then the dynamic photon proliferation is being calculated to help demonstrate the quantity of solar energy captured by the panels. Therefore, the calculation is performed using Eqs. (9.40). To confirm the net capture of the photon energy, the real condition $j(\omega)$, which is the solar panel transportation vehicle, is calculated. The quantity is used to define the density of states (DOS) through the release of the magnitude $V(\omega)$ in the active solar panel cell [4, 11, 26]. As a result, the solar energy is captured by considering different dimensions of the solar panel transportation vehicle, such as 1D, 2D, and 3D [31], as shown in Table 9.1:

Here, the ID, 2D, and 3D transportation PV panels require determining the sharp frequency cutoff at Ω_d, which is used to enhance the photon energy generation of the DOS, as shown in Fig. 9.7. Therefore, the two major variables that are dilogarithmic are $\mathrm{Li}_2(x)$ and $\mathrm{erfc}(x)$. The eigenfunctions of Maxwell's rule and photon eigenfrequencies are used in determining the DOS values of the other panels, which are denoted as $\varrho_{PC}(\omega)$ [27, 38]. The DOS is also utilized in 1D solar cells, and it is given by the equation $\varrho_{PC}(\omega) \propto \frac{1}{\sqrt{\omega-\omega_e}}\Theta(\omega-\omega_e)$, where $\Theta(\omega-\omega_e)$ is known as the Heaviside step function. The frequency of the PBE in a DOS is represented by the quantity ω_e.

Here, the DOS is a critical factor to accurately calculate the qualitative state of the photon energy of the photon cell in the vehicle solar panel; thus, the DOS is being used classically next to an anisotropic PBE on the 1D, 2D, and 3D solar panel [11, 35]. Thus, the DOS of the photon energy is derived here as $\varrho_{PC}(\omega) \propto \frac{1}{\sqrt{\omega-\omega_e}}\Theta(\omega-\omega_e)$, which is further clarified in relation to the electromagnetic field factor [4, 7, 19, 38]. The photon energy is approximated as $\varrho_{PC}(\omega) \propto -[\ln|(\omega-\omega_0)/\omega_0|-1]\Theta(\omega-\omega_e)$ in 1D and 2D solar panels, where ω_e is the center of the DOS distribution.

Simply, here, $J(\omega)$ is clarified as the DOS forces generated in the cell of the solar panel; thus, here, the net solar energy release magnitude $V(\omega)$ within the acting PV cell of the vehicle solar panel is calculated as [4, 11, 26]:

$$J(\omega) = \varrho(\omega)|V(\omega)|^2. \tag{9.41}$$

Hereafter, this paper defines the proliferative photon dynamics $\langle a(t)\rangle = u(t,t_0)\langle a(t_0)\rangle$ and the PB frequency ω_c, where the function $u(t,t_0)$ describes the photon structure. $u(t,t_0)$ is derived by the dissipative integral–differential equation as in Eq. (9.18):

$$u(t,t_0) = \frac{1}{1-\Sigma'(\omega_b)}e^{-i\omega(t-t_0)} + \int_{\omega_e}^{\infty} d\omega \frac{J(\omega)e^{-i\omega(t-t_0)}}{[\omega-\omega_c-\Delta(\omega)]^2+\pi^2J^2(\omega)}, \tag{9.42}$$

Table 9.1 The table above shows different photon structures of the density of states (DOS) that can be accommodated in different dimensions of the solar panel. The photon dynamics are used in determining the quantity that shows the unit area (ω) as well as the self-energy induction in the reservoir. The two may differ depending on the structure. The variables that determine the forces between solar breaks include C, η, and χ, which are used in all three dimensions

Photon	Unit area $J(\omega)$ for different DOS*	Reservoir-induced self-energy correction $\Sigma(\omega)$*
1D	$\frac{C}{\pi}\frac{1}{\sqrt{\omega-\omega_e}}\Theta(\omega-\omega_e)$	$\varrho_{PC}(\omega)\propto\frac{1}{\sqrt{\omega-\omega_e}}\Theta(\omega-\omega_e),-\frac{C}{\sqrt{\omega_e-\omega}}$
2D	$\Omega_d-\eta\left[\ln\left\|\frac{\omega-\omega_0}{\omega_0}\right\|-1\right]\Theta(\omega-\omega_e)\Theta(\Omega_d-\omega)$	$\eta\left[\mathrm{Li}_2\left(\frac{\Omega_d-\omega_0}{\omega-\omega_0}\right)-\mathrm{Li}_2\left(\frac{\omega_0-\omega_e}{\omega_0-\omega}\right)-\varrho_{PC}(\omega)\propto-[\ln\|(\omega-\omega_0)/\omega_0\|-1]\Theta(\omega-\omega_e)\right]$
3D	$\chi\sqrt{\frac{\omega-\omega_e}{\Omega_C}}\exp\left(-\frac{\omega-\omega_e}{\Omega_C}\right)\Theta(\omega-\omega_e)$	$\chi\left[\pi\sqrt{\frac{\omega_e-\omega}{\Omega_C}}\exp\left(-\frac{\omega-\omega_e}{\Omega_C}\right)\mathrm{erfc}\sqrt{\frac{\omega_e-\omega}{\Omega_C}}-\sqrt{\pi}\right]$

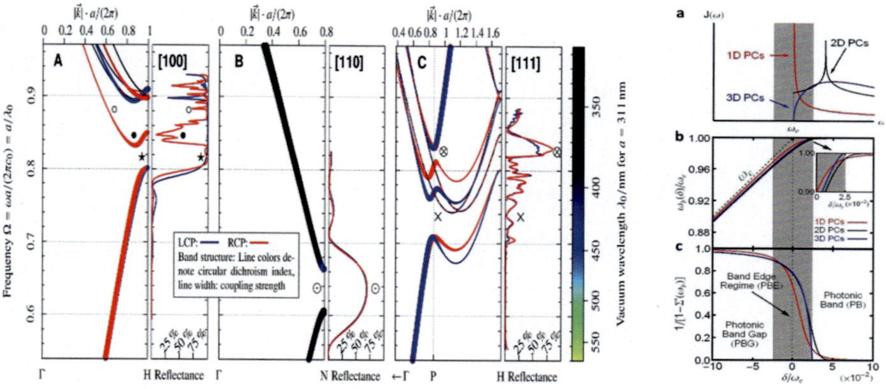

Fig. 9.7. (1) Represents the energy conversion modes and the photonic band structure. In (2), the diagram shows the relationship between the units and the frequency at several DOSs such as 1D, 2D, and 3D on the solar panel in subsection (2)-a. In subsection (2)-b, there is a representation of the frequencies of the photovoltaic modes for functional tuning. In sub figure (2)-c, the different levels of photonic modes necessary for the release of energy. Here, the solar panels 1D and 2D are the photonic modes assuming the crossover into solar panels, and 3D is depicted as the complex model which is related to the transformation from the PBG to the photonic band (PB) area for creating net solar energy [6, 35]

where $\Sigma'(\omega_b) = [\partial\Sigma(\omega)/\partial\omega]_{\omega=\omega_b}$ and $\Sigma(\omega)$ is the self-energy correction of the PB photon introduced in the reservoirs.

$$\Sigma(\omega) = \int_{\omega_e}^{\infty} d\omega' \frac{J(\omega')}{\omega - \omega'}. \tag{9.44}$$

The solar energy frequency mode in the PBG ($0 < \omega_b < \omega_e$) is represented by the frequency ω_b in Eq. (9.44). This energy can be calculated under the pole condition $\omega_b - \omega_c - \Delta(\omega_b) = 0$, where the principal-value integral is given by $\lesssim \Delta(\omega) = \mathcal{P}\left[\int d\omega' \frac{J(\omega')}{\omega - \omega'}\right]$.

Figure 9.7a shows the cooling photonic dynamics of the proliferation energy $|u(t, t_0)|$, and the energy is determined in 3D, 2D, and 1D photon cells with different measures of the detuning parameter δ [8, 17, 20, 21]. The proliferation magnitude is then integrated into the PB to the PBG area and shows a plot of the rates of solar energy generation dynamics $\kappa(t)$. From the results, it is shown that dynamic photons are generated rapidly once ω_c has crossed PB from the PBG area. Since the $u(t, t_0)$ range is$1 \geq |u(t, t_0)| \geq 0$, the crossover area is defined to satisfy $0.9 \gtrsim |u(t \to \infty, t_0)| \geq 0$. This represents $-0.025\omega_e \lesssim \delta \lesssim 0.025\omega_e$, with solar energy generation rate $\kappa(t)$ within PBG ($\delta < -0.025\omega_e$) and next to PBE($-0.025\omega_e \lesssim \delta \lesssim 0.025\omega_e$).

More specifically, the PB was first considered as the solar energy determination, n_0, i.e., $\rho(t_0) = |n_0\rangle\langle n_0|$, derived theoretically by real-time quantum feedback control

[36, 38], solving Eq. (9.44), in consideration of the state of energy state photon generation at time t:

$$\rho(t) = \sum_{n=0}^{\infty} \mathcal{P}_n^{(n_0)}(t)|n_0\rangle\langle n_0| \tag{9.45}$$

$$\mathcal{P}_n^{(n_0)}(t) = \frac{[v(t,t)]^n}{[1+v(t,t)]^{n+1}}[1-\Omega(t)]^{n_0} \times \sum_{k=0}^{\min\{n_0,n\}} \binom{n_0}{k} \times \binom{n}{k}\left[\frac{1}{v(t,t)}\frac{\Omega(t)}{1-\Omega(t)}\right]^k, \tag{9.46}$$

where $\Omega(t) = \frac{|u(t,t_0)|^2}{1+v(t,t)}$. The findings show that energy state solar photons are introduced into dynamic states $\mathcal{P}_n^{(n_0)}(t)$ of $|n_0\rangle$ and graphs of the proliferation of photon dissipation $\mathcal{P}_n^{(n_0)}(t)$ in the primary state $|n_0 = 5\rangle$ as well as in the limit of steady states $\mathcal{P}_n^{(n_0)}(t \to \infty)$. Therefore, the proliferation of the produced energy state photons will ultimately reach a non-equilibrium state of energy that supplies enough energy to run the transportation vehicle.

Solar Energy Conversion

Since the magnitude of solar irradiance is usually considered as the solar energy transformed into electrical energy for the transportation vehicle, the static light quanta are being counted in relation to the range of ν_r to $\nu_r + d\nu_r$ [14, 15, 38]. The optimal solar irradiation is being attained here at 1.4 eV and thus the energy value of 27.77 mW/m^2 eV has been configured mathematically, and thus a solar panel installed in a transportation vehicle shall produce 277,700 kW annually or 760.08 kW daily on an average of 5 h of solar irradiance during peak levels [19, 30]. The amount of energy that is simply produced by a solar panel of approximately 1 m^2 on a peak level day is equivalent to 23 gallons of gas that can be used to run any standard transportation vehicle for up to 690 miles. Considering the waste factor based on physical principles, the energy is lost during the conversion of solar energy to direct current (DC) power and finally to alternating current (AC) is approximately 0.8 [8, 11, 20, 21]. Since solar panels have an efficiency of up to 80%, the transformation of energy by a solar panel may be up to 125% higher, and thus the net energy will remain the same which is equivalent to (277,700 $\times$ 1.25 $\times$ 0.8) = 277,700 kW annually or 760 kW daily (Fig. 9.8).

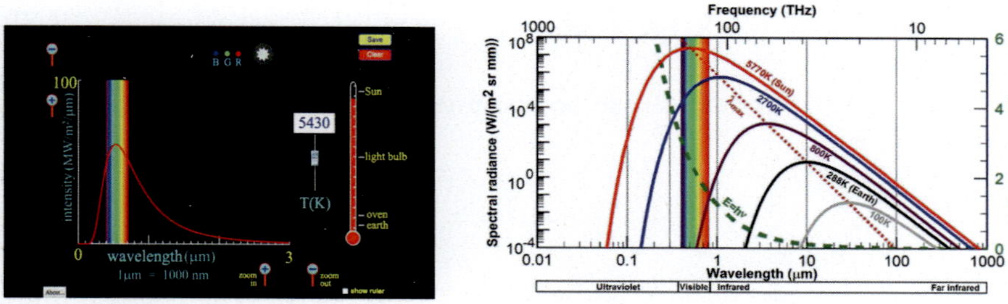

Fig. 9.8 An illustration of blackbody solar irradiation in different temperature at 5770 K. Power is 6.31×107 (W/m^2); Peak E is 1.410 (eV); Peak λ is 0.88 (μm); Peak μ is 2.81×107 (W/m^2 eV)

Electrical Subsystem

Once both wind and solar energy conversion modeling have been completed, then two renewable energy systems have been integrated using MATLAB 9.0. The utilization of this integrated hybrid renewable energy has been configured via the electrical subsystem. Hence, the functional mechanism of the electrical subsystem is important in transforming the electric energy produced into DC before it is finally converted into AC. The functional mechanism is quantified by the *d–q* synchronous voltage equation below:

$$V_q = -R_s i_q - L_q \frac{di_q}{dt} - \omega L_d i_d + \omega \lambda_m \tag{9.47}$$

The electrical energy produced can therefore be simplified as an equation that demonstrates the dynamic modeling of active energy, as shown below:

$$\begin{bmatrix} V_{qs} \\ V_{ds} \\ V_{qr} \\ V_{dr} \end{bmatrix} = \begin{bmatrix} R_s + pL_s & 0 & pL_m & 0 \\ 0 & R_s + pL_s & 0 & pL_m \\ pL_m & -\omega_r L_m & R_r + pL_r & -\omega_r L_r \\ \omega_r L_m & pL_m & \omega_r L_r & R_r + pL_r \end{bmatrix} \begin{bmatrix} i_{qs} \\ i_{ds} \\ i_{qr} \\ i_{dr} \end{bmatrix} \tag{9.48}$$

Here, electricity energy generation is précised by further analysis of stator control and expressed as:

$$V_{qs} = R_s i_{qs} + \frac{d}{dt} \lambda_{qs} \tag{9.49}$$

which can be further simplified as

$$V_{q\mathrm{r}} = R_{\mathrm{r}} i_{q\mathrm{r}} + \frac{d}{dt}\lambda_{q\mathrm{r}} - \omega_{\mathrm{r}}\lambda_{d\mathrm{r}} \tag{9.50}$$

In case an air gap flux leakage occurs in generation, the equations are further corrected to be

$$\lambda_{q\mathrm{r}} = L_m\left(i_{qr} + i_{qs}\right) \tag{9.51}$$

where R_{s}, R_{r}, L_{m}, L_{ls}, L_{lr}, ω_{r}, i_d, i_q, V_d, V_q, λ_d, and λ_q represent the resistance of the stator, current resistance, current inductance, stator leakage inductance, current leakage inductance, electrical conductance, current, voltage, and fluxes of the d–q model, respectively [38]. Then, the net hybrid electricity energy (T_{t}) produced is calculated as:

$$P_{\mathrm{w}} = \frac{1}{2}\rho A C_p(\lambda, \beta)\left(\frac{R\omega_{\mathrm{opt}}}{\lambda_{\mathrm{opt}}}\right)^3 \tag{9.52}$$

$$T_{\mathrm{t}} = \frac{1}{2}\rho A C_{\mathrm{p}}(\lambda, \beta)\left(\frac{R}{\lambda_{\mathrm{opt}}}\right)^3 \omega_{\mathrm{opt}} \tag{9.53}$$

The equation below is used in determining the power coefficient (C_{p}) as a nuclear function:

$$C_{\mathrm{p}}(\lambda, \beta) = c_1\left(c_2\frac{1}{\lambda} - c_3\beta - c_4\right)e^{-c_5\frac{1}{\lambda_i}} + c_6\lambda \tag{9.54}$$

where

$$\frac{1}{\lambda_i} = \frac{1}{\lambda + 0.08\beta} - \frac{0.035}{\beta^3 + 1} \tag{9.55}$$

The net hybrid solar and wind energy generation module produced was then determined using an energy conversion circuit and a converter available in MATLAB 9.0 used in powering the transport vehicle (Fig. 9.9).

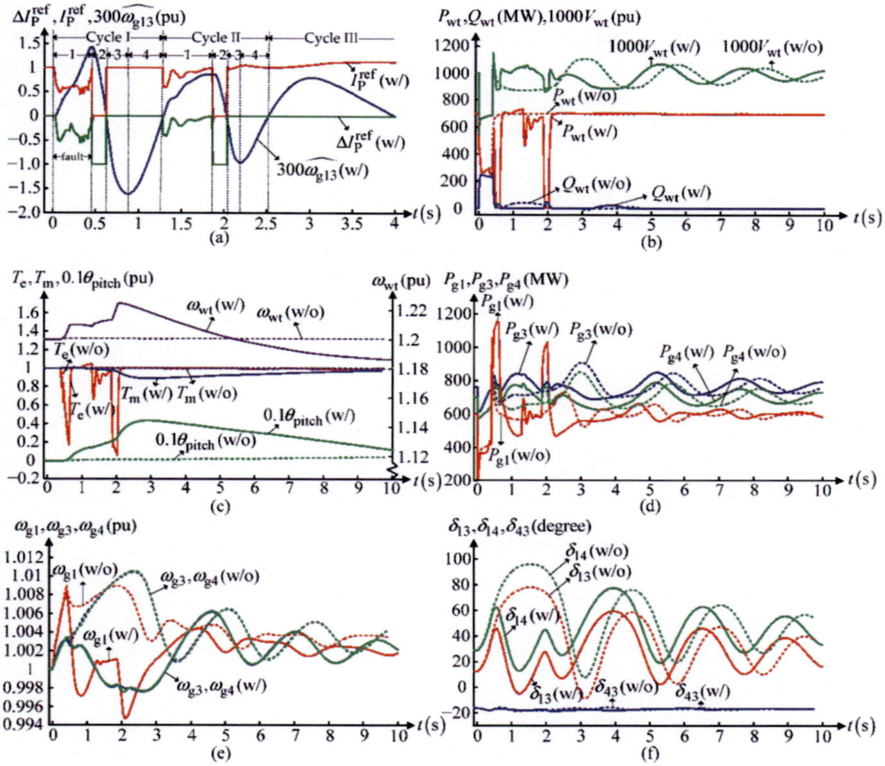

Fig. 9.9 An illustration of how current output (MW) by an installed wind turbine that is connected to the electrical subsystem circuit can be determined at different angles during the rotation of the turbines

Conclusions

Fossil fuels are the most commonly used sources of energy in most transportation vehicles; however, these sources of energy are expensive and cause environmental vulnerability. To meet the total energy needs in the global transportation sector, conventional energy technologies are being used, which places tremendous stress on fossil fuels, and thus fossil fuel supplies are becoming finite. It is well established that the transportation sector consumes 30% of the global energy, which is 5.6×10^{20} J/year (560 EJ/year) that is captured from burning fossil fuel and causes environmental vulnerability. Interestingly, the implementation of hybrid wind and solar energy into the transportation vehicle would be an important finding once it is utilized by the vehicles, which will increase the use of wind and solar power to run transportation vehicles globally. Concretely, it can be said that once the vehicles start to run by air and light, it indeed would be the gateway of new technology to eradicate the energy cost for the transportation sectors, which will also play a key factor in mitigating climate change.

Acknowledgments This research was sponsored by the Green Globe Technology under grant RD-02020-01 in effort to developing environmental conservation. However, the hypothesis, findings, and conclusions of this research are solely carried out by the author. The author confirms that there is no conflict of interest in publishing this research paper in a standard journal.

References

1. M. Abulizi, L. Peng, B. Francois, Y. Li, Performance analysis of a controller for doubly-fed induction generators based wind turbines against parameter variations. Int. Rev. Electr. Eng. **9**, 264–269 (2014). https://doi.org/10.15866/iree.v9i2.1797
2. N. Bahri, W. Ouled Amor, Intelligent power supply management of an autonomous hybrid energy generator. Int. J. Sustain. Eng. **12**, 312–332 (2019). https://doi.org/10.1080/19397038.2019.1581852
3. F.I. Bakhsh, D.K. Khatod, A novel method for grid integration of synchronous generator based wind energy generation system, in *2014 IEEE International Conference on Power Electronics, Drives and Energy Systems (PEDES)*, (IEEE, Mumbai, 2014), pp. 1–6
4. F. Bento, A.J.M. Cardoso, A comprehensive survey on fault diagnosis and fault tolerance of DC-DC converters. Chin. J. Electr. Eng. **4**, 1–12 (2018). https://doi.org/10.23919/CJEE.2018.8471284
5. A. Boumassata, D. Kerdoun, M. Madaci, Grid power control based on a wind energy conversion system and a flywheel energy storage system, in *IEEE EUROCON 2015—International Conference on Computer as a Tool (EUROCON)*, (IEEE, Salamanca, 2015), pp. 1–6
6. A. Darvish Falehi, Augment dynamic and transient capability of DFIG using optimal design of NIOPID based DPC strategy. Environ. Prog. Sustain. Energy **37**, 1491–1502 (2018). https://doi.org/10.1002/ep.12811
7. A. Elmansouri, J.E. El-mhamdi, A. Boualouch, Wind energy conversion system using DFIG controlled by back-stepping and RST controller, in *2016 International Conference on Electrical and Information Technologies (ICEIT)*, (IEEE, Tangiers, 2016), pp. 312–318
8. K. Elyaalaoui, M. Ouassaid, M. Cherkaoui, Supervision system of a wind farm based on squirrel cage asynchronous generator, in *2016 International Renewable and Sustainable Energy Conference (IRSEC)*, (IEEE, Marrakech, 2016), pp. 403–408
9. A. Gaillard, P. Poure, S. Saadate, Reactive power compensation and active filtering capability of WECS with DFIG without any over-rating. Wind Energy **13**, 603–614 (2009). https://doi.org/10.1002/we.381
10. K. Ghedamsi, D. Aouzellag, Improvement of the performances for wind energy conversions systems. Int. J. Electr. Power Energy Syst. **32**, 936–945 (2010). https://doi.org/10.1016/j.ijepes.2010.02.012
11. F. Hamoud, M.L. Doumbia, A. Cheriti, Hybrid PI-sliding mode control of a voltage source converter based STATCOM, in *2014 16th International Power Electronics and Motion Control Conference and Exposition*, (IEEE, Antalya, 2014), pp. 661–666

12. M. Heydari, K. Smedley, Comparison of maximum power point tracking methods for medium to high power wind energy systems, in *2015 20th Conference on Electrical Power Distribution Networks Conference (EPDC)*, (IEEE, Zahedan, 2015), pp. 184–189
13. M.F. Hossain, Design and construction of ultra-relativistic collision PV panel and its application into building sector to mitigate total energy demand. J. Build. Eng. **9**, 147–154 (2017). https://doi.org/10.1016/j.jobe.2016.12.005
14. M.F. Hossain, Flying transportation technology, in *Sustainable Design and Build: Building, Energy, Roads, Bridges, Water and Sewer Systems*, ed. by M. F. Hossain, (Butterworth-Heinemann, Cambridge, 2019a), pp. 282–300
15. M.F. Hossain, Sustainable technology for energy and environmental benign building design. J. Build. Eng. **22**, 130–139 (2019b). https://doi.org/10.1016/j.jobe.2018.12.001
16. M.F. Hossain, N. Fara, Integration of wind into running vehicles to meet its total energy demand. Energy Ecol. Environ. **2**, 35–48 (2016). https://doi.org/10.1007/s40974-016-0048-1
17. A. Junyent-Ferré, O. Gomis-Bellmunt, Wind turbine generation systems modeling for integration in power systems, in *Handbook of renewable energy technology*, ed. by A. Zobaa, R. Bansal, (World Scientific, Singapore, 2011), pp. 53–68
18. A. Junyent-Ferré, O. Gomis-Bellmunt, A. Sumper, M. Sala, M. Mata, Modeling and control of the doubly fed induction generator wind turbine. Simul. Model. Pract. Theory **18**, 1365–1381 (2010). https://doi.org/10.1016/j.simpat.2010.05.018
19. R. Karthikeyan, A.K. Parvathy, Peak load reduction in micro smart grid using non-intrusive load monitoring and hierarchical load scheduling, in *2015 International Conference on Smart Sensors and Systems (IC-SSS)*, (IEEE, Bangalore, 2015), pp. 1–6
20. K. Kerrouche, A. Mezouar, K. Belgacem, Decoupled control of doubly fed induction generator by vector control for wind energy conversion system. Energy Procedia **42**, 239–248 (2013a). https://doi.org/10.1016/j.egypro.2013.11.024
21. K. Kerrouche, A. Mezouar, L. Boumedien, A simple and efficient maximized power control of DFIG variable speed wind turbine, in *3rd International Conference on Systems and Control*, (IEEE, Algiers, 2013b), pp. 894–899
22. P. Lap-Arparat, T. Leephakpreeda, Real-time maximized power generation of vertical axis wind turbines based on characteristic curves of power coefficients via fuzzy pulse width modulation load regulation. Energy **182**, 975–987 (2019). https://doi.org/10.1016/j.energy.2019.06.098
23. H. Ligang, W. Xiangdong, Y. Kang, Optimal speed tracking for double fed wind generator via switching control, in *The 27th Chinese Control and Decision Conference (2015 CCDC)*, (IEEE, Qingdao, 2015), pp. 2638–2643
24. Y. Ling, A fault ride through scheme for doubly fed induction generator wind turbine. Aus. J. Electr. Electron. Eng. **15**, 71–79 (2018). https://doi.org/10.1080/1448837X.2018.1525172
25. M. Loucif, A. Boumédiène, Modeling and direct power control for a DFIG under wind speed variation, in *2015 3rd International Conference on Control, Engineering & Information Technology (CEIT)*, (IEEE, Tlemcen, 2015), pp. 1–6
26. S.N. Mahato, S.P. Singh, M.P. Sharma, Dynamic behavior of a single-phase self-excited induction generator using a three-phase machine feeding single-phase dynamic load. Int. J. Electr. Power Energy Syst. **47**, 1–12 (2013). https://doi.org/10.1016/j.ijepes.2012.10.067
27. P. Mani, J.H. Lee, K.W. Kang, Y.H. Joo, Digital controller design via LMIs for direct-driven surface mounted PMSG-based wind energy conversion system. IEEE Trans. Cybern. **50**, 3056–3067 (2020). https://doi.org/10.1109/TCYB.2019.2923775
28. A.A.B. Mohd Zin, H.A.M. Pesaran, A.B. Khairuddin, L. Jahanshaloo, O. Shariati, An overview on doubly fed induction generators′ controls and contributions to wind based electricity generation. Renew. Sustain. Energy Rev. **27**, 692–708 (2013). https://doi.org/10.1016/j.rser.2013.07.010
29. J. Mwaniki, H. Lin, Z. Dai, A condensed introduction to the doubly fed induction generator wind energy conversion systems. J. Eng. **2017**, 1–18 (2017). https://doi.org/10.1155/2017/2918281

30. M. Ouassaid, K. Elyaalaoui, M. Cherkaoui, Reactive power capability of squirrel cage asynchronous generator connected to the grid, in *2015 3rd International Renewable and Sustainable Energy Conference (IRSEC)*, (IEEE, Marrakech, 2015), pp. 1–4
31. T.A.M. Pugh, A.R. MacKenzie, J.D. Whyatt, C.N. Hewitt, Effectiveness of green infrastructure for improvement of air quality in urban street canyons. Environ. Sci. Technol. **46**, 7692–7699 (2012). https://doi.org/10.1021/es300826w
32. M.A.V. Rad, R. Ghasempour, P. Rahdan, S. Mousavi, M. Arastounia, Techno-economic analysis of a hybrid power system based on the cost-effective hydrogen production method for rural electrification, a case study in Iran. Energy **190**, 116421 (2020). https://doi.org/10.1016/j.energy.2019.116421
33. M.D. Rani, M.S. Kumar, Development of doubly fed induction generator equivalent circuit and stability analysis applicable for wind energy conversion system, in *2017 International Conference on Recent Advances in Electronics and Communication Technology (ICRAECT)*, (IEEE, Bangalore, 2017), pp. 55–60
34. E. Ribeiro, A. Monteiro, A.J.M. Cardoso, C. Boccaletti, Fault tolerant small wind power system for telecommunications with maximum power extraction, in *2014 IEEE 36th International Telecommunications Energy Conference (INTELEC)*, (IEEE, Vancouver, BC, 2014), pp. 1–6
35. M.S.R. Saeed, E.E.M. Mohamed, M.A. Sayed, Design and analysis of dual rotor multi-tooth flux switching machine for wind power generation, in *2016 Eighteenth International Middle East Power Systems Conference (MEPCON)*, (IEEE, Cairo, 2016), pp. 499–505
36. B. Touaiti, H.B. Azza, M. Jemli, A MRAS observer for sensorless control of wind-driven doubly fed induction generator in remote areas, in *2016 17th International Conference on Sciences and Techniques of Automatic Control and Computer Engineering (STA)*, (IEEE, Sousse, 2016), pp. 526–531
37. B. Touaiti, H.B. Azza, M. Jemli, Control scheme for stand-alone DFIG feeding an isolated load, in *2019 10th International Renewable Energy Congress (IREC)*, (IEEE, Sousse, 2019), pp. 1–6
38. V.C. Upadhyay, K.S. Sandhu, Reactive power management of wind farm using STATCOM, in *2018 International Conference on Emerging Trends and Innovations in Engineering and Technological Research (ICETIETR)*, (IEEE, Ernakulam, 2018), pp. 1–5
39. C. Venkatesan, K. Sundararaman, M. Gopalakrishnan, Grid integration of PMSG based wind energy conversion system using variable frequency transformer, in *2017 International Conference on Intelligent Computing, Instrumentation and Control Technologies (ICICICT)*, (IEEE, Kannurs, 2017), pp. 1451–1456
40. A. Zohoori, A. Vahedi, M.A. Noroozi, S. Meo, A new outer-rotor flux switching permanent magnet generator for wind farm applications. Wind Energy **20**, 3–17 (2017). https://doi.org/10.1002/we.1986

Chapter 10
Flying Transportation Technology to Console Global Communication Crisis

Abstract To have pleasant road trips, avoid commute, and do less time on trips to arrive at the desired destination much faster, a model of ***Flying Transportation Technology*** has been proposed. The vision behind this technology is to design an economical, safe, and environmentally friendly mode of transportation. This study seeks to present a 3D numerical simulation of external flow for a flying automobile with well-designed rectangular NACA 9618 wings. To enhance its airborne capabilities, this car's aerodynamic traits have been professionally measured and adjusted, such that it utilizes minimal takeoff velocity. Besides, the vehicle will have an integrated 3D *k-omega* turbulence model, which captures a fundamental flow physics enhancing the performance during takeoff. This forms the theoretical basis of the flying car. The numerical aspect comprises limited edition Reynolds-Averaged Navier–Stokes equations (RANS) comprehensible schemes. Generally, the vehicle is being designed with highly functional wings that allow divergent deployments during takeoff to maximize its air performance. Because of the utilization of wind during the flying process, the model has integrated wind turbines that enables wind recodification to propel the car in the air. Considering the combination of technologies involved in the design of the flying car, it is one of the most sophisticated inventions that will not only facilitate safe transportation and save trillions of dollars annually but will also significantly help in saving the ecosystem.

Keywords 3D numerical simulation · 3D *k-omega* turbulence modeling · Flying car design · Wind energy modeling · Cleaner energy implementation · Smart transportation technology

Introduction

The construction of roads and other transportation infrastructure consumes 0.9% of the earth surface. In addition, 1.77 million dollars is spent on constructing every mile of the infrastructure. As a result, several forest covers are destroyed to pave the way for the infrastructure, thus contributing to an approximated 6% of climate change. This costs about \$5575 trillion (100,000,000 mi^2 × 5280 sf × 5280 sf × \$200 per SF cost) as the heat is reflected back into space. The current statistics indicate that 2% of

M. F. Hossain, *Sustainable Design for Global Equilibrium*,
https://doi.org/10.1007/978-3-030-94818-4_10

the existing infrastructure undergoes annual repairs totaling to $12 trillion, with another $55.75 trillion spent in new infrastructure development annually. The rate of infrastructure development is estimated at 1% annually. Other than the infrastructure development and rehabilitation processes, road transport results in frequent traffic jams that cost trillions annually [14, 15]. The fact that the current traditional transport system heavily relies on fossil fuels, which results in undesirable emissions in the atmosphere, also makes it unfriendly to the environment. Compared to the flying transportation technology, the new model is much more safe, convenient, and economical. A lot of research has been conducted in designing the flying car, and it might help save both the environment and the economy upon its completion.

The engineering disciplines have of late seen diverse integration of computational fluid dynamic (CFD) models on a large-scale basis. This is a commendable step that will significantly improve the affordability of the models and enhance their power. The CFD model can also be adopted in the development of flying cars. It, however, requires much more precision as the flying cars are much more complicated than the regular race cars and aircraft currently in use primarily due to the additional features of its wings that support its deployment during takeoffs. This study, therefore, aims to introduce a propulsive and levitative force model that can be used in addressing the aerodynamics takeoff velocity and drag force computational fluid problems to enhance the performance of the flying cars. The research will adopt the use of MATLAB software to illustrate the model.

Materials, Methods, and Simulation

The flying vehicle's performance in the air is mainly determined by takeoff velocity, the propelling forces, drag force, stability, and proper control capabilities. These are the areas that require significant attention, especially during the wings designing phase. When creating the flying car's flanks, the shape, the aspect ratio, the cross-sections, and surface areas must be considered as they will determine the stability, control, and takeoff forces required for the maximum performance of the car. This research involves an illustration of the Mach number contours and the physical model of a complete 3D CFD, a *k-omega* turbulence mode required in a flying vehicle (Fig. 10.1). To validate the reliability of a CFD design in designing flying cars, the study will review various deployment histories involving its use.

Solving the Concept Numerically

As discussed earlier, the 3D standard *k-omega* turbulence model is based on the transport velocity model as an empirical model in assuring accurate dissipation rates and for turbulence kinetic energy and indicated in Eqs. (10.2) and (10.45).

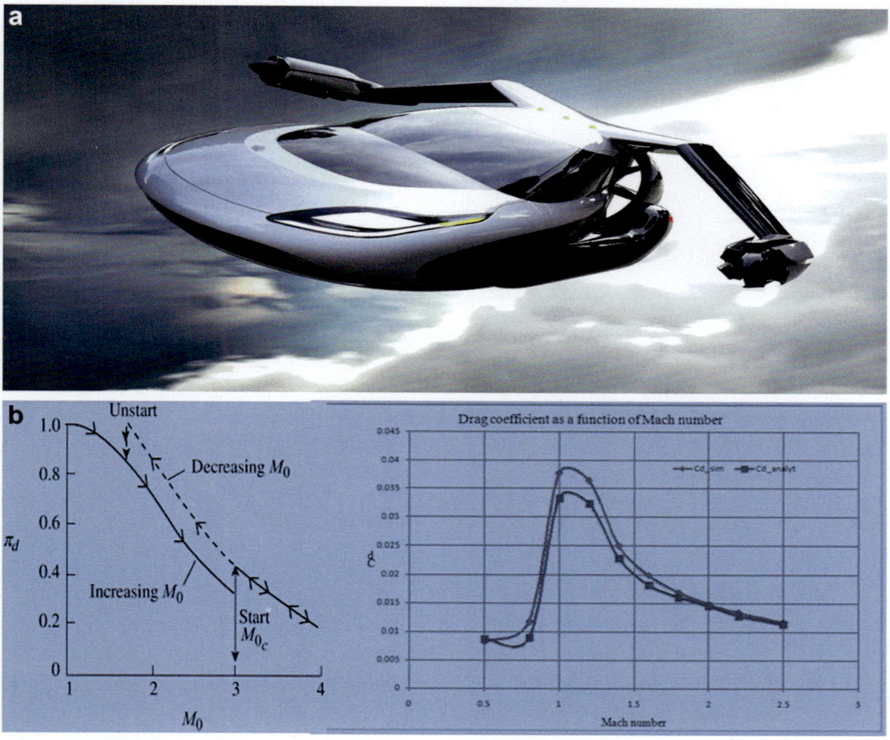

Fig. 10.1 (**a**) Indicates the flying car's conceptual model, (**b**) the proposed flying vehicle's Mach number contours

$$\psi = \left[-\left(\frac{v_\infty r_o^2}{4}\right)\frac{r_o}{r} + \left(\frac{3v_\infty r_o}{4}\right)r - \left(\frac{v_\infty}{2}\right)r^2\right]\sin^2\theta \tag{10.1}$$

$$v_r = v_\infty\left[1 - \frac{3}{2}\left(\frac{r_o}{r}\right) + \frac{1}{2}\left(\frac{r_o}{r}\right)^3\right]\cos\theta \tag{10.2}$$

$$v_\theta = -v_\infty\left[1 - \frac{3}{4}\left(\frac{r_o}{r}\right) - \frac{1}{4}\left(\frac{r_o}{r}\right)^3\right]\sin\theta \tag{10.3}$$

The equation utilizes second coupled hierarchy formulation to adequately find the solution to standard *k-omega* turbulence with shear flow corrections. The numerical solution will, therefore, be considered as an entirely limited volume scheme of the Reynolds-Averaged, Navier-Stokes force compressible that neutralizes the flying car's total pressure and wall temperatures. To achieve reliable codes that can be adopted in invalidating the baseline solutions to the flying car's wing designs 9618 NACA series airfoils can be incorporated to achieve maximum aerodynamic performance (Fig. 10.2).

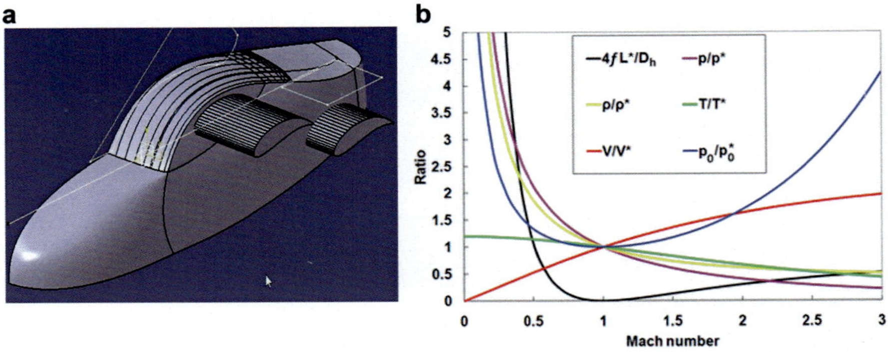

Fig. 10.2 (**a**) Indicates half of the flying car's 3D idealized model. (**b**) Aerodynamic traits in relation to the velocity and the Mach number

The Flying Car's Wind Energy Modeling Sequence

Originally, the installed single wind turbines in the cars help in the generation of energy necessary for powering it, thus satisfying the high-energy demands of flying cars. It is also essential that a doubly fed induction generator also is installed to facilitate electricity production (1,4,1,3). The equation below summarizes the entire processes involved in energy production using wind turbines.

$$P_{\mathrm{w}} = \frac{1}{2} C_{\mathrm{p}}(\lambda, \beta) \rho A V^3 \tag{10.4}$$

where

$V =$ Average speed of wind (m/s)
$P =$ Air density (kg/m^3)
$C =$ Betz's coefficient (maximum value of 0.593)
$\lambda =$ tip speed ratio
$A =$ rotor blades intercepting areas (m^2)

Here the tip speed ration equation is rewritten as

$$\lambda = \frac{R\omega}{V} \tag{10.5}$$

In the above equation

$\omega =$ is the angular speed in rad/s
$V =$ wind speed average (m/s)
$R =$ turbine radius (m)

To calculate the wind-generated energy the equation below is utilized

$$Q_w = P \times (\text{Time}) \ [\text{kWh}] \tag{10.6}$$

The direct measurements from a given motion cannot be used in obtaining velocity; thus, a lower motion should first be established using the equation below.

$$v(z) \ln\left(\frac{z_r}{z_0}\right) = v(z_r) \ln\left(\frac{z}{z_0}\right) \tag{10.7}$$

with

z = reference height (m)
z_0 = surface roughness measurement (in crop lands = 0.1–0.25)
$V(z)$ = speed of wind at z height (m/s)
$V(z_r)$ = speed of wind at z (m/s)

The equation below represents the estimation of power output based on the wind speed:

$$P_w(v) = \begin{cases} \dfrac{v^k - v_C^k}{v_R^k - v_C^k} \cdot P_R & v_C \le v \le v_R \\ P_R & v_R \le v \le v_F \\ 0 & v \le v_C \text{ and } v \ge v_F \end{cases} \tag{10.8}$$

In the above equation

P_R = represents the rated power
v_C = wind speed cut-in
v_R = total wind rated
v_F = cut-out speed rated
k = shape factor of Weibull

The primary purpose of altering the angular speed is to enable the extraction of optimum power levels, a technique commonly known as maximum power point tracking (MPPT). In scenarios where the blade pitch is at a zero angle, the optimum TRS is obtained by maximizing the power coefficients.

$$\omega_{opt} = \frac{\lambda_{opt}}{R} V_{wn} \tag{10.9}$$

resulting in

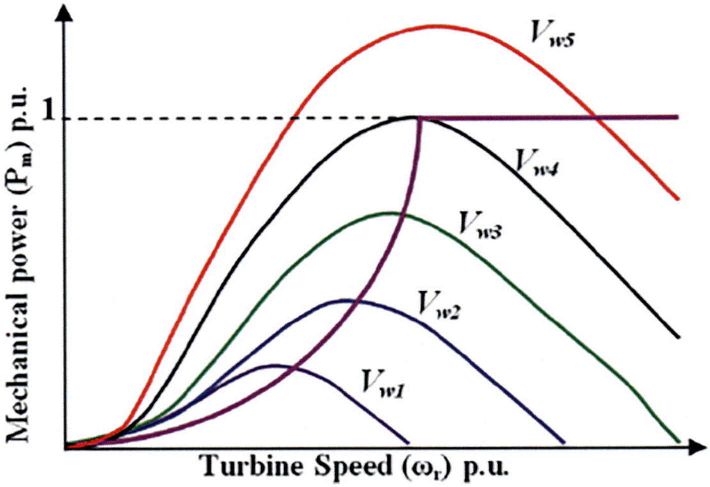

Fig. 10.3 Correlation between turbine speed and the generated power

$$V_{\mathrm{wn}} = \frac{R\omega_{\mathrm{opt}}}{\lambda_{\mathrm{opt}}} \tag{10.10}$$

with ω_{opt} as the optimum speed of the rotor angle measured in rad/s

λ_{opt} represents the ratio of optimum tip speed
R = turbine radius (m)
while V indicates the speed of wind (m/s)

Figure 10.3 indicates the changes in the rates of speed wind and power when the car is active [10, 12]. From the analysis of Fig. 10.3, it is clear that quality results in the car's performance can only be achieved in steady winds. On the other hand, a rapid increment in the speed of wind results in the rise in energy levels.

Conversion of Wind Energy

The conditions above require adequate airflow monitoring and regulation at specific kinetic energy levels. This is usually indicated by a wind energy conversion system (WECS) used in DFIG, such as the speed of the wind, gearbox, and wind turbine, among others (Fig. 10.4).

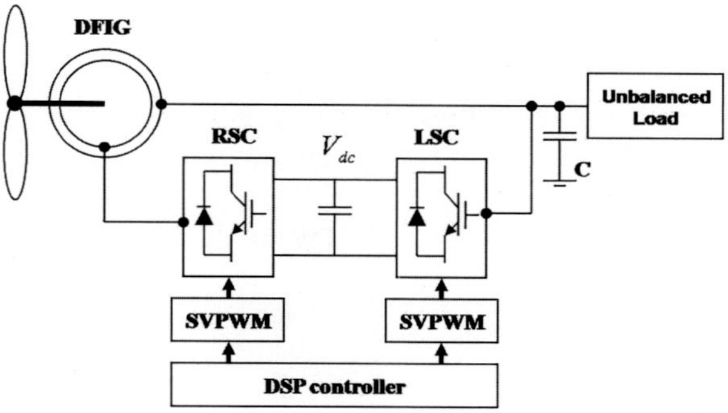

Fig. 10.4 The process of wind conversion to usable energy through the use of LSC, RSC, DFIG, DSP, and SVPWM

Modeling the Generator

Two options are always available when energy is required for the wind turbines. This includes the use of induction generators or synchronous models. The use of an asynchronous model drastically increases reliability. However, it results in a significant reduction of the nacelle weight [7, 8]. The equation below indicates the synchronous model principles based on a *d*–*q* referencing framework,

$$V_q = -R_s i_q - L_q \frac{di_q}{dt} - \omega L_d i_d + \omega \lambda_m \tag{10.11}$$

$$V_d = -R_s i_d - L_d \frac{di_q}{dt} + \omega L_q i_q \tag{10.12}$$

in which electronic torque is indicated by:

$$T_e = 1.5\rho\left[\lambda i_q + \left(L_d - L_q\right) i_d i_q\right] \tag{10.13}$$

In the equation, *q*-axis inductance is represented by L_q, while L_d indicates *d*-axis inductance. i_q is the current from the *q*-axis while i_d current from *d*-axis.

V_q represents voltage from the *q*-axis, and the energy from the *d*-axis is indicated by V_d

ω_r represents the velocity of the angular rotor

λ the induced flux amplitude

p as the number of poles

The squirrel cage induction generator can also be used in modeling the generator using the equation below.

$$\begin{bmatrix} V_{qs} \\ V_{ds} \\ V_{qr} \\ V_{dr} \end{bmatrix} = \begin{bmatrix} R_s + pL_s & 0 & pL_m & 0 \\ 0 & R_s + pL_s & 0 & pL_m \\ pL_m & -\omega_r L_m & R_r + pL_r & -\omega_r L_r \\ \omega_r L_m & pL_m & \omega_r L_r & R_r + pL_r \end{bmatrix} \begin{bmatrix} i_{qs} \\ i_{ds} \\ i_{qr} \\ i_{dr} \end{bmatrix} \tag{10.14}$$

Beginning from the stator position:

$$\begin{aligned} \lambda_{ds} &= L_s i_{ds} + L_m i_{dr} \\ \lambda_{qs} &= L_s i_{qs} + L_m i_{dr} \\ L_s &= L_{is} + L_m \\ L_r &= L_{lr} + L_m \\ V_{ds} &= R_s i_{ds} + \frac{d}{dt}\lambda_{ds} \\ V_{qs} &= R_s i_{qs} + \frac{d}{dt}\lambda_{qs} \end{aligned} \tag{10.15}$$

Beginning from the rotor position:

$$\begin{aligned} \lambda_{dr} &= L_r i_{dr} + L_m i_{ds} \\ \lambda_{qr} &= L_r i_{qr} + L_m i_{qs} \\ V_{dr} &= R_r i_{dr} + \frac{d}{dt}\lambda_{dr} + \omega_r \lambda_{qr} \\ V_{qr} &= R_r i_{qr} + \frac{d}{dt}\lambda_{qr} - \omega_r \lambda_{dr} \end{aligned} \tag{10.16}$$

To determine the air flux gaps:

$$\begin{aligned} \lambda_{dm} &= L_m(i_{ds} + i_{dr}) \\ \lambda_{qr} &= L_m\left(i_{qr} + i_{qs}\right) \end{aligned} \tag{10.17}$$

In the above equation, R_s, R_r, ω_r, i_d, i_q, V_d, V_q, λ_d, λ_q, L_m, L_{ls}, *and* L_{lr} indicate the resistance from wind by the stator and the fluxes. Energy conversion diagram-implemented inverter is often used in the preparation of the wind energy output. This is done through the calculation in Simulink-MATLAB.

Battery Modeling

In situations where the car is nonfunctional or where the engine power is exceeded by the power produced, the battery modeling is adopted in order to conserve the extra power [9, 13]. This process can be calculated by using the formula below.

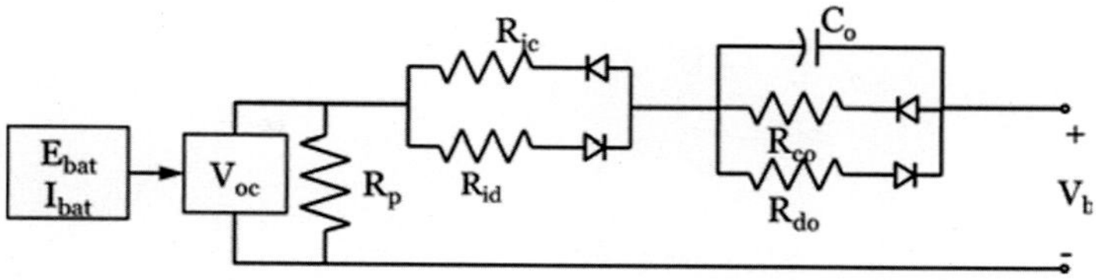

Fig. 10.5 The power backup model in a motionless car's battery for the ignition of the vehicle and to provide required energy while the transport is not in a motion

$$t_{\text{discharge}} = H\left(\frac{C}{IH}\right)^k \tag{10.18}$$

In the formula, the discharge time of the battery is represented by *t*. *C* shows the capacity of the battery in Ampere hour value. The drawn current is indicated by *I*, while *H* represents the time taken to discharge the power. The Peukert's coefficient is shown by *k* and can be calculated by using the equation below.

$$k = \frac{\log T_2 - \log T_1}{\log I_1 - \log I_2} \tag{10.19}$$

in which I_1 and I_2 represent the rates of current discharge. T_1 and T_2 show the duration of discharges. Because of the changes that take place after particular recharge cycles, including a decrement in the capacity of the battery, the *k* value must be redefined by adopting 1.3–1.4. To calculate the exact time taken to discharge a battery fully, the equation below is used (Fig. 10.5)

$$t_{\text{charging}} = \frac{\text{Ampere hour of battery}}{-\text{Charging current}} \tag{10.20}$$

Results, Optimization, and Discussion

In relation to proposed models, the fly car has been identified to be able to produce a satisfactory lift with a proper takeoff velocity within various wing positions and attack angles. Also, an in-depth analysis of the study indicates that the car has the ability to attain an appropriate lift even in low takeoff velocity situations. This is because of the presence of the deployable high wings shown in Fig. 10.6. Mathematically, it is evident that the flying vehicle's stream velocity ranges and the various geometrical options are able to produce sufficient upward force, thus lifting the car even in low takeoff velocity and conditions and also across multiple wing positions.

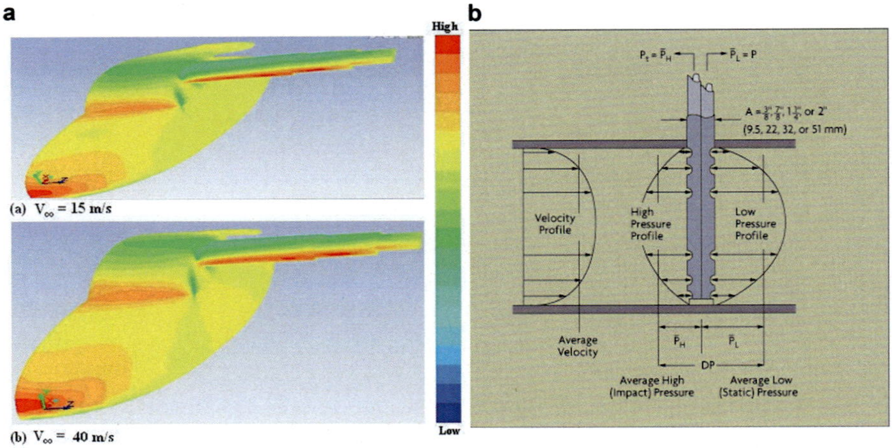

Fig. 10.6 (**a**) A flying vehicle's contours at different speeds (15 and 40 m/s), (**b**) a short high-wing flying car's model static pressure contours at a different velocity

The findings from the assessment of the aerodynamic characteristics and the features of the external flow of the flying car indicate that the car's speed heavily relies on the aerodynamic traits and the steadiness of the freestream velocity. This property enables the vehicle to initiate a takeoff from the ground, its flying ability, and proper control at different altitudes, as indicated in Fig. 10.7.

Through the adoption of these principles, this study adequately depicts the required standards and conditions necessary to facilitate the flying vehicle's takeoff and flying pattern. In order to achieve better performance and to keep the car on air, the body of the vehicle must be streamlined, and the wings must be made suitable for enhancing the lifting abilities of the vehicle. One way to achieve this is through the installation of NACA 9816 wings at various points of the vehicle. This will enable the vehicle to adequately launch attacks at various angles in the air, facilitating a smooth flying experience (Fig. 10.7).

From the robustness tests conducted in the simulation phase, the findings indicated that the flying car's capability of handling the conditions is remarkable. This test was conducted by adding voltage dips and the wind speed signals. Other suitable controls that can be added to enhance the performance of the flying car are DFIG and MPPT control. These controls can also be used in controlling the reactive powers and the stator active powers to facilitate a united power on the stator position. Also, after an increment in the wind speed, the generator shaft is able to detect an increased performance in angular momentum. In such situations, the reactive and the active bidirectional power transfer between the power and the rotor system can be calculated through the realization of the nominal potentials of the stator and the super synchronous operation. The stator power is reactive and is controlled by the converter on the load side to produce sufficient energy to power the car.

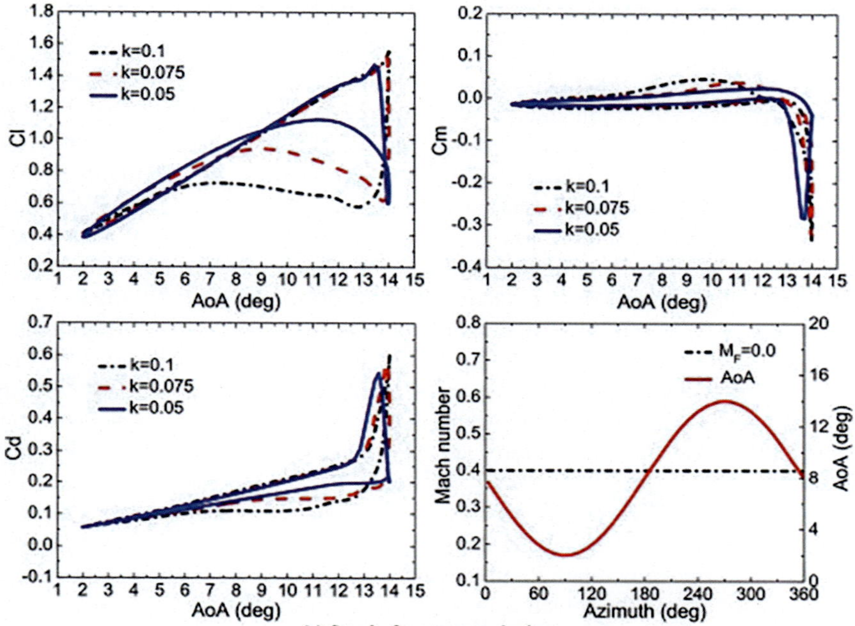

(a) Steady freestream velocity

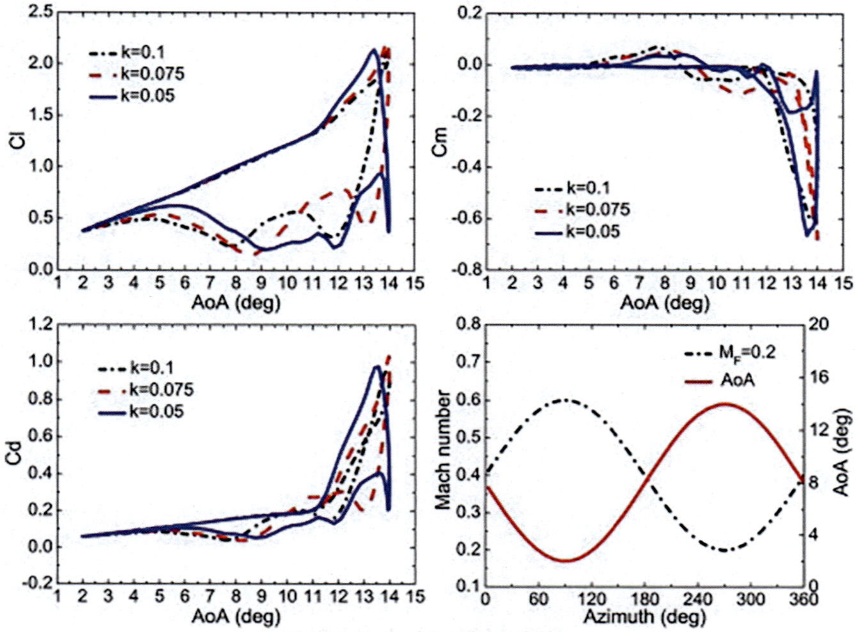

(b) Variational freestream velocity

Fig. 10.7 Indicates the car's speed as determined by (**a**) the freestream velocity variable in respect to the aerodynamics traits to facilitate the takeoff, fly, and optimization of the velocity, (**b**) steady freestream velocity

Wind Energy Modeling for the Flying Vehicles

This paper presented the flying car's complete wind turbine generating system, which facilitates the production of wind energy for the vehicle. Also, a control algorithm of a cascade has been meticulously molded to ensure that the system performs optimally, especially in the MPPT control system and the flux-orientation system [1, 11]. This mechanism has been successfully installed in a prototype sedan car to track the active energy power through the use of a rotor converter control computed by MATLAB Simulink. Subsequently, this strategy is also being used to control the pitch control through making adjustments on the power coefficient value in relation to the variation in the speed of wind with the purpose of extracting maximum wind power. The addition of voltage dip and the wind speed signals also enabled the robustness test to be efficiently conducted. From the test, the findings indicated the wind turbine's capabilities at various wind speeds in relation to the pitch angle, tip speed ratio, and the wind energy conversion chain DFIG represented by Fig. 10.8.

The increase in the speed of wind causes the generator shaft speed to achieve its extreme angular momentum by trailing the absolute powerpoint velocity. Subsequently, to affirm the workability of the control scheme, there was the measuring of disassociation among the constituents of the rotor current (Fig. 10.9). Consequently, the function of the generator with regard to the splendid synchronous action, insignificant station power, and the volatile power which is organized through a consignment-side transformer to acquire the component's power aspect to create energy was used in the calculation of the bidirectional dynamic and reactive power transmission among the rotor and electric scheme.

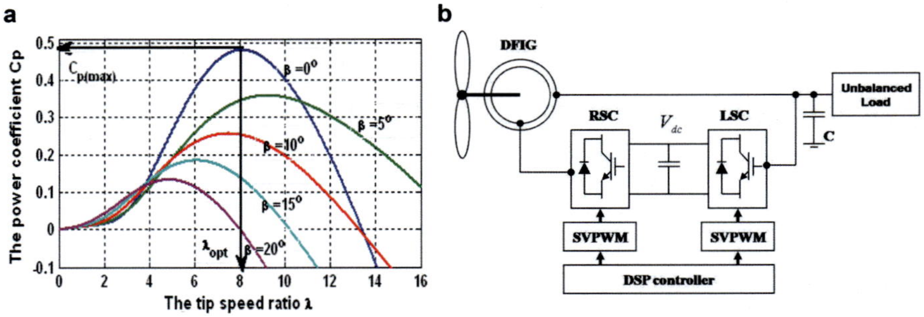

Fig. 10.8 (**a**) Indicates how to achieve the C_p's maximum values (0.5) in relation to the $\beta = 2°$. The value $\lambda_{opt} = 0.91$ indicates the maximum speed ratio (8 m/s) with a 10 m/s wind speed rating. The entire system testing was conducted under tight conditions with an approximated 50% (0.5–4.5) and 25% (6–6.5) and 50% (8–8.5), (**b**) For the DFIG control, the profile of the wind was considered as the wind speed signal, which permits the application of DSP control of the wind turbine to form the V_{dc} energy in the turbine. Thus, the wind turbine is considered to be working and in proper condition. In addition, to confirm a unity power factor at the stator position, the turbine got the reactive power as zero, where the stator active and reactive powers are regulated by the MPPT system [4, 6]

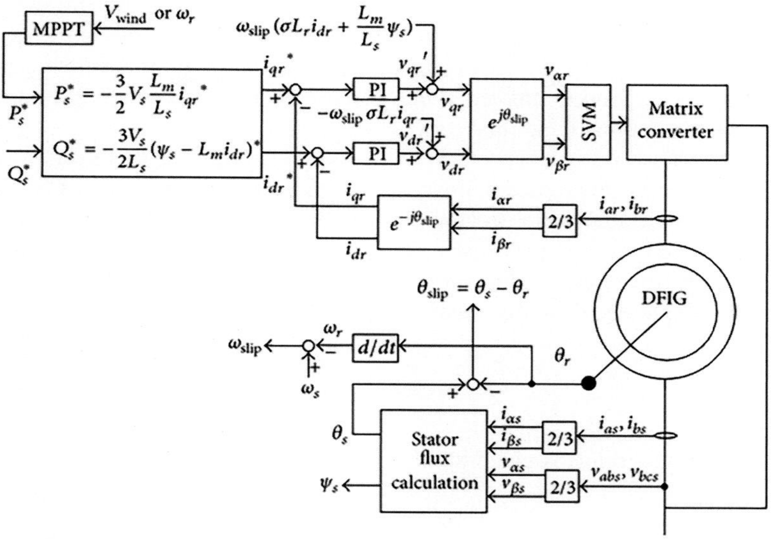

Fig. 10.9 The block diagram of the MPPT (V_{wind}) control bearing in mind the incarcerated DFIG velocity of the aerodynamic subsystem to find out the rate of wind intake and the transformation process of the wind energy into electric energy for operating the running vehicle by means of stator flux rotor initiation wind turbine of the shipping vehicle

Through an analysis of the electromagnetic torque, which implies that a predefined power-speed play the crucial role of trailing the extreme powerpoint into the rotor to convert the wind into energy computation of MPPT control has been determined. The calculation of MPPT, which was done, had the aim of confirming this stator active power. Thus, the achieving of the maximum power cofactor, which depicted that the optimum electromagnetic torque deduced from MPPT is by determination of the rapidity of the shaft of the wind-driven turbine. Since the control of the MPPT revealed the effectiveness of the wind-driven turbine, consequently the computation of the rotor dynamics is being determined by the referenced reactive and active powers that are expressed by the equation below

$$\begin{cases} i_{\text{qr}}^{*} = -\dfrac{L_{\text{s}}}{MV_{\text{s}}} P_{\text{s}}^{*} \\ i_{\text{dr}}^{*} = -\dfrac{L_{\text{s}}}{MV_{\text{s}}} \left(Q_{\text{s}}^{*} - \dfrac{V_{\text{s}}^{2}}{\omega_{\text{s}} L_{\text{s}}} \right) \end{cases} \tag{10.21}$$

The whole of this mechanism tries to imply that an accurate wind speed intake by the wind-driven turbine where the MPPT control tracked the velocity of the wind uninterruptedly and attuned the impose of the electromagnetic torque of the DFIG in to follow its aerodynamics accordingly to produce the energy. Therefore, the block diagram below of the clarified subsystem confirms the rate of wind intake and processed by this mechanism (Fig. 10.9).

What is being revealed in the MPPT control and DGIG block diagram is that a double rate of harvesting wind by the turbine is attained by controlling the speed of the generator. Thus the stator flux computation of the wind turbine, which is passed through the drive train to convert it into electric energy for operating the vehicle, determines the optimal wind intake rate.

Both induction and synchronous variable speed are produced by the drive train at the turbine, and they both had a direct impact on the synchronous gearbox. Consequently, the application of the gearbox in variable speed wind turbine shaft is being controlled to produce the maximum rate of energy being produced and given out. Therefore, the drive train model is being carried out with regard to the *d–q* synchronous and is expressed as in the equation below:

$$V_q = -R_s i_q - L_q \frac{di_q}{dt} - \omega L_d i_d + \omega \lambda_m \tag{10.22}$$

The electronic torque is given by

$$T_e = 1.5\rho\left[\lambda i_q + \left(L_d - L_q\right) i_d i_q\right] \tag{10.23}$$

In the above equations,

L_q represents the q axis mutual induction
L_d represents the d axis mutual induction
i_q represents the q axis flow of electricity
i_d represents the d axis flow of electricity
V_q represents the q axis energy
V_d represents the d axis energy
ω_r represents the rate of change of angular rotor position
λ refers to the maximum extent of oscillation of flux made
p represents the number of pairs of poles

The equation below can be applicable in case of dynamic modeling and is also pertinent in the instance of squirrel confine initiation generator (SCIG)

$$\begin{bmatrix} V_{qs} \\ V_{ds} \\ V_{qr} \\ V_{dr} \end{bmatrix} = \begin{bmatrix} R_s + pL_s & 0 & pL_m & 0 \\ 0 & R_s + pL_s & 0 & pL_m \\ pL_m & -\omega_r L_m & R_r + pL_r & -\omega_r L_r \\ \omega_r L_m & pL_m & \omega_r L_r & R_r + pL_r \end{bmatrix} \begin{bmatrix} i_{qs} \\ i_{ds} \\ i_{qr} \\ i_{dr} \end{bmatrix} \tag{10.24}$$

From the stator side, the equivalence is

$$\begin{aligned}
\lambda_{ds} &= L_s i_{ds} + L_m i_{dr} \\
\lambda_{qs} &= L_s i_{qs} + L_m i_{dr} \\
L_s &= L_{is} + L_m \\
L_r &= L_{lr} + L_m \\
V_{ds} &= R_s i_{ds} + \frac{d}{dt}\lambda_{ds} \\
V_{qs} &= R_s i_{qs} + \frac{d}{dt}\lambda_{qs}
\end{aligned} \tag{10.25}$$

From the rotor side, the equivalence is

$$\begin{aligned}
\lambda_{dr} &= L_r i_{dr} + L_m i_{ds} \\
\lambda_{qr} &= L_r i_{qr} + L_m i_{qs} \\
V_{dr} &= R_r i_{dr} + \frac{d}{dt}\lambda_{dr} + \omega_r \lambda_{qr} \\
V_{qr} &= R_r i_{qr} + \frac{d}{dt}\lambda_{qr} - \omega_r \lambda_{dr}
\end{aligned} \tag{10.26}$$

For the inflight gap flux connection, the equivalences are

$$\begin{aligned}
\lambda_{dm} &= L_m(i_{ds} + i_{dr}) \\
\lambda_{qr} &= L_m(i_{qr} + i_{qs})
\end{aligned} \tag{10.27}$$

In the equation

R_s = stator curving resistance
R_r = motor snaking resistance
L_m = magnetizing inductance
L_{ls} = stator seepage inductance
L_{lr} = dynamic rotor angular velocity
ω_r = current, energy
λ_q = fluidities

A substitution for the above equations can be expressed as

$$P_w = \frac{1}{2}\rho A C_p(\lambda, \beta)\left(\frac{R\omega_{opt}}{\lambda_{opt}}\right)^3 \tag{10.28}$$

and can be used to obtain the productivity power and rotating force of the turbine (T_t) in terms of gyratory rapidity.

$$T_t = \frac{1}{2}\rho A C_p(\lambda, \beta)\left(\frac{R}{\lambda_{opt}}\right)^3 \omega_{opt} \tag{10.29}$$

The power constant (C_p) is a counterclockwise principle articulated by the appropriate equivalence in the method.

$$C_p(\lambda, \beta) = c_1\left(c_2\frac{1}{\lambda_i} - c_3\beta - c_4\right)e^{-c_5\frac{1}{\lambda_i}} + c_6\lambda \tag{10.30}$$

with,

$$\frac{1}{\lambda_i} = \frac{1}{\lambda + 0.08\beta} - \frac{0.035}{\beta^3 + 1} \tag{10.31}$$

The value of constants c_1–c_6 has been explained in the later section. Consequently, the energy conversion circuit diagram implemented in Simulink systems by solicitation of wind energy alteration mechanism processes the productivity of the airstream energy generation module.

Wind Energy Conversion

Determination of the sturdiness of the Fuzzy Lucidity Checker (FLC) is done in consideration of its nonlinear processes and physiognomies to wind energy contrivance to convert it into electrical energy. To determine the operation of the conversation of wind by the presentation of the control rules of the fuzzy block diagram, the evaluation of the FLC mechanism by the fuzzy inference system is applicable. The fuzzification block reveals the calculated risk output of wind energy that is controlled by the rules of the defuzzification of the airstream turbine.

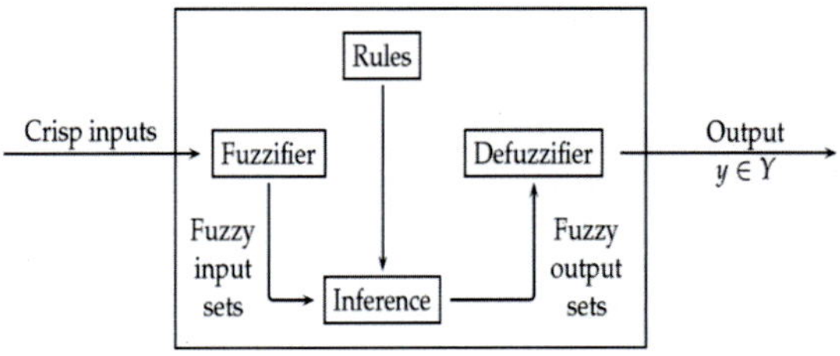

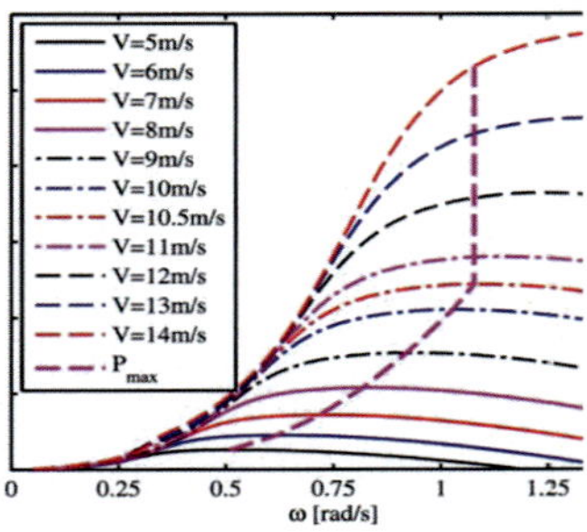

The determination of the airstream turbine and its DFIG subsystem to deal with the wind energy transformation mechanism to come up with electric energy is by utilization of this FLC as expressed below

$$\begin{cases} e_{\Omega_g(n)} = \Omega_g^*(n) - c_1(d) \\ \Delta e_{\Omega_g(n)} = \Omega_g^*(n) - \Omega_g(n-1) \end{cases} \tag{10.32}$$

The subsequent determination of the net realistic energy variants into the fuzzy controller is usually from the fuzzy controller. With regard to the above statement, it is evident that the net wind energy production affirms the triangle, trapezoidal, and symmetrical speed purposes of the wind-driven turbine. Analyzation of the formation of the net current of the turbines driven by wind is simply by the precise calculation of the FLC of DFIG by expression of the equation below

$$\begin{cases} e_{i_{dr}}(n) = i_{dr}^*(n) - i_{dr}(n) \\ e_{i_{dr}}(n) = i_{qr}^*(n) - i_{qr}(n) \end{cases} \tag{10.33}$$

The above equation can be further simplified as

$$\begin{cases} \Delta e_{i_{dr}}(n) = i_{dr}^*(n) - i_{dr}(n-1) \\ \Delta e_{i_{qr}}(n) = i_{qr}^*(n) - i_{qr}(n-1) \end{cases} \tag{10.34}$$

In conclusion, the input and output of the conversion of wind energy using the fuzzy controller have been quantized. Consequently, the effect of wind speed in altering the output power into the rotor of the airstream-driven turbine is used to determine the rate at which the wind is transforming. Consequently, there is the computation of the changes in the output power into the rotor to the trade-off between exactitude and intricacy of the conversion of wind energy in the electrical subsystem via simulation that is extremely thorough and careful to produce enteric energy that is useful and desirable for the operation of a vehicle.

Electrical Subsystem

Once the wind energy modeling has been analyzed, the implementation of this wind energy into the electrical subsystem has been calculated to run the vehicle. Therefore, both induction and synchronous wind energy driven by the rotor are being captured by a permanent synchronous generator (PMSG) to convert it into electric power [13, 14]. Hence, the functional mechanism of the gearbox of this PMSC plays a vital role to convert the electrical energy into DC current, and the *d–q* synchronous voltage equation quantifies it as:

Analysis of the wind energy modeling implies that the enactment of this wind energy into the electrical subsystem has been premeditated to operate the vehicle. It is thus clear that both induction and synchronous wind energy driven by the rotor are being captured by a permanent synchronous generator (PMSG) to convert it into electric energy. Hence, the efficient mechanism of the gearbox of this PMSC plays a vivacious role to convert the electrical energy into DC current, and it is quantified by the *d–q* synchronous voltage equation as

$$V_q = -R_s i_q - L_q \frac{di_q}{dt} - \omega L_d i_d + \omega \lambda_m \tag{10.35}$$

The electronic torque is given by

$$T_e = 1.5\rho \left[\lambda i_q + \left(L_d - L_q\right) i_d i_q\right] \tag{10.36}$$

In the above equation, L_q represents the q axis mutual induction, L_d considered as the d axis inductor, i_q represents the q axis flow of electricity, i_d represents the d axis flow of electricity, V_q represents the q axis energy, V_d denoted as the d axis energy, ω_r represents the range of speed of rotor, and λ represents the maximum extent of the oscillation of the induced flux. P is the quantity of pairs of poles. The demonstration of the dynamic modeling of wind simplifies the electronic torque, as illustrated in the equation below.

$$\begin{bmatrix} V_{qs} \\ V_{ds} \\ V_{qr} \\ V_{dr} \end{bmatrix} = \begin{bmatrix} R_s + pL_s & 0 & pL_m & 0 \\ 0 & R_s + pL_s & 0 & pL_m \\ pL_m & -\omega_r L_m & R_r + pL_r & -\omega_r L_r \\ \omega_r L_m & pL_m & \omega_r L_r & R_r + pL_r \end{bmatrix} \begin{bmatrix} i_{qs} \\ i_{ds} \\ i_{qr} \\ i_{dr} \end{bmatrix} \tag{10.37}$$

The equation that expresses the stator side is

$$\begin{aligned} \lambda_{ds} &= L_s i_{ds} + L_m i_{dr} \\ \lambda_{qs} &= L_s i_{qs} + L_m i_{dr} \\ L_s &= L_{is} + L_m \\ L_r &= L_{lr} + L_m \\ V_{ds} &= R_s i_{ds} + \frac{d}{dt}\lambda_{ds} \\ V_{qs} &= R_s i_{qs} + \frac{d}{dt}\lambda_{qs} \end{aligned} \tag{10.38}$$

A simplified form of the above equation is

$$
\begin{aligned}
\lambda_{dr} &= L_{r} i_{dr} + L_{m} i_{ds} \\
\lambda_{qr} &= L_{r} i_{qr} + L_{m} i_{qs} \\
V_{dr} &= R_{r} i_{dr} + \frac{d}{dt}\lambda_{dr} + \omega_{r}\lambda_{qr} \\
V_{qr} &= R_{r} i_{qr} + \frac{d}{dt}\lambda_{qr} - \omega_{r}\lambda_{dr}
\end{aligned}
\tag{10.39}
$$

If an air gas flux leaks into the generation, the equation is further simplified to

$$
\begin{aligned}
\lambda_{dm} &= L_{m}(i_{ds} + i_{dr}) \\
\lambda_{qr} &= L_{m}(i_{qr} + i_{qs})
\end{aligned}
\tag{10.40}
$$

where R_s, R_r, L_m, L_{ls}, L_{lr}, ω_r, i_d, i_q, V_d, V_q, λ_d, and λ_q represent the stator resistance from one end to another, motor resistance from one end to another, self-inductance of an inductor with magnetic core, stator leakage inductance, rotor outflow inductance, electrical rotor angular velocity, electric flow, energy, and fluxes of the *d–q* model correspondingly [5]. Then the net productivity power and rotating force of turbine (T_t) in terms of gyratory velocity can be attained by the following equations.

$$
P_{w} = \frac{1}{2}\rho A C_{p}(\lambda, \beta)\left(\frac{R\omega_{opt}}{\lambda_{opt}}\right)^{3} \tag{10.41}
$$

$$
T_{t} = \frac{1}{2}\rho A C_{p}(\lambda, \beta)\left(\frac{R}{\lambda_{opt}}\right)^{3}\omega_{opt} \tag{10.42}
$$

The fitting equation below expresses the power constant (C_p), which is a discipline that is not linear.

$$
C_{p}(\lambda, \beta) = c_{1}\left(c_{2}\frac{1}{\lambda_{i}} - c_{3}\beta - c_{4}\right)e^{-c_{5}\frac{1}{\lambda_{i}}} + c_{6}\lambda \tag{10.43}
$$

with,

$$
\frac{1}{\lambda_{i}} = \frac{1}{\lambda + 0.08\beta} - \frac{0.035}{\beta^{3} + 1} \tag{10.44}
$$

In conclusion, an energy transformation circuit diagram processes the output of the wind energy generator module through the implementation of the converter of Simulink Power Systems. Volt that assists in powering the vehicle is the result of what is calculated as the result of the circuit mode electricity generation (Fig. 10.10).

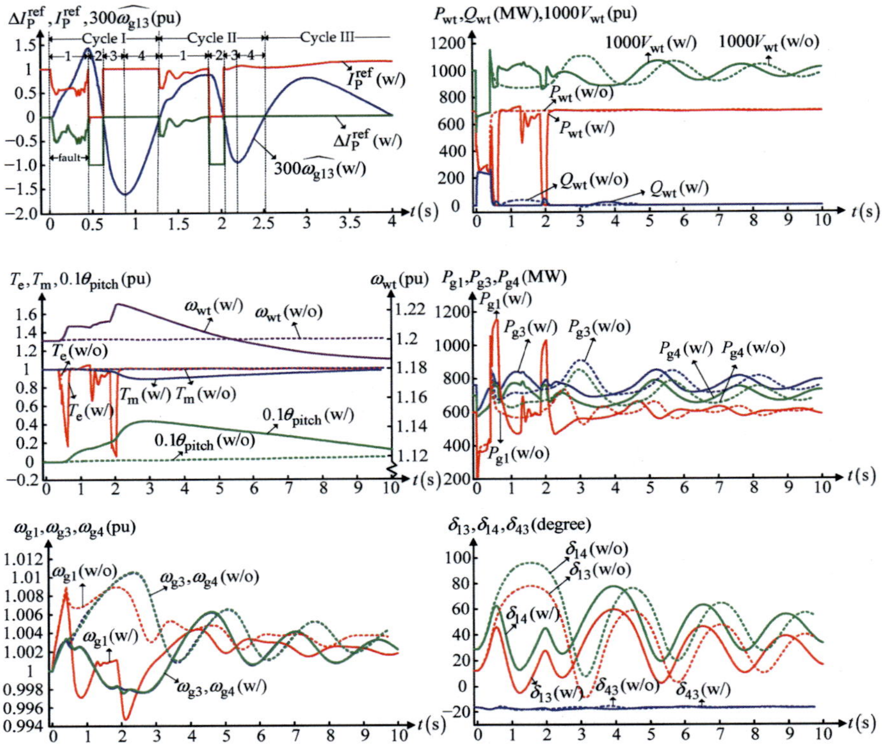

Fig. 10.10 Shows the aerodynamical power generation by an installed wind turbine, which is linked to the electrical subsystem circuit to determine the current generation (MW) at various degrees of the angle of the pitch of the wind turbine rotation

Generator Modeling

One of the two between initiation and asynchronous dynamo can be applicable in the WT schemes. Adjustable velocity straight-driven multipole perpetual magnet synchronous dynamos (PMSGs) are similarly expansively applicable in airstream-driven energy schemes since they have higher competence, subordinate weight, cheap upkeep, and stress-free workability and since they do not necessitate a responsive and attracting flow of electricity. The existence of a gearbox in variable-velocity WTs results in an additional encumbrance of fee and upkeep. Exhausting a straight-driven PMSG does both the increasing of consistency and decreasing of the mass of the nacelle.

The classical for a PMSG is its basis on a *d*–*q* synchronous situation frame. The equation of PMSG energy is given by

$$V_d = -R_s i_d - L_d \frac{di_q}{dt} + \omega L_q i_q$$

$$V_q = -R_s i_q - L_q \frac{di_q}{dt} - \omega L_d i_d + \omega \lambda_m \tag{10.45}$$

The automated torque is given by

$$T_e = 1.5\rho\left[\lambda i_q + \left(L_d - L_q\right) i_d i_q\right] \tag{10.46}$$

A WT system can accommodate both an induction and a synchronous generator. Because of their unique features and characteristics such as high efficiency, reduced cost of maintenance, and ease in controlling and since they can function without the use of reactive and magnetizing current, the variable velocity direct-driven multipole perpetual magnet synchronous dynamos (PMSGs) are also expansively being used in airstream-driven power schemes. An extra cost of maintenance and burden is created by the existence of a gearbox in variable-speed WTs generators. The decreasing of the weight of the nacelle is not the only impact created by the use of a direct-driven gearbox in variable-speed WTs but also the generation of an extra burden of cost. A *d–q* synchronous reference frame provides the basis for the model for a PMSG. The equivalence of the PMSG is

$$V_d = -R_s i_d - L_d \frac{di_q}{dt} + \omega L_q i_q \tag{10.47}$$

$$V_q = -R_s i_q - L_q \frac{di_q}{dt} - \omega L_d i_d + \omega \lambda_m \tag{10.48}$$

The electronic torque is given by

$$T_e = 1.5\rho\left[\lambda i_q + \left(L_d - L_q\right) i_d i_q\right] \tag{10.49}$$

In the above equation, the q axis mutual induction is L_q, d axis mutual induction is L_d, the q axis flow of electricity is i_q, i_d is the d axis flow of electricity, V_q is the q axis energy, V_d is the d axis energy, ω_r is the rate of change of angular position of the rotor, λ is the amplitude of flux prompted, and p is the quantity of pairs of poles. The equivalence that follows, which can be applicable in conjunction with a static *d–q* frame of reference for dynamic modeling, is relevant with regard to a squirrel cage induction dynamo (SCIG).

$$\begin{bmatrix} V_{qs} \\ V_{ds} \\ V_{qr} \\ V_{dr} \end{bmatrix} = \begin{bmatrix} R_s + pL_s & 0 & pL_m & 0 \\ 0 & R_s + pL_s & 0 & pL_m \\ pL_m & -\omega_r L_m & R_r + pL_r & -\omega_r L_r \\ \omega_r L_m & pL_m & \omega_r L_r & R_r + pL_r \end{bmatrix} \begin{bmatrix} i_{qs} \\ i_{ds} \\ i_{qr} \\ i_{dr} \end{bmatrix} \tag{10.50}$$

The equations that are applicable for the stator side are

$$\begin{aligned} \lambda_{ds} &= L_s i_{ds} + L_m i_{dr} \\ \lambda_{qs} &= L_s i_{qs} + L_m i_{dr} \\ L_s &= L_{is} + L_m \\ L_r &= L_{lr} + L_m \\ V_{ds} &= R_s i_{ds} + \frac{d}{dt}\lambda_{ds} \\ V_{qs} &= R_s i_{qs} + \frac{d}{dt}\lambda_{qs} \end{aligned} \tag{10.51}$$

The equations that are applicable for the rotor side are

$$\begin{aligned} \lambda_{dr} &= L_r i_{dr} + L_m i_{ds} \\ \lambda_{qr} &= L_r i_{qr} + L_m i_{qs} \\ V_{dr} &= R_r i_{dr} + \frac{d}{dt}\lambda_{dr} + \omega_r \lambda_{qr} \\ V_{qr} &= R_r i_{qr} + \frac{d}{dt}\lambda_{qr} - \omega_r \lambda_{dr} \end{aligned} \tag{10.52}$$

The equations that apply to the air gap flux connotation are

$$\begin{aligned} \lambda_{dm} &= L_m(i_{ds} + i_{dr}) \\ \lambda_{qr} &= L_m(i_{qr} + i_{qs}) \end{aligned} \tag{10.53}$$

where R_s, R_r, L_m, L_{ls}, L_{lr}, ω_r, i_d, i_q, V_d, V_q, λ_d, and λ_q are the stator winding resistance, motor winding resistance, fascinating inductance, stator leakage inductance, rotor leakage inductance, electrical rotor angular velocity, flow of electricity, energy, and fluxes, correspondingly, of the *d–q* model. Obtaining of the torque turbine (T_t) and the production power in terms of the consistent speed is through substitution of the formulas below.

$$P_w = \frac{1}{2}\rho A C_p(\lambda, \beta)\left(\frac{R\omega_{opt}}{\lambda_{opt}}\right)^3$$

$$T_t = \frac{1}{2}\rho A C_p(\lambda, \beta)\left(\frac{R}{\lambda_{opt}}\right)^3 \omega_{opt} \tag{10.54}$$

Battery Modeling

The application of battery modeling plays an essential role in the functioning of a vehicle. Some of the uses of application battery modeling in a car are: functioning as the starting power that ignites the car and operation as the source that provides backup whenever a car is not moving. Battery modeling functions as the source that provides backup whenever a car is not moving by integrating the Peukert's Law of battery charge, which is

$$t_{discharge} = H\left(\frac{C}{IH}\right)^k \tag{10.55}$$

In the above formula, the time for battery charge is represented by t, C represents the capacity of the battery, the time that is rated discharge is represented by H, the constant for Peukert is represented by K. Peukert's constant is calculated using the formula

$$k = \frac{\log T_2 - \log T_1}{\log I_1 - \log I_2} \tag{10.56}$$

In the above formula, the variance of the charge current flow rates is represented by I_1 and I_2, and T_1 and T_2 represent the corresponding time to charge the battery completely. The formula used to acquire the corresponding time to charge the battery entirely is (Fig. 10.11)

$$t_{charging} = \frac{\text{Ampere hour of battery}}{\text{Charging current}}$$

Conclusion

Coextensively, conventional transport infrastructure systems all over the world cause adverse environmental impacts and also result in a waste of funds annually. The solution for balancing the link amid traffic, price, energy, and atmosphere in this research may come from the introduction of vehicle technology. The installation of a wind turbine into the flying vehicle for the production of energy by itself when the car is functioning in consideration of the sophisticated energy demand and the necessity for a more conducive ecosystem. The conduction of an examination with

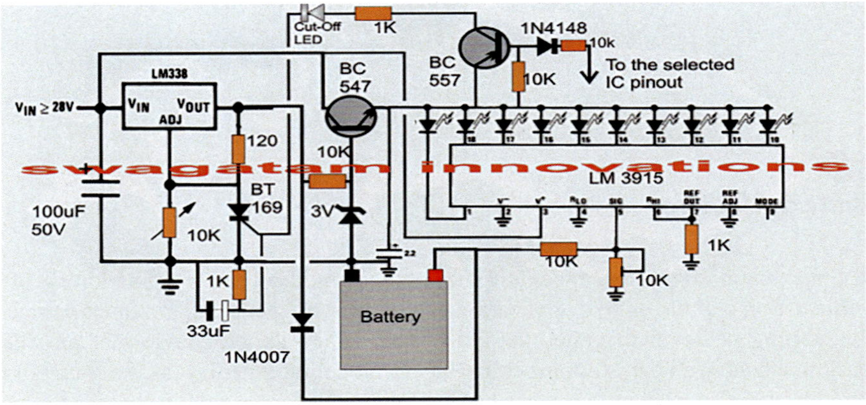

Fig. 10.11 A block diagram of the battery modeling of a car is being shown to determine the V_{oc} and SOC to control its voltage source to ignite the engine and run the car when it is not in motion

a sequence of arithmetical computations of the transfiguration procedure, regulator organization, and generator modeling in this study has the aim of enhancing the use of the turbine, the high extensile form of existence of profitable application of the flying vehicle by coming up with the dependability of getting off, soaring, security regulator of rapidity and sleekness, and match numbers is publicized by the replication outcomes of NACA 9618 model by means of MATLAB. In recent days, the construction of infrastructure costs trillions of dollars; the amount of fossil fuel consumed by the transport sector annually is 5.6×10^{20} J/yr (560 EJ/yr). Despite substructure building costing trillions of dollars annually and the transport segment consuming 5.6×10^{20} J/yr (560 EJ/yr) fossil fuel annually, the two are also accountable for approximately 34 percent of the total yearly change in climatic conditions. If the sectors involved in this technology take it incumbent upon themselves to develop this technology appropriately, a new epoch of science to dispatch unembellished transport, substructure, and energy glitches can be seen in this technology.

Acknowledgments This research was supported by Green Globe Technology under grant RD-02018-03. Any findings, conclusions, and recommendations expressed in this paper are solely those of the author and do not necessarily reflect those of Green Globe Technology.

References

1. A.M. Eltamaly, A.I. Alolah, M.H. Abdel-Rahman, Improved simulation strategy for DFIG in wind energy applications. Int. Rev. Model. Simul. **4**(2), 525–532 (2011)
2. T.B. Haines, *First Roadable Airplane Takes Flight* (Aircraft Owners and Pilots Association (AOPA), Frederick, MD, 2009). Retrieved 19 Mar 2009
3. M. Hossain, Faruque., Solar energy integration into advanced building design for meeting energy demand and environment problem. Int. J. Energy Res. **40**, 1293–1300 (2016)
4. E. Kamal, M. Koutb, A.A. Sobaih, B. Abozalam, An intelligent maximum power extraction algorithm for hybrid wind-diesel-storage system. Int. J. Electr. Power Energy Syst. **32**(3), 170–177 (2010)
5. M.V. Kazemi, M. Moradi, R.V. Kazemi, Minimization of powers ripple of direct power controlled DFIG by fuzzy controller and improved discrete space vector modulation. Electr. Power Syst. Res. **89**, 23–30 (2012)
6. J. Khodakarami, P. Ghobadi, Urban pollution and solar radiation impacts. Renew. Sust. Energ. Rev. **57**, 965–976 (2016)
7. L. Page, Terrafugia Flying Car Gets Road-Safety Exemptions, The Register, 4 Jul 2011. Retrieved 11 Jul 2011
8. G. Sivasankar, V. Suresh Kumar, Improving low voltage ride through capability of wind generators using dynamic voltage restorer. J. Electr. Eng. **65**(4), 235–241 (2014)
9. S.J. Thompson, Congressional Research Service, High Speed Ground Transportation (HGST): Prospects and Public Policy, 6 Apr 1989, p. 5
10. H. Tien, C. Scherer, J. Scherpen, V. Muller, Linear parameter varying control of doubly fed induction machines. IEEE Trans. Ind. Electron. **63**(1), 216–224 (2016)
11. G. Tsourakisa, B.M. Nomikosb, C.D. Vournasa, Effect of wind parks with doubly fed asynchronous generators on small-signal stability. Electr. Power Syst. Res. **79**, 190–200 (2009)
12. J.P.A. Vieira, A. Nunes, M. Vinicius, U.H. Bezerra, W. Barra Jr., New fuzzy control strategies applied to the DFIG converter in wind generation systems. IEEE Trans Am Latina **5**(3), 142–149 (2007)
13. H. Xu, X. Ma, D. Sun, Reactive current assignment and control for DFIG based wind turbines during grid voltage sag and swell conditions. J. Power Electron. **15**, 235–245 (2015)
14. X.J. Zheng, J.J. Wu, Y.H. Zhou, Numerical analyses on dynamic control of five-degree-of-freedom maglev vehicle moving on flexible guideways. J. Sound Vib. **235**, 43–61 (1997)
15. X.J. Zheng, J.J. Wu, Y.H. Zhou, Effect of spring non-linearity on dynamic stability of a controlled maglev vehicle and its guideway system. J. Sound Vib. **279**, 201–215 (2005)

Part IV
Sustainable Society, and Environment

Chapter 11
Photophysical Reaction Technology to Eliminate Pathogens Naturally from Earth

Abstract Advanced building design technology is being proposed to kill all pathogens including COVID-19 inside the building naturally before it invades into the human body. Simply photon energy is being implemented using exterior glazing wall surface to form *Ultraviolet Germicidal Irradiation* (UVGI) from sun to release high-frequency ultraviolet light to kill all pathogens such as bacteria, protozoa, prion, viroid, fungus, molds, and virus including COVID-19 inside the buildings and houses everywhere in the world. Since killing pathogens with solar radiation will only require short-range wavelengths, 254–280 nanometers (nm) wavelength has been utilized to initiate UVGI to destroy the pathogenic nucleic acids DNA and/or RNA, leaving all pathogens inside buildings and houses unable to perform their vital cellular functions, and letting them die in seconds. Simply, usage of exterior glazing wall surface of the building and house design as an acting photophysical reaction technology will be an innovative field of science to kill all pathogens inside the buildings and houses naturally entire world.

Keywords Advanced building design · Photonic wavelengths · Ultraviolet germicidal irradiation (UVGI) · Killing pathogens · Global sustainability

Introduction

The average microorganisms in indoors that occupy building's room of $10' \times 10'$ are typically around 8000 types of virus, bacteria, molds, and fungus and the concentrations are 10^5 pathogens per m^3 [6, 12, 25]. Pathogens, the infectious microorganism such as bacterium, protozoan, prion, viroid, fungus, and virus including COVID-19, are indeed the dangerous beings that threaten entire human race to survive on earth in the near future. These tiny creatures can easily invade into the human body through absorbing, adsorbing, inhaling, ingesting, and producing mutants into the human body rapidly to form toxins which penetrate the tissues, hijack nutrients, immunosuppress the human body, and causes serious illness or death [1, 17, 26]. The recent threat of the COVID-19 is a deadly infectious disease that causes severe dreadful respiratory syndrome (SARS-CoV-2) which is first detected in November 2019 at Wuhan, China. Since then it has spread to

M. F. Hossain, *Sustainable Design for Global Equilibrium*,
https://doi.org/10.1007/978-3-030-94818-4_11

210 countries and the World Health Organization (WHO) has declared “the 2019–20 coronavirus outbreak” a pandemic [3, 20, 27]. Besides this current COVID-19 pandemic crisis, we are not prepared to defeat any forthcoming deadly outbreak which could be much more deadly than COVID-19 that probably could completely wipe out the entire human race from earth. The simple biochemistry is that some of the pathogens have the extremely brilliant ability to change their mutants and pathway frequently into the human body to deceive any vaccines to kill them into the host body which is indeed a clear and present danger for the entire human race [13, 28]. Thus, prevention technology is an urgent demand to eliminate these deadly pathogens before it invades into the human body as a first level of defense. Interestingly, photonic ultraviolet germicidal irradiation (UVGI) has the excellent functionality to eliminate these deadly pathogens in seconds regardless of their ability of changing mutants and pathways. Therefore, in this research, photonic irradiance structure has been studied which has been formed by using exterior curtain wall of a building or house to release UVGI at short-range electromagnetic spectrum of wavelengths of 185–254 nanometers (nm) to destroy the nucleic acids and DNA and/or RNA structure of all pathogens in order to kill them eventually. Simply, in this research, photophysical reaction application is being proposed to implement to design all buildings and homes to have the ability to create UVGI from sunlight in order to eradicate all pathogens inside the buildings and homes naturally before it invades into the human body.

Materials and Methods

UVGI Production by the Exterior Glazing Wall Skin

The quantum field of the sunlight has been clarified by using the outer glazing wall surface to emit UVGI light in between 254 and 280 nm for a limited time when nobody lives inside the building or house. Naturally photonic electromagnetic field generation has been analyzed considering the penetration of solar irradiance into the outer glazing wall surface and it has been clarified by the semiconductor connected to the glazing wall (Fig. 11.1).

Consequently, photon dynamics has been analyzed through the clarification of wavelength spectrum into the outer glazing wall surface in order to determine the photonic UVGI release rate from sunlight which is expressed by Hamiltonian [2, 14]

$$H = \sum \omega_{ci} a_i^\dagger a_i + \sum_K \omega_k b_k^\dagger b_k + \sum_{ik} \left(V_{ik} a^\dagger b_k + V_{ik}^* b_k^\dagger a_i \right) \tag{11.1}$$

where $a_i\left(a_i^\dagger\right)$ represents the kinetics of the photon energy mode, $b_k\left(b_k^\dagger\right)$ represents the kinematics of the photodynamic mode, and the cofactors V_{ik} denote the mode of photon energy in the curtain wall surface (Fig. 11.2).

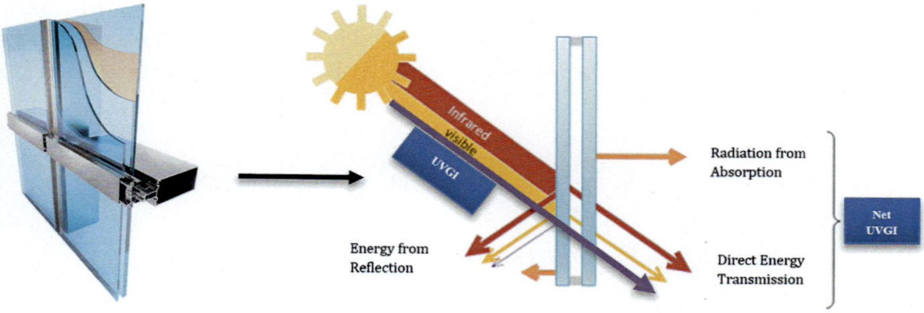

Fig. 11.1 Solar energy penetration into the exterior glazing wall surface has been analyzed by photon spectrum considering the penetration of solar irradiance

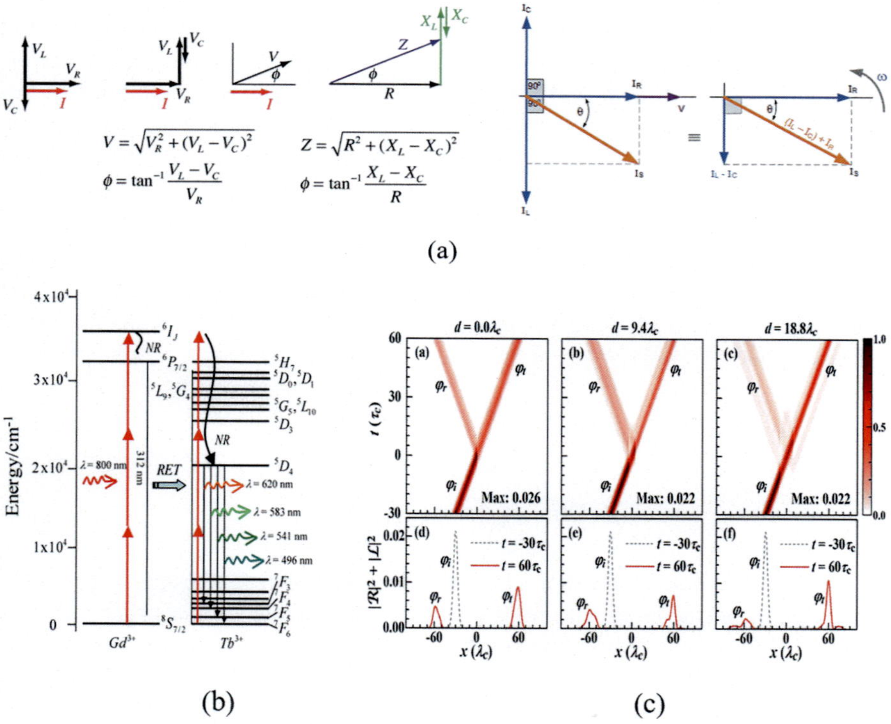

Fig. 11.2 (**A**) Photon energy mode considering the angle of irradiance penetration on the surface of outer glazing wall, (**B**) solar energy deliberation at various wavelength of the outer wall surface at per unit area, (**C**) (a–c) the photon density functions of x and t at time period, (d–f) is the net solar irradiance on the surface of the incident (φ_i), reflected (φ_r), and transmitted (φ_t) photons

Considering the initial photon energy dynamic proliferation, here, the whole electromagnetic photonic spectrum is being structured considering excited energy state photons in the outer glazing wall surface and expressed by the following equation [10, 13, 15]:

$$\rho(t) = -i\left[H'_{c}(t)\rho(t)\right] + \sum_{ij}\Big\{k_{ij}(t)\left[2a_{j}\rho(t)a_{i}^{\dagger} - a_{i}^{\dagger}a_{j}\rho(t) - \rho(t)a_{i}^{\dagger}a_{j}\right] + k_{ij}(t)\left[a_{i}^{\dagger}\rho(t)a_{j} + a_{j}\rho(t)a_{i}^{\dagger} - a_{i}^{\dagger}a_{j}\rho(t) - \rho(t)a_{j}a_{i}^{\dagger}\right]\Big\} \tag{11.2}$$

Here, $\rho(t)$ represents the attenuated density of photon energy states, $H'_{c}(t) = \sum_{ij}\omega'_{cij}(t)a_{i}^{\dagger}a_{j}$ represents re-standardized photon energy frequencies $\omega'_{cii}(t) = \omega'_{ci}(t)$, which is being factored induced photonic energy $\omega'_{cij}(t)$. The factors $\kappa_{ij}(t)$ and $\widetilde{\kappa}_{ij}(t)$ are being here clarified photon energy proliferation in the curtain wall connected to a semiconductor. Consequently, the wavelength frequencies, ω'_{cij}, and time-related function, $\kappa_{ij}(t)$ and $\widetilde{\kappa}_{ij}(t)$, are being clarified by integration vector-perturbative principle and, thus, the photonic energy dynamic is being denoted as $H_{I} = \sum_{k}\lambda_{k}xq_{k}$, where x and q_{k} represent the area of the photon energy reservoir on the glazing wall surface. Considering the quantum dynamics of the solar energy, the net area of the reservoir of the photon dynamic of the curtain wall skin is being modified as $H_{I} = \sum_{k}V_{k}\left(a^{\dagger}b_{k} + b_{k}^{\dagger}a + a^{\dagger}b_{k}^{\dagger} + ab_{k}\right)$ to determine the mode of the solar energy penetration within the curtain wall surface. Thus, the solar energy dynamics is being clarified by the dissipation of the solar energy function $\kappa(t)$ and $\widetilde{\kappa}(t)$ (sub-indices (i, j) in Eq. (11.2)), and therefore it can be modified as the following variable equation [4, 28]:

$$\omega'_{c}(t) = -\mathrm{Im}[u\,(t, t_{0})/u\,(t, t_{0})] \tag{11.3}$$

$$k(t) = -\,\mathrm{Re}\,[u\,(t, t_{0})/u\,(t, t_{0})] \tag{11.4}$$

$$\widehat{k}(t) = \dot{v}(t, t) + 2v(t, t)k(t). \tag{11.5}$$

where $u(t, t_0)$ represents the photonic energy area and $v(t, t)$ denotes the solar energy proliferation on the outer glazing wall surface. This equation is further modified using the following equilibrium photonic dynamic mechanism on the outer glazing wall surface [5, 19]:

$$\dot{u}(t, t_{0}) = -i\omega_{c}u(t, t_{0}) - \int_{t_{0}}^{t} dt' g(t - t')\, u\,(t', t_{0}) \tag{11.6}$$

$$v(t, t) = \int_{t_{0}}^{t} dt \int_{0}^{t} dt_{2}u^{*}\,(t_{1}, t_{0})\widehat{g}\,(t_{1} - t_{2})\, u\,(t_{2}, t_{0}) \tag{11.7}$$

Here v_c denotes the primary frequency of the photonic energy, and thus the scalar functions in Eqs. (11.6) and (11.7) are being determined by the back-up function within the photon energy, and the amount of photon energy formed on the outer glazing wall surface per *unit* area of $J(\varepsilon)$ considering the relations of: $g(t-t') = \int d\omega J(\omega) e^{-i\omega(t-t')}$ and $\widetilde{g}(t-t') = \int d\omega J(\omega)\overline{n}(\omega,T)e^{-i\omega(t-t')}$, here $\overline{n}(\omega,T) = 1/\left[e^{\hbar\omega/k_B T} - 1\right]$ denotes the photon proliferation in the outer glazing wall surface at temperature T. The *unit* area $J(\omega)$ is clarified in conjunction with the density of state (DOS) $\varrho(\omega)$ photon energy generation in the glazing wall surface at the magnitude of V_k between the photon dynamics and glazing wall skin,

$$J(\omega) = \sum_k |V_k|^2 \, \delta(\omega - \omega_k) = \varrho(\omega)|V(\omega)|^2 = [n * e(1+2n)]^4 \tag{11.8}$$

Finally, photon energy dynamic rate has been simplified into the glazing wall skin as $V_k \to V(\omega)$ and i of V_{ik} at the non-equilibrium photon capture $J(\omega)$, and being expressed as the following equation:

$$J(\omega) = [n * e(1+2n)]^4 \tag{11.9}$$

where $n = E = hf$ represents the amount of solar energy generation on the glazing wall surface and e denotes the photonic constant and n is the electromagnetic radiation of photon which has been further clarified to determine the required electromagnetic radiation (EM) release from sunlight to destroy the pathogens naturally inside the building or house.

Electromagnetic Radiation

Then, the electromagnetic radiation (EM radiation), the waves of photonic quanta of the electromagnetic field, has been clarified considering the electromagnetic radiant energy emitted by exterior glazing wall surface of the building or house using a computerized semiconductor. Since the EM is the linearly polarized sinusoidal electromagnetic wave, it has been propagating the direction +z through a homogeneous, isotropic, dissipationless medium of glazing wall skin [2, 9]. Therefore, the electric fields denoted with blue arrows will oscillate in the $\pm x$-directional factor, and the orthogonal magnetic fields denoted with red arrows will oscillate in the $\pm y$-directional factors in various phases considering the electric fields (Fig. 11.3).

Since electromagnetic waves are being transmitted by photon particles, these waves will certainly react with other photon charges [7, 10]. Here, the EM waves will carry photon energy and will interact with the photonic momentum of its solar irradiance [8, 29]. Consequently, the electromagnetic radiation will be associated with EM waves and, thus, it will be free to release radiation continuously without any perturbation of the photonic flow [21, 24]. Since the EM of visibility has the

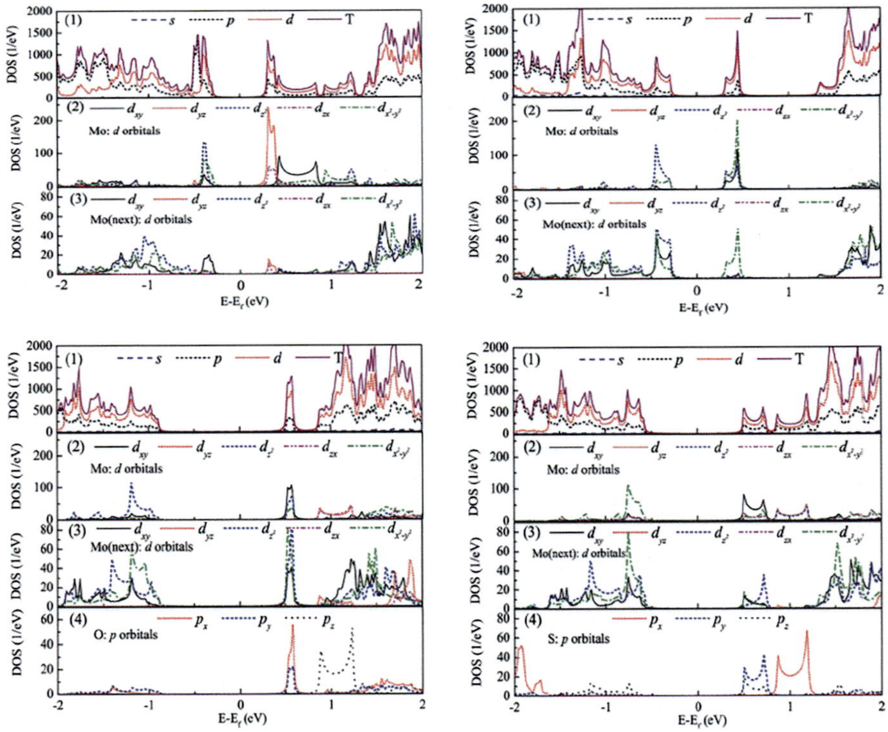

Fig. 11.3 The light electromagnetic radiation of the total density of states (DOS) solar energy transmitted on the glazing wall surface at various *s, p, d,* and *T* factors, of molecular (Mo) photon atomic (*S*) function

lower frequencies of *non-ionizing radiation*, its photonic bandgap will have enough energy to break biochemical bonds of the pathogens. Simply, the effects of these radiations on biochemical systems on pathogenic cell is solely the sunlight heating mechanism of the energy transfer from low frequency photon energy which will have enough energy to break chemical bonds and cause damage to pathogenic cell structure severely.

Killing Pathogens

Since the effects of UVGI on pathogens depend on the radiation's frequency and wavelength, the release of UVGI from sunlight is being measured and controlled by computerized PHOTOMETER connected with a semiconductor. Then the effects of EM on pathogens, including corona virus (COVID-19), is being determined using PCR analysis considering the radiation's frequency of wave lengths at the spectrum

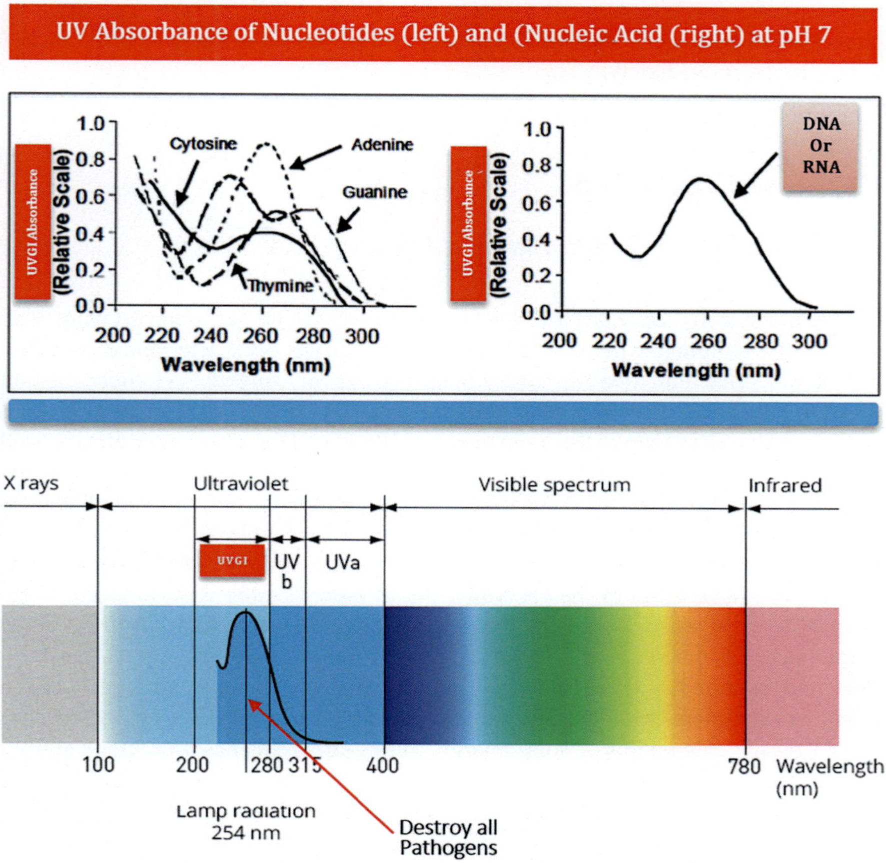

Fig. 11.4 The photophysics radiation application for eliminating pathogens inside the building or house which shows that, once UVc, which is UVGI radiation of 254–280 nm, is applied inside the house or building, it starts to disinfect all pathogens immediately

of 254–280 nm in order to break down the molecular bonds of microorganismal DNA and/or RNA producing Thymine dimers ("T" of ATGC). Simply, the high frequency of EM with short range of light spectrum 254–280 nm has been used to damage the pathogenic DNA and/or RNA thymine dimer. This damage to the pathogens will eliminate their ability to reproduce and biochemical function of their ability to cause harm to its hosts (Fig. 11.4).

Simply, the UVGI light-releasing mechanism using the exterior glazing wall skin will inactivate pathogens by destroying their deoxyribonucleic acid (DNA) and ribonucleic acid (RNA). Just because when DNA and/or RNA of the pathogens absorb UVGI lights, it will break down the dimers (covalent bonds between the same nucleic acids) of the pathogens and their failure dimers will cause the barrier in the transcription of information of DNA and/or RNA, which in turn will result in disruption of pathogens' biochemical function and replication and thus the pathogens will die immediately.

Results and Discussion

UVGI Production by the Exterior Glazing Wall Skin

To mathematically calculate the generation of UVGI by the outer glazing wall surface of a building or house, the photon dynamics has been determined considering the functional unit area $J(\omega)$ of the outer glazing wall surface [11]. Since the unit area $J(\omega)$ has a consistent solar irradiance emission rate, UVGI generation from sunlight will have a unique dynamic mode of 1D, 2D, 3D into the outer glazing wall surface. Simply, the generation of UVGI will ultimately evolve from the light spectrum equilibrium of the solar irradiance penetration per unit area in relation to photon reservoir-induced energy factor, as described in Table 11.1.

Consequently, a high-level solar energy frequency cutting-off Ω_C is being determined from bifurcation of the DOS in a 3D of outer glazing wall surface. Subsequently, a sharpened solar energy high-level frequency cutting-off has been employed at Ω_d maintaining the total DOS in 2D and 1D outer glazing wall surface. Since the function $Li_2(x)$ is being acted, here, as an alogarithm and $erfc(x)$ acted as an additional function, the DOS at outer glazing wall surface, $\varrho_{PC}(\omega)$, is being confirmed by the calculation of solar energy cutting-off frequencies on outer glazing wall surface [14, 22]. The determination of the combination of the total DOS on the outer glazing wall surface is being confirmed as $\varrho_{PC}(\omega) \propto \frac{1}{\sqrt{\omega - \omega_e}}\Theta(\omega - \omega_e)$, where $\Theta(\omega - \omega_e)$ denotes the variable function and ω_e denotes the solar energy frequency considering DOS on the outer glazing wall surface [4, 19, 25].

This DOS is, therefore, calculated from the glazing wall surface by conducting 3D analysis of the electromagnetic field (EMF) in order to confirm the accurate qualitative state of photon energy on the glazing wall surface as DOS: $\varrho_{PC}(\omega) \propto \frac{1}{\sqrt{\omega - \omega_e}}\Theta(\omega - \omega_e)$ [15, 23]. To conduct the 2D and 1D analysis of the glazing wall surface, the photonic DOS had been calculated from the pure algorithm of PBE, as $\varrho_{PC}(\omega) \propto -[\ln|(\omega - \omega_0)/\omega_0| - 1]\Theta(\omega - \omega_e)$, where ω_e denotes the peak algorithm, and thus the unit area $J(\omega)$ is determined from the calculation of DOS in the glazing wall surface by the fine photonic magnitude $V(\omega)$ and expressed as [16],

$$J(\omega) = \varrho(\omega)|V(\omega)|^2 \tag{11.10}$$

Here, it has been considered as the PB frequency ω_c and thus, the photon dynamics has been used as the function $u(t, t_0)$ of UVGI proliferation structure in the relation $\langle a(t)\rangle = u(t, t_0)\langle a(t_0)\rangle$. It is thus confirmed using the integral differential calculation given in Eq. (11.10) and confirmed as

Table 11.1 The formation of UVGI from the light spectrum at variables C, η, and χ functional energy emission rate on the surface curtain wall at 1D, 2D, and 3D directional modes which corresponded various DOS per unit area $J(\omega)$ and solar irradiance induction $\Sigma(\omega)$ at glazing wall surface

UVGI	Unit area $J(\omega)$ at various DOS	Solar irradiance induction $\Sigma(\omega)$
1D	$\frac{C-1}{\pi}\frac{1}{\sqrt{\omega-\omega_e}}\Theta(\omega-\omega_e)$	$-\frac{C-1}{\sqrt{\omega_e-\omega}}$
2D	$-\eta\left[\ln\left\|\frac{\omega-\omega_0}{\omega_0}\right\|-1\right]\Theta(\omega-\omega_e)\Theta(\Omega_d-\omega)$	$\eta\left[\mathrm{Li}_2\left(\frac{\Omega_d-\omega_0}{\omega-\omega_0}\right)-\mathrm{Li}_2\left(\frac{\omega_0-\omega_e}{\omega_0-\omega}\right)-\ln\frac{\omega_0-\omega_e}{\Omega_d-\omega_0}\ln\frac{\omega_e-\omega}{\omega_0-\omega}\right]$
3D	$\chi\sqrt{\frac{\omega-\omega_e}{\Omega_C}}\exp\left(-\frac{\omega-\omega_e}{\Omega_C}\right)\Theta(\omega-\omega_e)$	$\chi\left[\pi\sqrt{\frac{\omega_e-\omega}{\Omega_C}}\exp\left(-\frac{\omega-\omega_e}{\Omega_C}\right)\mathrm{erfc}\sqrt{\frac{\omega_e-\omega}{\Omega_C}}-\sqrt{\pi}\right]$

$$u(t,t_0) = \frac{1}{1-\Sigma'(\omega_b)} e^{-i\omega(t-t_0)} + \int_{\omega_e}^{\infty} d\omega \frac{J(\omega)e^{-i\omega(t-t_0)}}{[\omega-\omega_c-\Delta(\omega)]^2+\pi^2 J^2(\omega)} \quad (11.11)$$

where $\Sigma'(\omega_b) = [\partial\Sigma(\omega)/\partial\omega]_{\omega=\omega_b}$ and $\Sigma(\omega)$ denotes the solar irradiance induction,

$$\Sigma(\omega) = \int_{\omega_e}^{\infty} d\omega' \frac{J(\omega')}{\omega-\omega'} \quad (11.12)$$

Here, the frequency ω_b in Eq. (11.2) represents the UVGI frequency mode in the PBG ($0 < \omega_b < \omega_e$) and is calculated using the pole condition: $\omega_b - \omega_c - \Delta(\omega_b) = 0$, where $\lesssim\Delta(\omega) = \mathcal{P}\left[\int d\omega' \frac{J(\omega')}{\omega-\omega'}\right]$ is a principal-value integral.

Subsequently, the detailed UVGI spectrum, considering the emission magnitude $|u(t, t_0)|$, on the glazing wall surface has been calculated and is shown in Fig. 11.5a for 1D, 2D, and 3D considering the variable detuning δ integration of solar irradiance [18]. The photonic spectrum deliberation rate $\kappa(t)$ is shown in Fig. 11.5b, considering the function $\delta = 0.1\omega_e$. These results revealed that photonic UVGI is released at a modest rate once ω_c crosses on the surface of the glazing wall with a range of $u(t, t_0)$ as $1 \geq |u(t, t_0)| \geq 0$ as related to the condition $0.9 \gtrsim |u(t \to \infty, t_0)| \geq 0$. This corresponds to $-0.025\omega_e \lesssim \delta \lesssim 0.025\omega_e$, which revealed that UVGI release rate $\kappa(t)$ within the glazing wall surface ($\delta < -0.025\omega_e$) and in the vicinity of the glazing wall surface ($-0.025\omega_e \lesssim \delta \lesssim 0.025\omega_e$) is almost exponential for $\delta \gg 0.025\omega_e$ [11, 27]. In Fig. 11.5a it has been shown that the dash-doted black curves $\delta = 0.1\omega_e$ sharply increased photonic energy generation in the crossover area ($-0.025\omega_e \lesssim \delta \lesssim 0.025\omega_e$), and thus this UVGI proliferation of dynamics confirms that the DOS rate in the vicinity of the glazing wall surface is in an equilibrium modest rate [26].

Then, the UVGI spectrum dynamics in the curtain wall surface is being determined considering thermal variation with respect to the photon correlation function $v(t, t)$ by determining the non-equilibrium photon scattering theorem [18],

$$v(t,t) = \int_{t_0}^{t} dt_1 \int_{t_0}^{t} dt_2 u^*(t_1,t_0)\widetilde{g}(t_1,t_2)u(t_2,t_0) \quad (11.13)$$

Here, the time-related function $\widetilde{g}(t_1,t_2) = \int d\omega J(\omega)\overline{n}(\omega,T)e^{-i\omega(t-t')}$ reveals the photonic UVGI release is initiated by the surface relativistic condition of the glazing wall, where $\overline{n}(\omega,T) = 1/\left[e^{\hbar\omega/k_B T} - 1\right]$ is the proliferation of the UVGI emission in the glazing wall surface at temperature T and expressed as

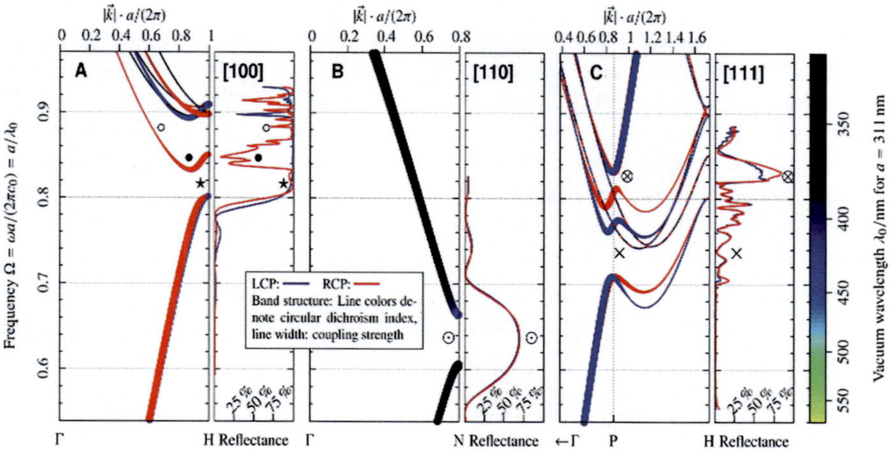

(a)

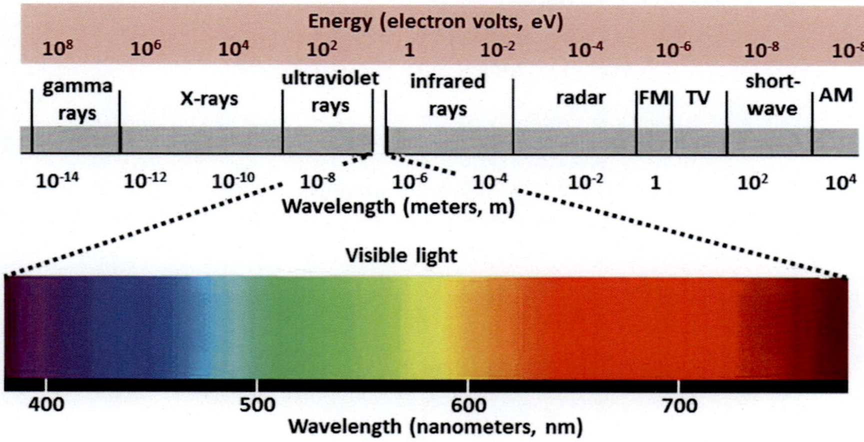

(b)

Fig. 11.5 (**a**) The formation of various photonic band structure and reflectance of the glazing wall surface. (**b**) The EM radiation shows that as frequency increases into the UVGI, the photons electron volts gotten to excite that damage the DNA and/or RNA of the pathogens at the visible light range for 254–280 nm

$$v(t, t \to \infty) = \int_{\omega_e}^{\infty} d\omega \mathcal{V}(\omega) \quad \text{with} \quad \mathcal{V}(\omega) = \overline{n}(\omega, T)[\mathcal{D}_l(\omega) + \mathcal{D}_d(\omega)] \tag{11.14}$$

Here, is simplified to determine the non-equilibrium condition: $\mathcal{V}(\omega) = \overline{n}(\omega, T)\mathcal{D}_{\rm d}(\omega)$. Under low-temperature conditions, Einstein's photon fluctuation dissipation is not functionally viable at the photonic band (PB), but it connects the within bandgap to release UVGI which is measurable as; $n(t) = \langle a^{\dagger}(t)a(t)\rangle = |u(t, t_0)|^2 n(t_0)v(t, t)$, where $n(t_0)$ represents the primary PB [7, 15, 25]. Therefore, in Fig. 11.5b, it has been plotted the rate of release of UVGI versus temperature to confirm non-equilibrium proliferated photon energy, as shown by the solid-blue curve. Therefore, it has been considered as the PB as the Fock state is the UVGI release rate n_0, i.e., $\rho(t_0) = |n_0\rangle\langle n_0|$, at time t and expressed as:

$$\rho(t) = \sum_{n=0}^{\infty} \mathcal{P}_n^{(n_0)}(t)|n_0\rangle\langle n_0| \tag{11.15}$$

$$\mathcal{P}_n^{(n_0)}(t) = \frac{[v(t,t)]^n}{[1+v(t,t)]^{n+1}}[1-\Omega(t)]^{n_0} \times \sum_{k=0}^{\min\{n_0,n\}} \binom{n_0}{k} \times \binom{n}{k}\left[\frac{1}{v(t,t)}\frac{\Omega(t)}{1-\Omega(t)}\right]^k \tag{11.16}$$

where $\Omega(\mathrm{t}) = \frac{|u(t,t_0)|^2}{1+v(t,t)}$.

Simply, this result revealed that a Fock state photon is being acted as the catalyst to release photon energy at various of $|n_0\rangle$ is $\mathcal{P}_n^{(n_0)}(t)$ and thus, the UVGI release has been determined as $\mathcal{P}_n^{(n_0)}(t)$ as the primary state $|n_0 = 5\rangle$ and steady-state limit, $\mathcal{P}_n^{(n_0)}(t \to \infty)$ and expressed as:

$$\mathcal{P}_n^{(n_0)}(t \to \infty) = \frac{[\overline{n}(\omega_{\rm c}, T)]^n}{[1+\overline{n}(\omega_{\rm c}, T)]^{n+1}} \tag{11.17}$$

To probe this UVGI production in curtain wall surface, it has been further broken down through extreme coherent states considering the proliferation state of UVGI, and expressed as

$$\rho(t) = \mathcal{D}[\alpha(t)]\rho_{\rm T}[v(t,t)]\mathcal{D}^{-1}[\alpha(t)] \tag{11.18}$$

where $\mathcal{D}[\alpha(t)] = \exp\{\alpha(t)a^{\dagger} - \alpha^*(t)a\}$ denotes the displacement function of $\alpha(t) = u(t, t_0)\alpha_0$ and

$$\rho_{\rm T}[v(t,t)] = \sum_{n=0}^{\infty} \frac{[v(t,t)^n]}{[1+v(t,t)]^{n+1}} \mid n\rangle\langle n| \tag{11.19}$$

Here, $\rho_{\rm T}$ denotes a standard state photon particle $v(t, t)$, where Eq. (11.19) confirms the net UVGI generation from the glazing wall surface electromagnetic radiation (EM).

Electromagnetic Radiation

EM radiation is being determined by the wavelength of UVGI sinusoidal monochromatic waves, which in turn is being classified from the EM spectrum of the photon particles. Thus, the EM waveform is being calculated using the spectral analysis of photonic frequency considering their *power* content which is the spectral density of the solar energy. Thus, the random electromagnetic radiation of UVGI is being determined from the wideband forms of solar radiation from the unique sinusoidal wave field of the photon energy.

Killing Pathogens

The EM field of the solar energy is being thus clarified the outer glazing wall surface in order to emit light between 254 and 280 nm which is released by electromagnetic field in order to allow to add a mass-term sunlight energy transform into UVGI by a connected semiconductor. Since UVGI has the extraordinary ability to destroy the DNA and/or RNA of the fatal pathogens in seconds regardless of the pathogens' frequent changing mechanism of mutants and pathways, the UVGI light wavelength implantation suggested that EM field has been paved for the solar irradiance to penetrate uniformly into the pathogen bodies which is determined by using semiconductor connected with a controlled circuit system into house and building. Since DNA and/or RNA of the pathogens is composed of double- or single-strand bonds consisting of chemically coded nucleotides adenine (A), thymine (T), guanine (G), and cytosine (C) that are very sensitive to UBGI, UVGI light interfered directly to these bonds between the nucleotides in the DNA and/or RNA and completely damaged their cyclobutane pyrimidine dimers (CPD) and 6-4 pyrimidine pyrimidone photoproducts (6-4PPs), and its Dewar isomers (Fig. 11.6c). The results suggested that CPDs are being formed when two adjacent pyrimidine bases (thymine—TT or cytosine—CC) become covalently linked producing a cyclic ring structure, and thus 6-4PPs resulted from a single covalent bond formed between the 5′ end of C6 and 3′ end of C4 of adjacent pyrimidines [12, 25]. This leads to the formation of an unstable oxetane or azetidine intermediate end base of the thymine or cytosine dimers. Consequently, spontaneous rearrangement of these intermediate ends gave rise to 6-4PP and eventually, the pyrimidine dimers caused a kink into the DNA strands, halting transcription and protein synthesis (Fig. 11.6a). 6-4 Pyrimidine pyrimidone adducted isomerization and caused severe damage to cellular function (Fig. 11.6a). Subsequently, UVGI radiation does cause indirect damage to DNA and/or RNA via absorption of photons by non-DNA chromophores which eventually generates the reactive oxygen charge that oxidizes the DNA bases causing prohibition of mutations just because most of the common mutations occurred at the G-T transversion process where guanine gets oxidized into 8-oxo-7,8-dihydroguanine (8-oxoG) hindering its pairing with cytosine. Interestingly, UVGI changes all these

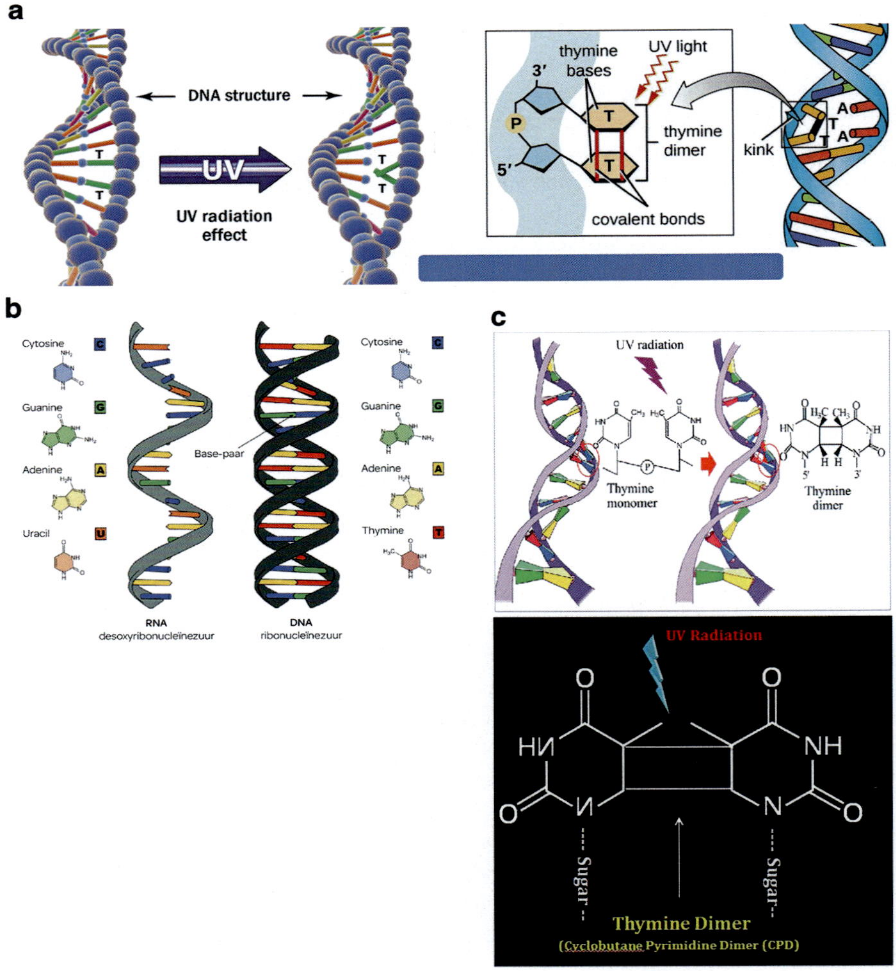

Fig. 11.6 (**a**) Showing how UV light affects DNA structure, (**b**) detail RNA and DNA structure of pathogens, (**c**) UV radiation for disinfection involvement to generate desired germicidal mechanism to destroy the thymine dimer of the pathogens to ultimately disable their biochemical function and reproduction

biochemical functions of the pathogens during this replication process especially when the 8-oxoG attempted to pair with adenine during the synthesis of the strand, and 8-oxoG is being replaced with a thymine in order to G-T transversion and eventually completely stop the mutation process.

Conclusions

Solar energy has been implemented using advanced design technology of exterior glazing wall surface of a building or house to create *Ultraviolet Germicidal Irradiation* (UVGI) to release short-range ultraviolet light to make extremely effective destroying DNA and/or RNA of the pathogens such as bacteria, viroid, bacteria, protozoa, fungus, molds, and virus including COVID-19 naturally. Since some of the pathogens are extremely dangerous to the human being, which also surprisingly can change their mutants and pathway to deceive the functionality of the vaccines, the prevention technology has been studied to defend these deadly pathogens rather than treatment technology. Therefore, in this research, implementation of photonic irradiance has been conducted to release short-range wavelength 254–280 nm using the outer glazing wall surface of a building or house to form UVGI to eliminate deadly pathogens inside the building and house in seconds regardless their ability to change the mutants and pathways. Simply using the UVGI to eliminate all pathogens inside the building by damaging their DNA and/or RNA structure and cellular function naturally would be an innovative technology to eliminate these deadly pathogens before it invades into the human body and will mitigate future global outbreak challenges.

Acknowledgments Thanks to the GGT, Inc. for furnishing required fund (RD-02020-03) to conduct this research on time. Author confirms that this research does have any conflicts of interest and welcome any suitable journal to publish this paper.

References

1. J. Douglas, H. Habibian, C. Hung, A. Gorshkov, H. Kimble, D. Chang, Quantum many-body models with cold atoms coupled to photonic crystals. Nat. Photonics **23**, 45–49 (2015)
2. D. Englund, Resonant excitation of a quantum dot strongly coupled to a photonic crystal nanocavity. Phys. Rev. Lett. **104**, 904–909 (2010)
3. H. Faruque, Solar energy integration into advanced building design for meeting energy demand and environment problem. Int. J. Energy Res. **78**, 671–679 (2016)
4. H. Faruque, Green science: independent building technology to mitigate energy, environment, and climate change. Renew. Sustain. Energy Rev. **45**, 223–229 (2017a)
5. H. Faruque, Design and construction of ultra-relativistic collision PV panel and its application into building sector to mitigate total energy demand. J. Build. Eng. **36**, 270–278 (2017b)
6. H. Faruque, Transforming dark photons into sustainable energy. Int. J. Energy Environ. Eng. **11**, 38–45 (2018a)
7. H. Faruque, Green science: advanced building design technology to mitigate energy and environment. Renew. Sustain. Energy Rev. **56**, 87–93 (2018b)
8. H. Faruque, Photon energy amplification for the design of a micro PV panel. Int. Energy Res. **66**, 340–349 (2018c)
9. H. Faruque, Sustainable technology for energy and environmental benign building design. J. Build. Eng. **44**, 123–132 (2019)
10. J. Hou, H. Wang, P. Liu, Applying the blockchain technology to promote the development of distributed photovoltaic in China. Int. J. Energy Res. **29**, 450–459 (2018)

11. J. Huang, F. Shi, T. Sun, Controlling single-photon transport in waveguides with finite cross section. Phys. Rev. A **88**, 13836–13842 (2013)
12. C. Lei, U. Zhang, A quantum photonic dissipative transport theory. Ann. Phys. **327**, 1408–1412 (2012)
13. J. Liao, Q. Law, Correlated two-photon transport in a one-dimensional waveguide side-coupled to a nonlinear cavity. Phys. Rev. A **82**, 538–545 (2010)
14. P. Longo, P. Schmitteckert, Few-photon transport in low-dimensional systems. Phys. Rev. A **83**, 638–647 (2011)
15. H. Min, C. Veronis, Subwavelength slow-light waveguides based on a plasmonic analogue of electromagnetically induced transparency. Appl. Phys. Lett. **99**, 143–151 (2011)
16. D. O'Shea, C. Junge, Fiber-optical switch controlled by a single atom. Phys. Rev. Lett. **111**, 193–201 (2013)
17. K. Oulton, R. Zhang, Nonlinear quantum optics in a waveguide: distinct single photons strongly interacting at the single atom level. Phys. Rev. Lett. **106**, 113–119 (2011)
18. T. Pregnolato, E. Song, Single-photon non-linear optics with a quantum dot in a waveguide. Nat. Commun. **6**, 86–95 (2015)
19. M. Reed, L. Maxwell, Connections between groundwater flow and transpiration partitioning. Science **353**, 377–380 (2015)
20. A. Reinhard, Strongly correlated photons on a chip. Nat. Photonics **6**, 2–4 (2011). https://doi.org/10.1038/nphoton.2011.321
21. D. Roy, Two-photon scattering of a tightly focused weak light beam from a small atomic ensemble: an optical probe to detect atomic level structures. Phys. Rev. A **87**, 638–6419 (2013)
22. E. Saloux, Explicit model of photovoltaic panels to determine voltages and currents at the maximum power point. Sol. Energy **66**, 450–459 (2015)
23. C. Sayrin, Real-time quantum feedback prepares and stabilizes photon number states. Nature **477**, 73–76 (2011)
24. T. Shi, S. Fan, Two-photon transport in a waveguide coupled to a cavity in a two-level system. Phys. Rev. A **84**, 638–643 (2011)
25. C. Song, Fano resonance analysis in a pair of semiconductor quantum dots coupling to a metal nanowire. Opt. Lett. **37**, 978–980 (2012)
26. C. Wang, R. Zhang, J. Xiao, Multiple plasmon-induced transparencies in coupled-resonator systems. Opt. Lett. **37**, 5133–5135 (2012)
27. Y. Wang, Y. Zhang, Q. Zhang, B. Zou, U. Schwingenschlogl, Dynamics of single photon transport in a one-dimensional waveguide two-point coupled with a Jaynes-Cummings system. Sci. Rep. **98**, 451–458 (2016)
28. Y. Xiao, C. Meng, P. Wang, Y. Ye, Single-nanowire single-mode laser. Nano Lett. **48**, 340–348 (2011)
29. G. Yan, H. Lu, A. Chen, Single-photon router: implementation of information-holding of quantum states. Int. J. Theor. Phys. **24**, 78–83 (2016)

Chapter 12
Air Pollution and the Survival Period of Earth

Abstract Global total CO_2 emission and sequestration are being analyzed from 1960 to 2029 reports interpreted from DEP, DOE, IPCC, CFC, CDIAC, IEA, UNEP, NOAA, and NASA. Consequently, these reports have been transcribed into each 10-year period data set by using MATLAB software to accurately calculate the decadal emission and sequestration rate of total CO_2 within the world. Then these data were further analyzed to determine the final annual increasing rate (yr^{-1}) of CO_2 accumulation into the atmosphere. The study revealed that total CO_2 emissions throughout the world since the 1960s have been increasing rapidly and the recent year the net CO_2 increasing rate is 2.11% annually. If the current annual CO_2 growth rate is not copped now, the atmospheric CO_2 accumulation shall indeed reach a toxic level of 1200 ppm concentration of CO_2 into the atmosphere in 53 years. Consequently, the entire human race will become extinct on Earth due to the toxic level of CO_2 presence in the air.

Keywords Global CO_2 emissions · Total CO_2 sequestration · Toxic level of CO_2 · Environmental vulnerability · Global Public Health Crisis

Introduction

Since the 1960s, massive development of industrialization and the misuse of the natural resources throughout the world quickening to accumulation of atmospheric CO_2 concentration heavily will certainly be dangerous for mankind to take fresh breath in the near future [3, 7]. Numerous studies revealed that currently accumulation of CO_2 into the atmosphere is 400 ppm and it is increasing in such a rapid rate that it will reach soon a the toxic level of 1200 CO_2; this will result in severe respiratory problem in human beings and many people will possibly die throughout the world [4, 8, 13]. So, it is the time without a doubt to make the global environment green by reducing CO_2 emissions which will confirm the versatility, adaptability, and manageability of our mother Earth, which won't result in maladjustment simultaneously, but will be presentable as a sustainable world for our future generation to take a fresh breath. Thus, in the research, a detail calculation of global CO_2 emission from all sources on Earth and sequestration of CO_2 by all sinks on this planet has

M. F. Hossain, *Sustainable Design for Global Equilibrium*,
https://doi.org/10.1007/978-3-030-94818-4_12

been estimated to evaluate the net increasing rate of CO_2 into the atmosphere to give an advance warning to the mankind for forthcoming environmental vulnerability due to the heavy accumulation CO_2 into the atmosphere. Simply, this study will help the global scientific community, policy makers, and leaders to take this forthcoming danger seriously to mitigate global CO_2 immediately to console the forthcoming deadly respiratory problem for mankind on Earth.

Methods and Simulation

CO_2 Emissions

The decadal increasing rate in CO_2 emissions due to all industrial development globally was estimated from the difference between two consecutive years decades from the period 1960s, 1970s, 1980s, 1990s, 2000s, 2010s, and 2020s, and then it was converted into yearly growth rate divided by past year emission to the current year emissions by using the following equation:

$$\mathrm{FF} = \left[\frac{E_{\mathrm{FF}\ (t_{0+1})} - E_{\mathrm{FF}(t_0)}}{E_{\mathrm{FF}\ (t_0)}}\right] * 100\% \ \ \mathrm{yr}^{-1} \tag{12.1}$$

Here, this simple calculation is being analyzed to determine per year CO_2 emissions increasing rate. To precisely estimate the CO_2 increasing rate considering each decadal period, a leap-year factor is also being applied to determine net yearly increasing rate of CO_2 (E_{Ff}) by using its logarithm equal to the below equation:

$$\mathrm{Ff} = \frac{1}{E_{\mathrm{FF}}} \frac{d\left(\ln E_{\mathrm{FF}}\right)}{dt} \tag{12.2}$$

Here, the net CO_2 emission increasing rates have been calculated accounting multi-decadal time scales by integrating a nonlinear function into $\ln(E_{\mathrm{FF}})$ in Eq. (12.2) to calculate eventually yearly increasing rate of CO_2 into the atmosphere [1, 5, 11]. Thus, the algorithm of E_{FF} of this equation is being fitted into MATLAB algorithm E_{FF} to confirm the precise increasing rate of CO_2 yearly.

Similarly, the CO_2 emission calculated here (E_{LUC}) due to the misuse of all natural resources throughout the world were calculated by implementing dynamic global environmental modeling (DGVM) simulations in MATLAB considering the difference between two consecutive decadal periods of 1960s, 1970s, 1980s, 1990s, 2000s, 2010s, and 2020s [2, 18, 19]. Then, a time series is being implemented in this simulation by allocating the dynamic emission of CO_2 due to the misuse of all natural resources throughout the world within two consecutive years, and then it was converted into yearly growth rate divided by past year emission to the current year emissions by using the following equation:

$$\text{LUC} = \left[\frac{E_{\text{LUC}(t_{0+1})} - E_{\text{LUC}(t_0)}}{E_{\text{LUC}(t_0)}}\right] * 100\% \ \text{yr}^{-1} \tag{12.3}$$

Here, the equation is being calculated in yearly CO_2 emissions growth rate [12, 15, 19]. However, to precisely determine the increasing rate of CO_2 in multiple decades, a leap-year factor is also being applied to ensure the net yearly increasing rate of carbon di oxide (E_{LUC}) which is expressed by the following equation:

$$\text{LUC} = \frac{1}{E_{\text{LUC}}} \frac{d\left(\ln E_{\text{LUC}}\right)}{dt} \tag{12.4}$$

Here, the CO_2 emission increasing rates have been estimated corresponding to all decadal time scales by applying a nonlinear function in ln(E_{LUC}) in Eq. (12.4) to determine annual CO_2 emission into the atmosphere [16, 17]. Thus, the algorithm of E_{FF} is being integrated into MATLAB to confirm the precise emission rate of CO_2 from all natural resources misuses.

Finally, the global total CO_2 emission from all industrial development and misuse of natural resources per year has been calculated by combining all four equations (Eqs. 12.1–12.4) as follows:

$$\begin{aligned} \text{FL} = & \left[\frac{E_{\text{FF}(t_{0+1})} - E_{\text{FF}(t_0)}}{E_{\text{FF}(t_0)}}\right] * 100\%\text{yr}^{-1} + \frac{1}{E_{\text{FF}}} \frac{d\left(\ln E_{\text{FF}}\right)}{dt} \\ & + \left[\frac{E_{\text{LUC}(t_{0+1})} - E_{\text{LUC}(t_0)}}{E_{\text{LUC}(t_0)}}\right] * 100\%\text{yr}^{-1} + \frac{1}{E_{\text{LUC}}} \frac{d\left(\ln E_{\text{LUC}}\right)}{dt} \end{aligned} \tag{12.5}$$

Thereafter, the global CO_2 sequestration considering all (1) ocean sink and (2) terrestrial sink available throughout the world has been calculated from 1960 to 2029 by conducting 10-year period each experimental data set and then converted it into the time period for an average annual rate.

CO_2 Sink

Consequently, the CO_2 sequestered by the ocean is being calculated for the past years and the next years from the decadal set of 1960s, 1970s, 1980s, 1990s, 2000s, 2010s, and 2020s by implementing all oceans' carbon sink cycle models [10, 14]. This approach is being implemented to accurately analyze the physio-biological processes of global oceans directly involved in CO_2 sequestering by the ocean surfaces and its fauna [6, 9]. Thus, the oceans' CO_2 sink is being determined accurately by dividing the individual yearly values with the previous year's value; therefore, the oceanic CO_2 sequestration per year (t) in GtC yr^{-1} is being calculated as follows:

$$S_{\text{OCEAN}}(t) = \frac{1}{n}\sum\nolimits_{m=1}^{m=n} \frac{S^m_{\text{OCEAN}}(t)}{S^m_{\text{OCEAN}}(t10 - t1)} \tag{12.6}$$

Here n is the number of oceans; m is the factors involving CO_2 sequestration; and t represents the period.

Then the absorption of CO_2 per year by terrestrial vegetation and the Earth is also being determined to determine the total CO_2 sequestration by the land (S_{LAND}) similarly from the decadal set of 1960s, 1970s, 1980s, 1990s, 2000s, 2010s, and 2020s and convert it into annual rate. Here, the net CO_2 sink by land is being calculated as follows:

$$S_{\text{LAND}} = E_{\text{FF}} + E_{\text{LUC}} - (G_{\text{ATM}} + S_{\text{OCEAN}}) \tag{12.7}$$

Here, S_{LAND} is calculated from the remainder of the estimates where G_{ATM} is the presence of CO_2 into the atmosphere, (E_{FF}) is the carbon from industrial development, and the E_{LUC} is the CO_2 from the misuse of all natural resources throughout the world [2, 19].

Then, the computation of S_{LAND} in Eq. (12.7) is being utilized to determine E_{LUC} by subtracting the (G_{ATM} + S_{OCEAN}) CO_2.

Subsequently, the total CO_2 sequestration in a year period has been calculated by combining these two equations (Eqs. 12.6 and 12.7) as follows:

$$\begin{aligned} S_{\text{OCEAN}}\ (t) + S_{\text{LAND}} &= \frac{1}{n}\sum\nolimits_{m=1}^{m=n} \left(\frac{S^m_{\text{OCEAN}}\ (t)}{S^m_{\text{OCEAN}(t10-t1)}} \right) + (E_{\text{FF}} + E_{\text{LUC}}) \\ &\quad - (G_{\text{ATM}} + S_{\text{OCEAN}}) \end{aligned} \tag{12.8}$$

Atmospheric CO_2 Concentration (G_{ATM}) Increasing Rate

Finally, the net yearly increasing rate of the atmospheric CO_2 concentration is being determined yearly from the variation of the total CO_2 emission and total CO_2 sequestration each year.

Results and Discussion

CO_2 Emission

The average global CO_2 emissions from 1960 to 2029 during this time scale showed that total CO_2 emissions from combined industrial development and the misuse of all natural resources throughout the world are at an annual average of 1.7 GtC yr^{-1} of 1.7 ± 0.7 GtC yr^{-1} per decade in the 1960s (1960–1969); annual average of

2.2 GtC yr^{-1} of 1.7 ± 0.8 GtC yr^{-1} per decade in the 1970s (1970–1979); annual average of 1.5 GtC yr^{-1} of 1.6 ± 0.8 GtC yr^{-1} per decade in the 1980s (1980–1989); annual average of 2.45 GtC yr^{-1} of 2.6 ± 0.8 GtC yr^{-1} per decade in the 1990s (1990–1999); annual average of 2.45 GtC yr^{-1} of 2.6 ± 0.8 GtC yr^{-1} per decade in the 2000s (2000–2009); annual average of 3.26 GtC yr^{-1} of 3.26 ± 0.5 GtC yr^{-1} per decade in the 2010s (2010–2019); and expected to be increased to annual average of 3.26 GtC yr^{-1} of 3.26 ± 0.5 GtC yr^{-1} per decade in the 2020s (2020–2029) (Table 12.1).

CO_2 Sink

Subsequently, the results of CO_2 sequestration by ocean and the terrestrial vegetation and land suggested that the average global CO_2 sink from 1960 to 2029 during this time scale showed that total CO_2 emissions from combined industrial development and the misuse of all natural resources throughout the world are at an annual average of 1.5 GtC yr^{-1} of 1.5 ± 0.2 GtC yr^{-1} per decade in the 1960s (1960–1969); annual average of 1.3 GtC yr^{-1} of 1.3 ± 0.5 GtC yr^{-1} per decade in the 1970s (1970–1979); annual average of 1.4 GtC yr^{-1} of 1.4 ± 0.6 GtC yr^{-1} per decade in the 1980s (1980–1989); annual average of 1.4 GtC yr^{-1} of 1.6 ± 0.4 GtC yr^{-1} per decade in the 1990s (1990–1999); annual average of 1.15 GtC yr^{-1} of 1.15 ± 0.5 GtC yr^{-1} per decade in the 2000s (2000–2009); annual average of 1.15 GtC yr^{-1} of 1.15 ± 0.5 GtC yr^{-1} per decade in the 2010s (2010–2019); and expected to be increased to annual average of 1.15 GtC yr^{-1} of 1.15 ± 0.5 GtC yr^{-1} per decade in the 2020s (2020–2029) (Table 12.1).

Atmospheric CO_2 Concentration (G_{ATM}) Increasing Rate

Then, the rate of growth of the atmospheric CO_2 concentration is being calculated by comparing the decadal and individual annual values for 10-year periodical set which suggested that the average global CO_2 annual growth from 1960 to 2029 are 0.2% at the decade 1960s; 0.9% at the decade 1970s; 0.1% at the decade 1980s; 1.15% at the decade 1990s; 1.3% at the decade 2000s; 2.11% at the decade 2010s; and expected to be 2.11% at the decade 2020s. The projected growth rate of atmospheric CO_2 concentration is presumably suggested that the increased at rate of CO_2 will remain same 2.11% per year for next several decades if we do not curb this acceleration of CO_2 emissions (Table 12.1).

Since the current CO_2 concentration into the atmosphere is 400 ppm and is being growing at a rate of 2.11% per year, following equations confirmed that it will attain a toxic level of 1200 ppm in 53 years.

Table 12.1 The results from DGVM simulation in MATLAB, implemented from the data of DEP, DOE, IPCC, CFC, CDIAC, IEA, UNEP, NOAA, and NASA to confirm the yearly increasing rate of atmospheric CO_2 (%). The results described the variation of the total CO_2 emissions from industrial development and misuse of all natural resources throughout the world and the total CO_2 sink (Ocean and Land) from the years of 1960–1969, 1970–1979, 1980–1989, 1990–1999, and 2000–2009, 2010–2019, and 2020–2029 shown in GtC yr^{-1}

Mean (GtC yr^{-1})	1960–1969	1970–1979	1980–1989	1990–1999	2000–2009	2010–2019	2020–2029
Total CO_2 emission (industrial development and misuse of natural resources)							
DGVM simulations and the mean average data of DEP, DOE, IPCC, CFC, CDIAC, IEA, UNEP, NOAA, and NASA for each decadal period of CO_2 emissions	1.7 ± 0.7	1.7 ± 0.8	1.6 ± 0.8	2.6 ± 0.8	2.6 ± 0.8	3.26 ± 0.5	3.26 ± 0.5
Net CO_2 emissions rate (%) per year	1.7	2.2	1.5	2.45	2.45	3.26	3.26
Total CO_2 sink (ocean and terrestrial vegetation and land)							
DGVM simulations and the mean average data of DEP, DOE, IPCC, CFC, CDIAC, IEA, UNEP, NOAA, and NASA for each decadal period of CO_2 sequestration	1.5 ± 0.5	1.3 ± 0.5	1.4 ± 0.6	1.6 ± 0.4	1.15 ± 0.5	1.15 ± 0.5	1.0 ± 0.5
Net CO_2 sequestration rate (%) per year	1.5	1.3	1.4	1.4	1.15	1.15	1.15
Annual increasing rate of atmospheric CO_2 (%)							
G_{ATM}	0.2	0.9	0.1	1.05	1.3	2.11	2.11

$$1200 = 400(1 + 0.0211)^{\text{Year}} \tag{12.9}$$

$$3 = (1 + 0.0211)^{\text{Year}} \tag{12.10}$$

$$\text{Log } 3 = \text{Year Log}(1.0211) \tag{12.11}$$

$$\text{Year} = 52.61 = 53 \text{ (round figure)} \tag{12.12}$$

Consequently, all human being on earth will be in serious breathing problem due to the toxic level of CO_2 into the atmosphere. Simply, it is an urgent demand to reduce the CO_2 emission globally to mitigate the forthcoming deadly breading problem for mankind as well as secure a better planet for our next generation.

Conclusion

The total global CO_2 emissions due to the industrial development and the misuse of all natural resources throughout the world estimated for the past several decades as well as total CO_2 sink by ocean and land were also calculated to determine the increasing rate of CO_2 into the atmospheric each year. The yearly increasing rate of the atmospheric CO_2 accumulation over the last several years was confirmed by simulated estimate which revealed that it is increasing at a rate of 2.11% yearly. If the current annual CO_2 growth rate is not curbed now, the atmospheric CO_2 accumulation shall indeed reach a toxic level of 1200 ppm in 53 years, which can lead to perishing of entire human race from Earth in the future. This will be the end of human civilization on Earth.

Acknowledgment This research was supported by Green Globe Technology under the grant RD-02021-03. Any findings, conclusions, and recommendations expressed in this paper are solely those of the author and do not necessarily reflect those of Green Globe Technology.

References

1. F. Achard et al., Determination of tropical deforestation rates and related carbon losses from 1990 to 2010. Glob. Chang. Biol. **20**, 2540–2554 (2014)
2. A.P. Ballantyne, C.B. Alden, J.B. Miller, P.P. Tans, J.W.C. White, Increase in observed net carbon dioxide uptake by land and oceans during the past 50 years. Nature **488**, 70–72 (2012)
3. J.E. Bauer et al., The changing carbon cycle of the coastal ocean. Nature **504**, 61–70 (2013)
4. R.A. Betts, C.D. Jones, J.R. Knight, R.F. Keeling, J.J.E. Kennedy, Nino and a record CO_2 rise. Nat. Clim. Change **6**, 806–810 (2016)
5. J.G. Canadell et al., Contributions to accelerating atmospheric CO_2 growth from economic activity, carbon intensity, and efficiency of natural sinks. Proc. Natl. Acad. Sci. USA **104**, 18866–18870 (2007)
6. F. Chevallier, On the statistical optimality of CO_2 atmospheric inversions assimilating CO_2 column retrievals. Atmos. Chem. Phys. **15**, 11133–11145 (2015)

7. S.J. Davis, K. Caldeira, Consumption-based accounting of CO_2 emissions. Proc. Natl. Acad. Sci. USA **107**, 5687–5692 (2010)
8. K.-H. Erb et al., Bias in the attribution of forest carbon sinks. Nat. Clim. Change **3**, 854–856 (2013)
9. M.F. Hossain, Theory of global cooling. Energy Sustain. Soc. **6**, 24 (2016)
10. M.F. Hossain, Green science: independent building technology to mitigate energy, environment, and climate change. Renew. Sustain. Energy Rev. **73**, 695–705 (2017)
11. R. Houghton, Balancing the global carbon budget. Annu. Rev. Earth Planet. Sci. **35**, 313–347 (2007)
12. A.K. Jain, P. Meiyappan, Y. Song, J.I. House, CO2 emissions from land-use change affected more by nitrogen cycle, than by the choice of land-cover data. Glob. Chang. Biol. **19**, 2893–2906 (2013). https://doi.org/10.1111/gcb.12207
13. W. Li et al., Reducing uncertainties in decadal variability of the global carbon budget with multiple datasets. Proc. Natl. Acad. Sci. USA **113**, 13104–13108 (2016)
14. Z. Liu et al., Reduced carbon emission estimates from fossil fuel combustion and cement production in China. Nature **524**, 335–338 (2015)
15. J. Mason Earles, S. Yeh, K.E. Skog, Timing of carbon emissions from global forest clearance. Nat. Clim. Change **2**, 682–685 (2012)
16. J. Prietzel, L. Zimmermann, A. Schubert, D. Christophel, Organic matter losses in German Alps forest soils since the 1970s most likely caused by warming. Nat. Geosci. **9**, 543–548 (2016)
17. S. Schwietzke et al., Upward revision of global fossil fuel methane emissions based on isotope database. Nature **538**, 88–91 (2016)
18. M. Shinji et al., Spatio-temporal variations of the atmospheric greenhouse gases and their sources and sinks in the Arctic region. Pol. Sci. **27**, 100553 (2021)
19. B.B. Stephens et al., Weak northern and strong tropical land carbon uptake from vertical profiles of atmospheric CO_2 science. Science **316**, 1732 (2007)

Chapter 13
Modeling of Climate Control to Secure Global Environmental Equilibrium

Abstract A natural mechanism is being proposed to control global climate change in order to leave all living being in a steady comfort temperature condition throughout the seasons. To achieve this, photon particle has been remodeled by implementing Bose–Einstein (*B–E*) dormant photonic dynamics of the earth surface plane. Simply, the proposed decoded *B–E* photons will be induced by the photonic bandgap of earth surface to convert the solar photons into cooling-state photons here named as the *Hossain Cooling Photon* (HcP^-) which eventually will cool the earth surface. Interestingly, this HcP^- could an also be converted into the thermo-state photons, named as the *Hossain Thermal Photon* (HtP^-) by implementing Higgs bosons ($H \rightarrow \gamma\gamma^-$) electromagnetic quantum fields utilized by earth's electromagnetic force. Because $H \rightarrow \gamma\gamma^-$ quantum field of earth surface plane has the extreme small length weak force which will enforce the electrically charged HcP^- quantum to get voracious to convert it into HtP^- in order to heat the earth surface naturally. The formation of HcP^- from the photon particles and then the conversion of HcP^- to HtP^- has being proved by a set of mathematical tests in this research which reveals the feasibility of the deformed photons (HcP^- and HtP^-) are actively doable into the earth surface to cool and heat the Earth naturally.

Keywords Reformation of Bose · Einstein photon dynamics · Higgs boson BR ($H \rightarrow \gamma\gamma^-$) quantum fields · Hossain cooling photon (HcP^-) · Hossain thermal photon (HtP^-) · Control of earth temperature

Highlights

- Photon energy mechanism has been remodeled to control the earth temperature.
- Photonic bandgap has been implemented on earth surface to reform the photonic energy into cooling-state photons to cool the earth naturally.
- Electromagnetic quantum fields of Earth have been modeled to reform cooling-state photons into thermal state photons to heat the earth surface naturally.

M. F. Hossain, *Sustainable Design for Global Equilibrium*,
https://doi.org/10.1007/978-3-030-94818-4_13

Introduction

Global heating and cooling mechanism during the sessions varied tremendously due to the anthropogenic activities throughout the world which causes adverse effect on all living being on earth to live comfortably. Consequently, trillions of dollars are being spent each year to make our houses, offices, and premises comfortable which is in fact creating adverse environmental and atmospheric vulnerability. This is because conventional heating mechanism devours fossil energy that delivers CO_2, which is the primary player to climate change. Consequently, it causes severe environmental and atmospheric non-equilibrium, resulting in tremendous fluctuation of temperature throughout the session [3, 16, 56]. At the same time, traditional cooling systems form chlorofluorocarbons (CFCs), which creates the holes in the ozone layer, a protection shield that defends the maximum high-frequency ultraviolet rays of the sun in order to reduce the cause of skin cancer and severe reproduction problem to all mammalians [1, 5, 18]. To mitigate these dangerous environmental and atmospheric vulnerability, in this study, I propose a natural cooling and heating mechanism for Earth surface by implementing Bose–Einstein photon dynamics and the implementation of Higgs bosons ($H \to \gamma\gamma^{-}$) quantum. Simply this novel mechanism will decode the photons (solar energy) into cooling-state photons once photon flux from solar irradiance penetrates Earth surface. This is because the solar photons will be absorbed by nano-point breaks waveguides of photon band edges (PBEs) by utilizing the quantum electrodynamics (QED) to form cooling-state photons in order to cool the earth surface naturally [13, 44]. Mediated by Earth surface, the cooling-state photons can then be converted into the heating-state photons via the photonic radiations (PR) irradiated by the quantum of Higgs bosons ($H \to \gamma\gamma^{-}$) by utilizing the electromagnetic fields of Earth to heat the Earth surface naturally [38, 42, 49]. This cooling and heating conversion mechanism of Earth surface is indeed a noble science to control global temperature naturally and mitigate the global energy, environment, and climate change perplexity dramatically for securing a sustainable Earth eventually.

Methods and Simulation

Cooling Mechanism of Earth Surface

Once the solar photons penetrate to the Earth surface it would be deformed into the cooling-state photons by the cliques of nano-point breaks waveguide of earth surface plane by creating point defects in the photon particle [9, 46, 50]. Simply, the cluster of photons' bandgap (PBG) waveguides will be defected once the photon particles hit the earth plane surface [7, 43, 47]. These nano-point defect and PBG waveguide will deform the quantum dynamic mechanism of photon in order to transform it into cooling-state photons into earth surface to cool the earth naturally. The present

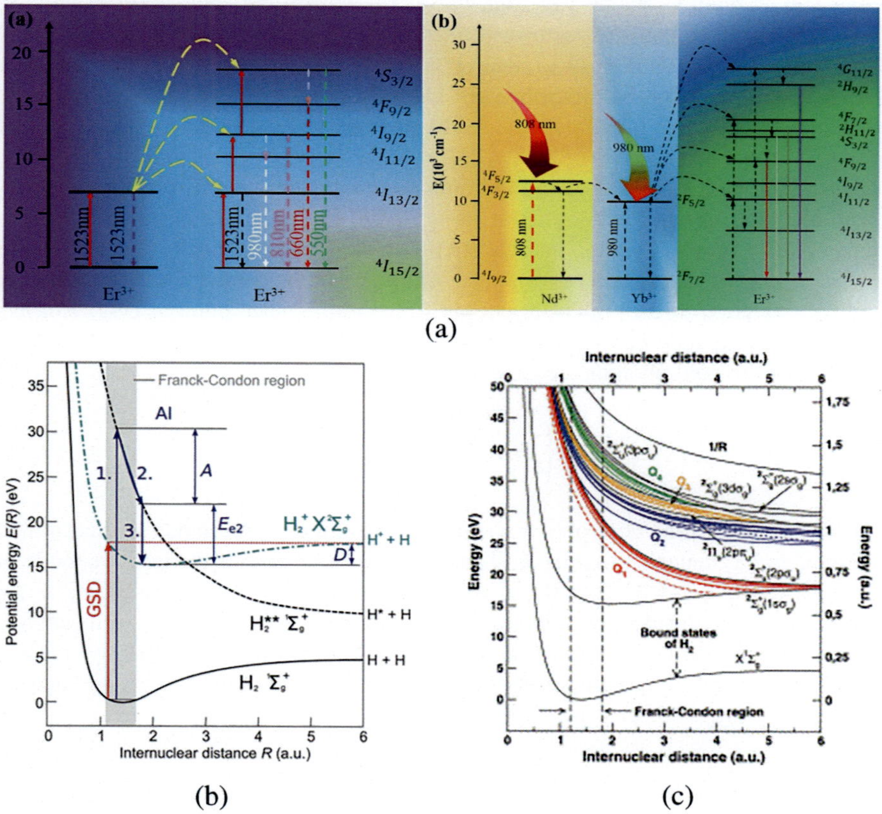

Fig. 13.1 (**A–C**) The array of the photonic probable density (standardized to its pick value) while the functions of x and t acted on the occurrence of the probable distributions of the reflection and transmission pulse of heating photons

research work outs the mechanism to form cooling-state photons using the transformation of thermal photons by using MATLAB software. The calculation revealed that the thermal waveguide that is being embedded in the earth surface plane acts as photons' reservoirs in order to conduct electrodynamics of the cool-state photon which can be calculated by the following *Hossain* equation [2, 10, 14]:

$$H = C + \sum \omega_{ci} a_i^\dagger a_i + \sum_K \omega_k b_k^\dagger b_k + \sum_{ik} \left(V_{ik} a^\dagger b_k + V_{ik}^* b_k^\dagger a_i \right), \tag{13.1}$$

where $a_i \left(a_i^\dagger \right)$ and $b_k \left(b_k^\dagger \right)$ present the operator of the nano-point breaks modules and the photo-dynamic modules of the photons' nano-structures, and the co-efficient, V_{ik}, presents the magnitude of the photon modules within the nano-breakpoint and photon nano-structures. Figure 13.1 shows the transmission contours and the spectrum of photons plane wave and pulses to confirm the calculation of cooling state of photons to cool the earth surface.

Simply the proposed cooling-state photonic modules that form HcP^{-} from solar irradiance to cool the earth surface are shown here considering the inter-nuclear distances, where the Q_1 denotes dissociation of $H(n = 1) + H(n = 2,...,\infty)$ and the Q_2 denotes the dissociation of $H(n = 2, l = 1) + H(n = 2,...,\infty)$, where n and l are, respectively, the principal and angular momentum quantum numbers of the cooling-state photons. The following photon energy-level diagram considering time (min), temperature (°C), and photon intensity (a.u.) by the module which is converted from the solar photons and mechanism of dissociative photons confirms the photon point breaks activation to cool the solar photons into cooling-state photons (Fig. 13.2).

This cooling-state photon that can be clarified by current and volt (I–V) characteristics of the cool photons are expressed by

$$I = I_L - I_o \left\{ \exp^{\left[\frac{q(V+I_{RS})}{AkT_c}\right]} - 1 \right\} - \frac{(V + I_{RS})}{R_s}, \tag{13.2}$$

where I_L presents the cool photons' current generation, I_o presents the unsaturated current, and R_s denotes the series resistances. Here A represents the inactive mode of the diodes, k ($= 1.38 \times 10^{-23}$ W/m^2 K) is the Boltzmann constant, q ($= 1.6 \times 10^{-19}$C) considers the function of the active charges of the electrons, and T_c defines the dynamic function of the cell thermal condition. Consequently, the I–q function in the photonic cells varies within the diode, which can be expressed by [12, 39, 51]

$$I_o = I_{RS} \left(\frac{T_c}{T_{ref}}\right)^3 \exp \left[\frac{qEG\left(\frac{1}{T_{ref}} - \frac{1}{T_c}\right)}{KA}\right]. \tag{13.3}$$

In Eq. (13.3), I_{RS} defines the unsaturation current, which relies on the dynamic function of temperature mode and photon speed, and qEG defines the photonic bandgap energy of the electron per unit area of the cool photons into the cell (Fig. 13.3).

In this cooling photons mode, the I–V relations integrate the I–V curve of all cells in the photon emissions on earth surface and are related to V–R relationship that can be expressed by

$$V = -IR_s + K \log \left[\frac{I_L - I + I_o}{I_o}\right], \tag{13.4}$$

where K represents the constant $\left(= \frac{AkT}{q}\right)$ and I_{mo} and V_{mo} are the current and voltage in earth surface. Therefore, the correlation between I_{mo} and V_{mo} is similar to the I–V relationship on the earth surface plane:

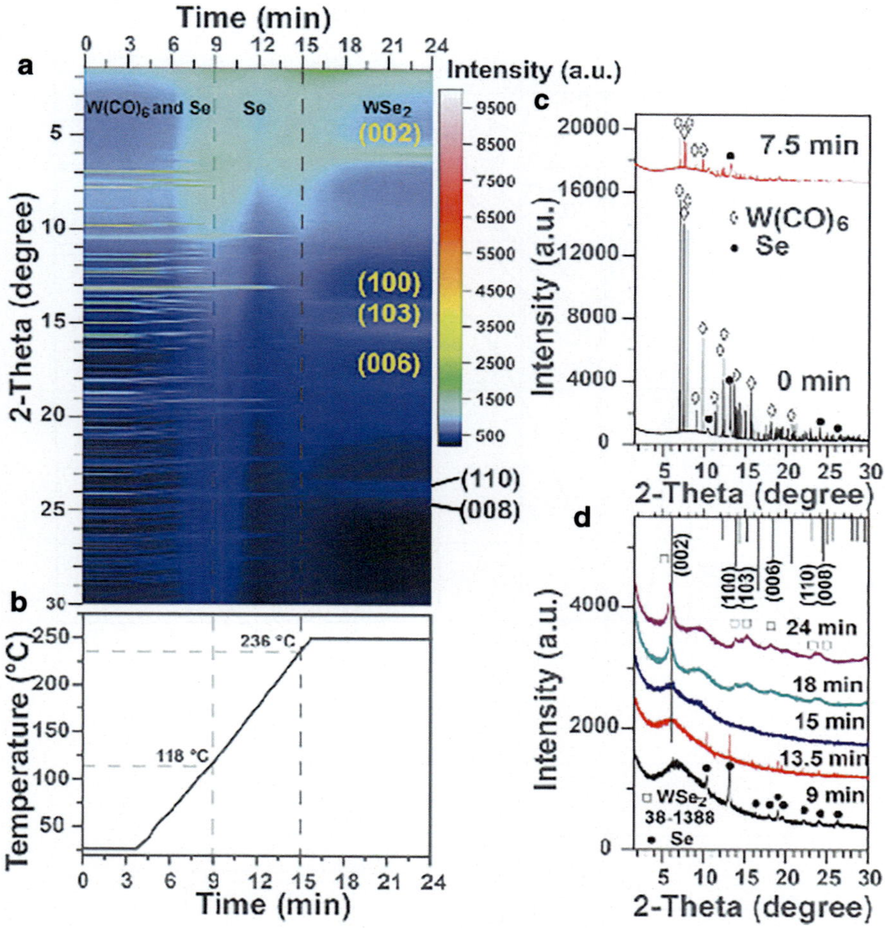

Fig. 13.2 Transformation mechanism of photon energy level diagram and dissociative photons. (**a**) Total cooling photonic energy systems as a function of internuclear distance (a.u., atomic units). Red and blue are the cooling series of excited states of cooling photon with $^1\Pi_u$ symmetry. (**b–d**) Semiclassical pathways for dissociative cooling ionization by absorption of cooling photons. (**b**) Direct cooling ionization leading to cooling-state photons. (**c**) Direct cooling photons leading to cooling energy through the lowest cooling excited states leading to cool the Earth surface. (*Source:* F. Martin et al. 2007. Science)

$$V_{\text{mo}} = -I_{\text{mo}}R_{\text{Smo}} + K_{\text{mo}} \log\left(\frac{I_{\text{Lmo}} - I_{\text{mo}} + I_{\text{Omo}}}{I_{\text{Omo}}}\right), \tag{13.5}$$

where I_{Lmo} defines the currents from the photon generation, I_{Omo} is the unsaturated current into the diode, R_{Smo} represents the series resistances, and K_{mo} represents the constant. The resistance from all non-series (NS) earth surface is being connected into the series, the total resistances expressed as $R_{\text{Smo}} = N_s \times R_s$, and the constant is

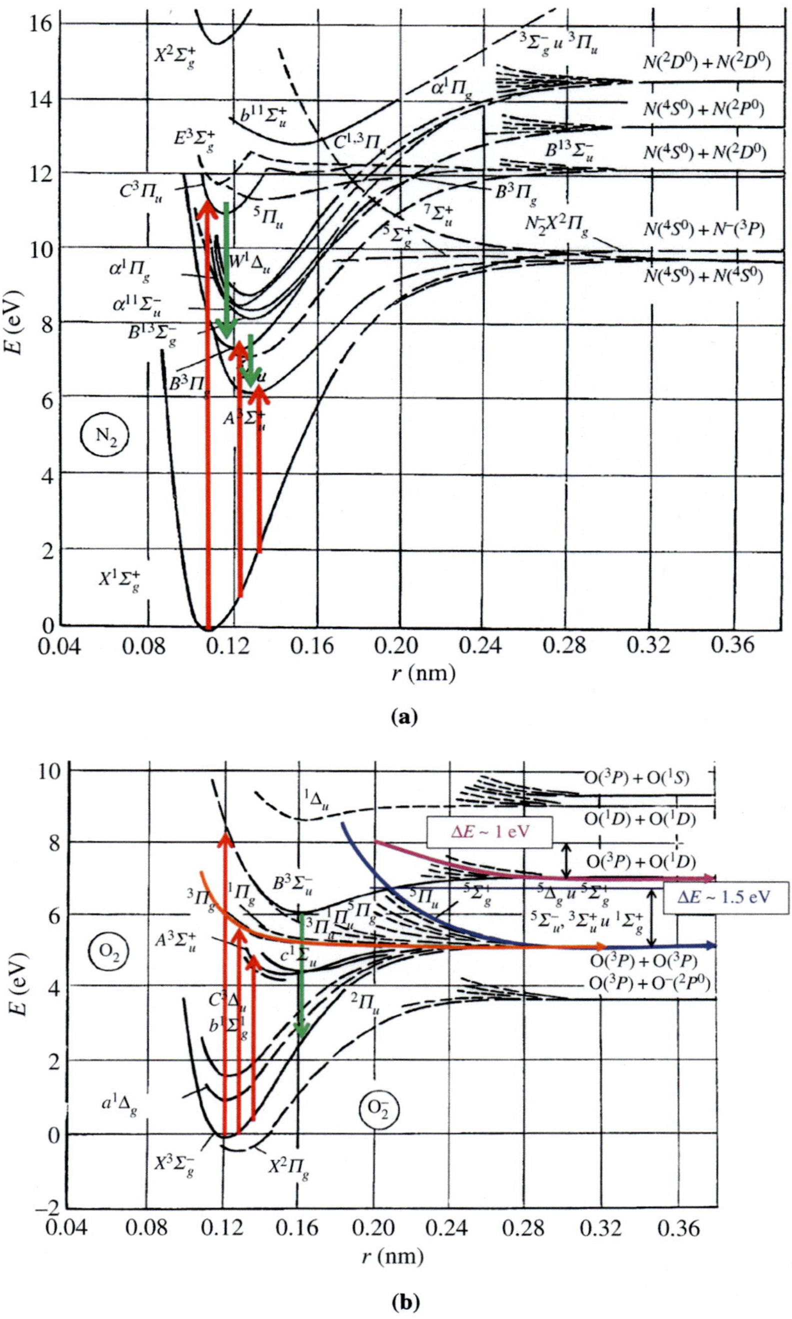

Fig. 13.3 Cooling-state photons' *I–V* characteristic of the single diode cooling energy dynamic formation: (**a**) earth surface plane in the normal state, (**b**) earth surface plane is not in the normalized condition

$K_{\text{mo}} = N_s \times K$. The current dynamics into the series connected earth surface is the same in each component, i.e., $I_{\text{Omo}} = I_o$ and $I_{\text{Lmo}} = I_L$. Thus, the I_{mo}–V_{mo} relationship in the N_s connected cells is given by

$$V_{\text{mo}} = -I_{\text{mo}} N_s R_s + N_s K \log\left(\frac{I_L - I_{\text{mo}} + I_o}{I_o}\right). \tag{13.6}$$

Accordingly, once all N_p are interconnected in parallel of earth surface, the I_{mo}–V_{mo} relationship is considered as [13, 48]

$$V_{\text{mo}} = -I_{\text{mo}} \frac{R_s}{N_p} + K \log\left(\frac{N_{\text{sh}} I_L - I_{\text{mo}} + N_p I_o}{N_p I_o}\right). \tag{13.7}$$

Just because the photon-generated cooling thermal conductivity relies mainly on photon flux and relativistic thermal state of the photon emission on earth surface, the thermal conductivity could be expressed as follows:

$$I_L = G[I_{\text{sc}} + K_I (T_{\text{cool}})] V_{\text{mo}} \tag{13.8}$$

$$T_{\text{cool}} = \left(\frac{I_L}{(G * V_{\text{mo}}) \times (I_{\text{sc}} + K_I)}\right),$$

where I_{sc} is the photonic currents per unit area at 25 °C, K_I defines the relativistic photonic co-efficient, T_{cool} is the cooling thermal of the photons, and G is the solar thermal conductivity per unit area (Fig. 13.4])

Heating Mechanism of Earth Surface

To convert cooling-state photon into heating-state photon, the local Higgs quantum field on the surface of earth plane has been used. Thus, I have simulated Abelian local symmetries by implementing Higgs boson electromagnetic field of earth to create comfortable heat on earth surface once is needed [37, 53, 55]. Simply, due to the penetration of solar light into the earth surface, it will break the gauge field symmetries of earth plane, and the Goldstone scalar particles would become the longitude modes of the vector Boson [27, 48, 60], where the local symmetries of each are spontaneously broken down to the particle T^α in respect of gauge field of $A^\alpha_\mu(x)$. These Higgs quantum fields will begin to form into local $U(1)$ phase symmetries and will create heat [28, 29]. Therefore, this mechanism can comprise the perplex scalar fields $\Phi(x)$ of electrically charged q coupled with the electromagnetic field $A^\mu(x)$, which is defined by the below Lagrangian equation:

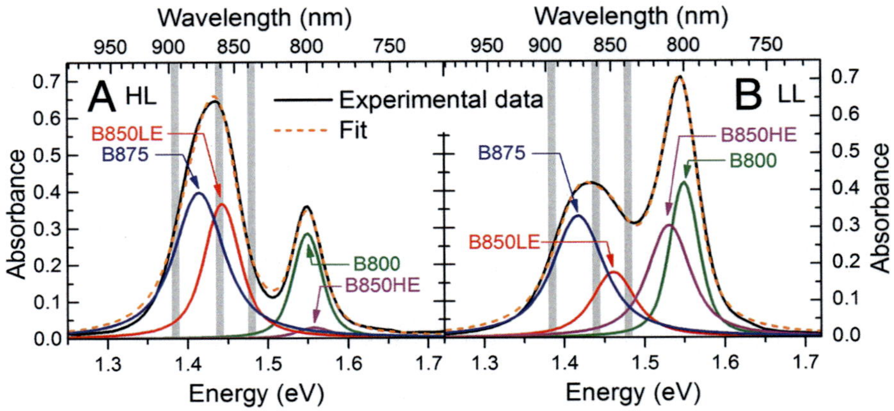

Fig. 13.4 Heating energy-level photons illustration of activation. (1) Total energy of the eV as a function of photon absorbance, (2) heating energy-level photons in relation to normal earth surface plane

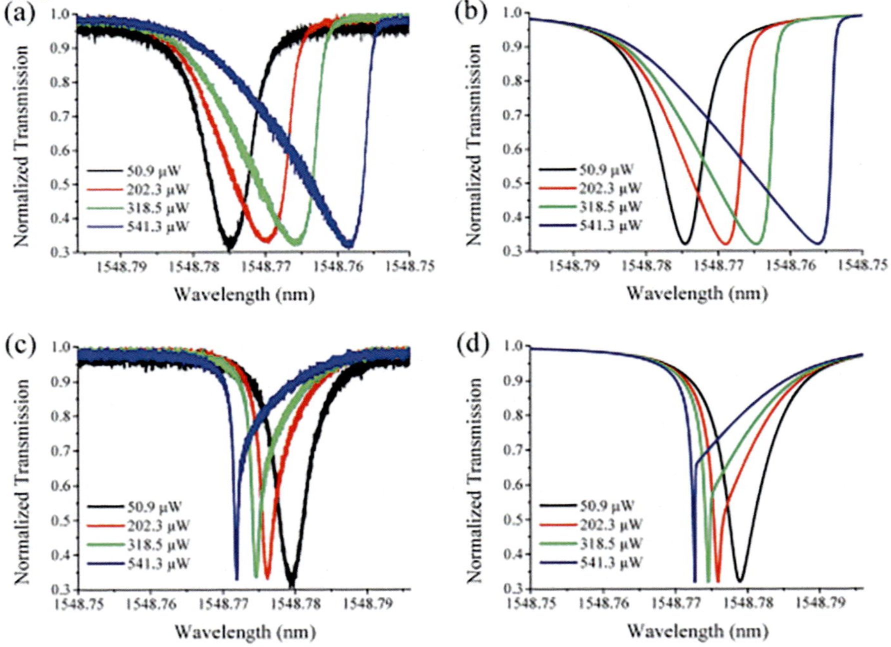

Fig. 13.4 (continued)

$$\mathcal{L} = -\frac{1}{4}F_{\mu\nu}F^{\mu\nu} + D_{\mu}\Phi^{*}D^{\mu}\Phi - V(\Phi^{*}\Phi), \tag{13.9}$$

where

$$D_\mu \Phi(x) = \partial_\mu \Phi(x) + iqA_\mu(x)\Phi(x)$$
$$D_\mu \Phi^*(x) = \partial_\mu \Phi^*(x) - iqA_\mu(\mathrm{x})\Phi^*(\mathrm{x}), \tag{13.10}$$

and

$$V(\Phi^*\Phi) = \frac{\lambda}{2}\,(\Phi^*\Phi)^2 + m^2(\Phi^*\Phi). \tag{13.11}$$

Here $\lambda > 0$ but $m^2 < 0$; therefore, $\Phi = 0$ is a local highest value of the scalar potentials, and the minimum form of deformed circle is $\Phi = \frac{v}{\sqrt{2}} * e^{i\theta}$ with respect to the following equation.

$$v = \sqrt{\frac{-2m^2}{\lambda}} \quad \text{for any real } \theta. \tag{13.12}$$

Subsequently, the scalar fields Φ will develop a non-zero expected value $\langle\Phi\rangle \neq 0$, which instinctually will create the $U(1)$ symmetries into electromagnetic field of earth. The deformation of this symmetries will thus create massless Goldstone scalars from the phase of the perplex fields $\Phi(x)$. Nevertheless, in local U-(1) symmetries, the phase of $\Phi(x)$ is sx-dependent phase of the dynamic $\Phi(x)$ fields rather than the phase of the expected value of earth surface $\langle\Phi\rangle$.

To determine this heating strategy, I have expressed the scalar field spaces with respect to the polar coordinate:

$$\Phi\,(x) = \frac{1}{\sqrt{2}}\,\Phi_\mathrm{r}(x) * e^{i\Theta(x)}, \quad \text{real } \Phi_\mathrm{r}(x) > 0, \text{real } \Phi(x). \tag{13.13}$$

Since the scalar field in this mechanism is singular at $\Phi(x) = 0$, it would be indeed applicable in theory of $\langle\Phi\rangle \neq 0$ since it is considered as the very much adequate for instinctually broken theory, where $\Phi\langle x\rangle \neq 0$ is expectation for most everywhere of the earth surface plane. Considering the real field of $\phi_\mathrm{r}(x)$ and $\Theta(x)$, the scalar potential which relies duly on the radial field ϕ_r,

$$V(\phi) = \frac{\lambda}{8}\,\left(\phi_r^2 - v^2\right)^2 + \text{const}. \tag{13.14}$$

Once the radical fields are shifted by the variable scalars, $\Phi_\mathrm{r}(x) = v + \sigma(x)$, then it can be expressed as

$$\phi_\mathrm{r}^2 - v^2 = (v+\sigma)^2 - v^2 = 2v\sigma + \sigma^2 \tag{13.15}$$

$$V = \frac{\lambda}{8}\left(2v\sigma - \sigma^2\right)^2 = \frac{\lambda v^2}{2} * \sigma^2 + \frac{\lambda v}{2} * \sigma^3 + \frac{\lambda}{8} * \sigma^4. \tag{13.16}$$

Meanwhile, the covariant derivative $D_\mu \phi$ becomes

$$\begin{aligned} D_\mu \phi &= \frac{1}{\sqrt{2}}\left(\partial_\mu\left(\phi_r e^{i\Theta}\right) + iqA_\mu * \phi_r e^{i\Theta}\right) \\ &= \frac{e^{i\Theta}}{\sqrt{2}}\left(\partial_\mu \phi_r + \phi_r * i\partial_\mu \Theta + \phi_r * iqA_\mu\right). \end{aligned} \tag{13.17}$$

$$\begin{aligned} \left|D_\mu \phi\right|^2 &= \frac{1}{2}\left|\partial_\mu \phi_r + \phi_r * i\partial_\mu \Theta + \phi_r * iqA_\mu\right|^2 \\ &= \frac{1}{2}\left(\partial_\mu \phi_r\right) + \frac{\phi_r^2}{2} * \left(\partial_\mu \Theta q A_\mu\right)^2 \\ &= \frac{1}{2}\left(\partial_\mu \sigma\right)^2 + \frac{(v+\sigma)^2}{2} * \left(\partial_\mu \Theta + qA_\mu\right)^2. \end{aligned} \tag{13.18}$$

The Lagrangian is then given by

$$\mathcal{L} = \frac{1}{2}\left(\partial_\mu \sigma\right)^2 - v(\sigma) - \frac{1}{4} F_{\mu\nu}F^{\mu\nu} + \frac{(v+\sigma)^2}{2} * \left(\partial_\mu \Theta + qA_\mu\right)^2. \tag{13.19}$$

To conduct the heat ($\mathcal{L}_{\text{heat}}$) from the magnetic field of earth surface into these Lagrangian, I have expanded $\mathcal{L}_{\text{heat}}$ as the energy series of the field into the earth and then I have extracted the quadratic parts by defying the free particles:

$$\mathcal{L}_{\text{heat}} = \frac{1}{2}\left(\partial_\mu \sigma\right)^2 - \frac{\lambda v^2}{2} * \sigma^2 - \frac{1}{4} F_{\mu\nu}F^{\mu\nu} + \frac{v^2}{2} * \left(qA_\mu + \partial_\mu \Theta\right)^2. \tag{13.20}$$

Thus, it is obviously will initiate to form comfortable heating photons into the quantum field of the earth plane, the free particle (with Lagrangian $\mathcal{L}_{\text{free}}$) shall act as the real scalar particles with positive $m^2 = \lambda v^2$ to confirm to deliver heat on the earth surface (where m represents as the particle mass; see Fig. 13.4).

Results and Discussion

Cooling Mechanism of Earth Surface

To mathematically confirm the deliberation of cool state photons by the earth surface plane, I have determined the dynamics of photonic proliferations by merging Eqs. (13.15) and (13.16). Due to the cool unit area requirement $J(\omega)$ and the steady weak coupling limits, the earth surface plane is to absorb the proliferated photons [13, 32]. Therefore, $J(\omega)$ is the quantum field area which is defined by the density of states (DOS) photon formed into the earth surface plane by the cool photon magnitudes $V(\omega)$ within the photon band (PB) on to the earth surface plane. Besides, photon deliveration follows the Weisskopf–Winger theory [6, 22, 23]. Subsequently, all the proliferated *HcP*s shall pass the dynamic state mode (1D, 2D, and 3D) in the earth surface, as shown in Table 13.1 [17, 57, 59].

In the 3D earth surface plane, Ω_C is the standard frequency cutoff which can avoid the bifurcation of the DOS. Naturally, the 1D and 2D earth surface plane conduct a sharp frequency cutoff at Ω_d to avoid negative DOS (Fig. 13.5). Here, $Li_2(x)$ and erfc(x) are the bi-logarithmic and additional variable functions. Consequently, the DOS of various earth surface plane, here named as $\varrho_{PC}(\omega)$, is confirmed by estimating the photonic energy frequencies and energy functions of Maxwell's theory in the nano-structure [30–32]. In a 1D earth surface plane, the DOS is thus provided by $\varrho_{PC}(\omega) \propto \frac{1}{\sqrt{\omega-\omega_e}}\Theta(\omega-\omega_e)$, where $\Theta(\omega-\omega_e)$ where ω_e is denoted as the frequency of the PBE at the used DOS.

Then, the DOS has been used to precisely determine the qualitative states of the non-Weisskopf–Winger modes and the cool-state photons calculated by the 3D isotropic classification of the earth surface plane. The DOSs and expected DOSs (PDOS) have been shown in Fig. 13.6. In the 3D earth surface plane, the DOS is closer to the PBE and it is determined by $\varrho_{PC}(\omega) \propto \frac{1}{\sqrt{\omega-\omega_e}}\Theta(\omega-\omega_e)$. This DOS is thereafter determined considering the electromagnetic field vectors on earth surface [30–32]. In 2D and 1D earth surface plane, the cool photon DOS shows a real logarithmic divergence close to the PBE, which is $\varrho_{PC}(\omega) \propto -[\ln|(\omega-\omega_0)/\omega_0| - 1]\Theta(\omega-\omega_e)$, where ω_e defines the primary point of the peak of the DOS distribution.

Therefore, the density of state (DOS) and the expected density of state (EDOS) of deformed photon to convert into the cool state will confirm (1) the total reproduction of DOS (T) at orbitals, (2) EDOS of the fourth level of earth surface quantum atoms, and (3) EDOS of the earth surface quantum (Fig. 13.7). As defined above, $J(\omega)$ denotes the DOS field produced in the earth surface plane by the standard cooling photon magnitudes $V(\omega)$ in the PB and the earth surface plane [25, 34, 35]:

$$J(\omega) = \varrho(\omega)|V(\omega)|^2. \tag{13.21}$$

Therefore, I have considered, here, the PB as the frequency of ω_c and the proliferative cooling photon dynamic is $\langle a(t)\rangle = u(t,t_0)\langle a(t_0)\rangle$, where the function $u(t,t_0)$ describes the photon structure. $u(t,t_0)$ is determined by clarifying the integral–differential equation given in Eq. (13.18):

Table 13.1 Photon structure of the densities of states (DOS) in various dimension modes of the earth surface plane. The unit areas $J(\omega)$ and the proliferation of self-energy inductions into the earth surface reservoir $\Sigma(\omega)$, demonstrated by the photonic dynamic into the standard relativistic earth surface plane, vary among the photonic structures. The variables C, η, and χ function as coupled forces between the point break and earth surface plane in 1, 2, and 3 dimensions [8, 54, 58]

Photons	Unit area $J(\omega)$ for different DOS*	Reservoir-induced self-energy correction $\Sigma(\omega)$*
1D	$\frac{C}{\pi}\frac{1}{\sqrt{\omega-\omega_e}}\Theta(\omega-\omega_e)$	$-\frac{C}{\sqrt{\omega_e-\omega}}$
2D	$-\eta\left[\ln\left\lvert\frac{\omega-\omega_0}{\omega_0}\right\rvert-1\right]\Theta(\omega-\omega_e)\Theta(\Omega_d-\omega)$	$\eta\left[\mathrm{Li}_2\left(\frac{\Omega_d-\omega_0}{\omega-\omega_0}\right)-\mathrm{Li}_2\left(\frac{\omega_0-\omega_e}{\omega_0-\omega}\right)-\ln\frac{\omega_0-\omega_e}{\Omega_d-\omega_0}\ln\frac{\omega_e-\omega}{\omega_0-\omega}\right]$
3D	$\chi\sqrt{\frac{\omega-\omega_e}{\Omega_C}}\exp\left(-\frac{\omega-\omega_e}{\Omega_C}\right)\Theta(\omega-\omega_e)$	$\chi\left[\pi\sqrt{\frac{\omega_e-\omega}{\Omega_C}}\exp\left(-\frac{\omega-\omega_e}{\Omega_C}\right)\mathrm{erfc}\sqrt{\frac{\omega_e-\omega}{\Omega_C}}-\sqrt{\pi}\right]$

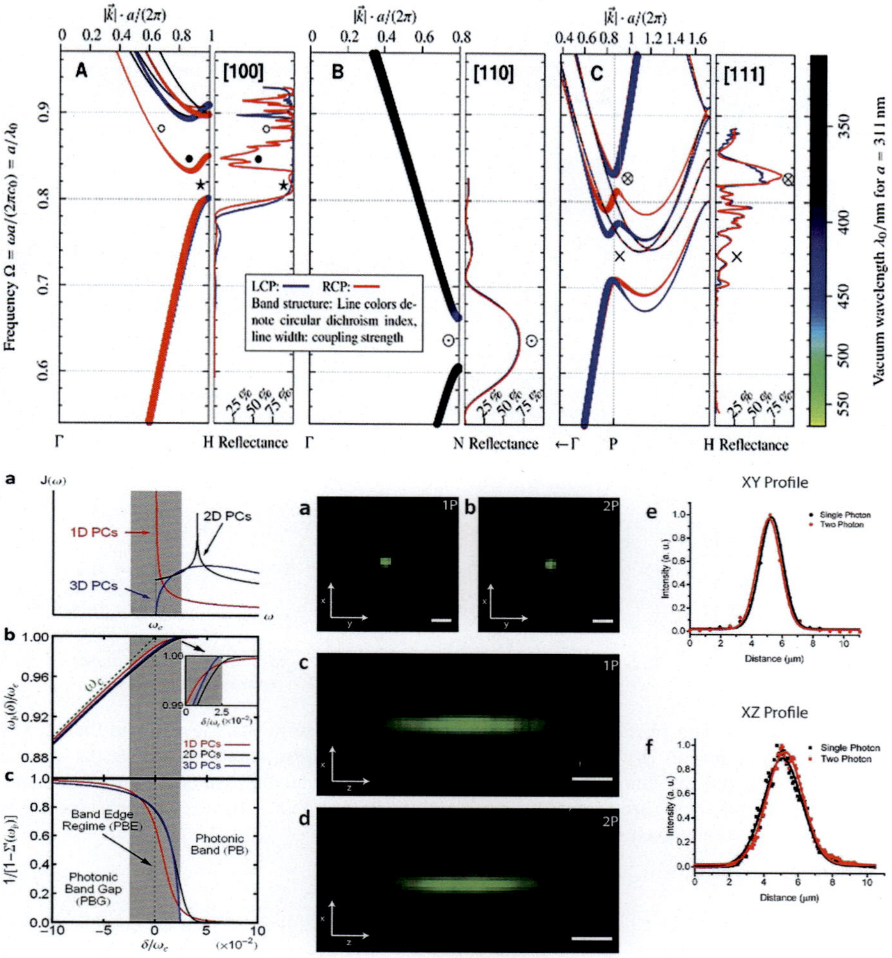

Fig. 13.5 (1) The photon band structures and photon transformation mode. (2) (a) Unit areas vs frequencies at different DOSs in 1D, 2D, and 3D earth surface plane. (b) The photon modes of frequency as the function tune. (c) The photon modes of magnitude to deliver the thermal energy (2). Photon modes show the earth surface plane

$$u(t, t_0) = \frac{1}{1 - \Sigma'(\omega_b)} e^{-i\omega(t-t_0)} + \int_{\omega_e}^{\infty} d\omega \frac{J(\omega)e^{-i\omega(t-t_0)}}{[\omega - \omega_c - \Delta(\omega)]^2 + \pi^2 J^2(\omega)}, \quad (13.22)$$

where $\Sigma'(\omega_b) = [\partial\Sigma(\omega)/\partial\omega]_{\omega=\omega_b}$ and $\Sigma(\omega)$ defines the PB photonic self-energy corrections inducted into the reservoir,

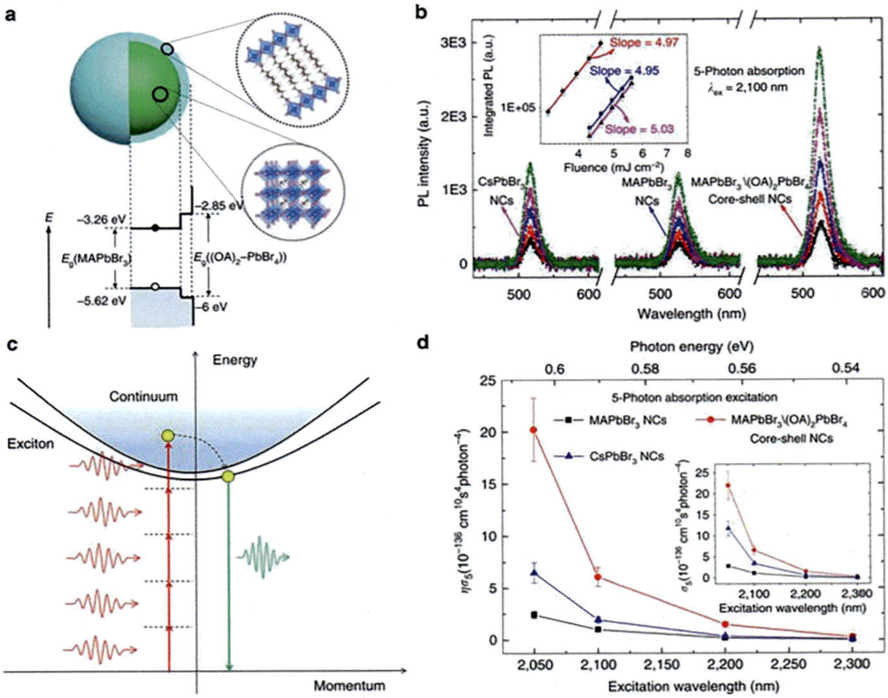

Fig. 13.6 (**a**) Cooling photons core–shell multidimensional energy state levels, and their type-I energy level alignment. (**b**) The photonic spectra in femtosecond excitation shows the quantic dependence on the excitation fluence of the spectrally integrated PL intensity. (**c**) Schematic illustrating the photonic spectral process. (**d**) Cooling photon action cross-section ($\eta\sigma_5$) spectra in relation to excitation wavelength

$$\Sigma(\omega) = \int_{\omega_e}^{\infty} d\omega' \frac{J(\omega')}{\omega - \omega'}. \tag{13.23}$$

Here, the frequency ω_b in Eq. (13.2) represents the cooling photonic frequency mode in the PBG ($0 < \omega_b < \omega_e$), calculated under the pole condition $\omega_b - \omega_c - \Delta(\omega_b) = 0$, where $\lesssim\Delta(\omega) = \mathcal{P}\left[\int d\omega' \frac{J(\omega')}{\omega - \omega'}\right]$ is a primary-value integral.

In detail, it can be explained that PB is referred as the Fock cooling determination n_0, i.e., $\rho(t_0) = |n_0\rangle\langle n_0|$, which is obtained mathematically from the real-time quantum field [54–56] by implementing the Eq. (13.21) to consider the cooling-state photons induction at time t:

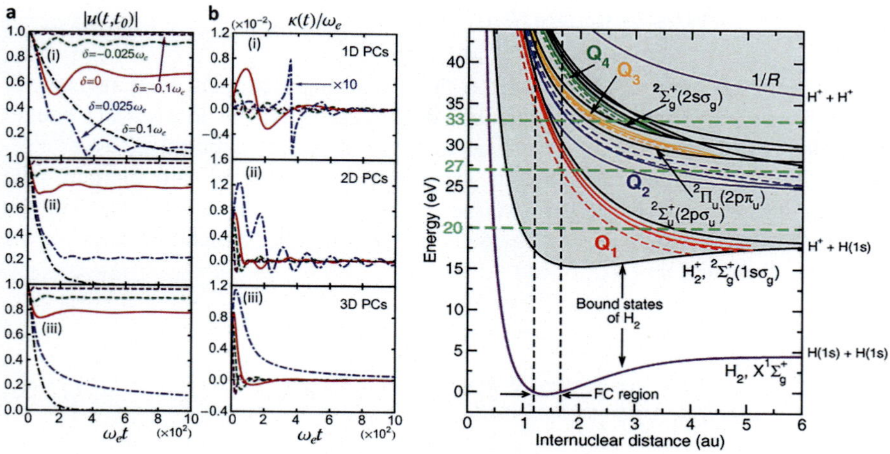

Fig. 13.7 (**a** Shows the plot for the cool photon dynamics $|u(t, t_0)|$, determined in 1D, 2D, and 3D photons for different values of the different parameters δ and accumulated from the PBG area [24, 33, 45]. The cooling photon dynamical rates $\kappa(t)$ are being plotted in Fig. 13.8b. The result shows that ω_c has crossed from the PBG to the PB area in order to produce the dynamic at a high rate. Because the $u(t, t_0)$ range is $1 \geq |u(t, t_0)| \geq 0$, it has been clarified as the crossover area to satisfy $0.9 \gtrsim |u(t \to \infty, t_0)| \geq 0$. This refers to $-0.025\omega_e \lesssim \delta \lesssim 0.025\omega_e$, with a cool photon inducing rate $\kappa(t)$ within the PBG ($\delta < -0.025\omega_e$) and near the PBE ($-0.025\omega_e \lesssim \delta \lesssim 0.025\omega_e$)

$$\rho(t) = \sum_{n=0}^{\infty} \mathcal{P}_n^{(n_0)}(t)|n_0\rangle\langle n_0| \tag{13.24}$$

$$\mathcal{P}_n^{(n_0)}(t) = \frac{[v(t,t)]^n}{[1+v(t,t)]^{n+1}}[1-\Omega(t)]^{n_0} \times \sum_{k=0}^{\min\{n_0,n\}} \binom{n_0}{k} \times \binom{n}{k}\left[\frac{1}{v(t,t)}\frac{\Omega(t)}{1-\Omega(t)}\right]^k, \tag{13.25}$$

where $\Omega(t) = \frac{|u(t,t_0)|^2}{1+v(t,t)}$. These results revealed that the Fock state cooling photons are indeed induced into dynamic states of earth plane $\mathcal{P}_n^{(n_0)}(t)$ of $|n_0\rangle$. Fig. 13.7 shows the plots of the proliferation of photon deliberation $\mathcal{P}_n^{(n_0)}(t)$ in the prime state $|n_0 = 5\rangle$ and in the steady-state limits $\mathcal{P}_n^{(n_0)}(t \to \infty)$. Thus, the deliberation of the generated cool photons shall eventually reach a non-equilibrium cooling state which will cool the earth surface finally.

Heating Mechanism of Earth Surface

In the proposed theory, I have utilized the electromagnetic field of Earth surface being formed by Higgs boson quantum dynamics. Here, the locals U(1) gauge-variable QED shall allow another mass in terms of gauge particles $\varnothing' \rightarrow e^{i\alpha(x)}\varnothing$, which are the cooling-state photons that could be converted into thermal state photons in order to heat the earth surface at a comfort level. This mechanism can be tested by the variable derivatives with the specific transformational rules for the scalar field, written by [20, 36, 61]

$$\begin{aligned} \partial_\mu \rightarrow D_\mu = \partial_\mu = ieA_\mu \quad &[\text{covariant derivatives}], \\ A'_\mu = A_\mu + \frac{1}{e}\,\partial_\mu\alpha \quad &[A_\mu \text{ derivatives}], \end{aligned} \tag{13.26}$$

Here, the local $U(1)$ gauge-invariant Lagrangian is considered as the perplex scalar field that is expressed by

$$\mathcal{L} = (D^\mu)^\dagger\,(D_\mu\varnothing) - \frac{1}{4}\,F_{\mu\nu}F^{\mu\nu} - V(\varnothing). \tag{13.27}$$

The term $\frac{1}{4}\,F_{\mu\nu}F^{\mu\nu}$ is the kinetic mechanism in the gauge field for considering thermal photons and $V(\varnothing)$ denotes the additional term written as $V(\varnothing^*\varnothing) = \mu^2(\varnothing^*\varnothing) + \lambda\,(\varnothing^*\varnothing)^2$.

As per equation of the Lagrangian $\mathcal{L}$, the perturbations of the quantum field initiate the production of massive scalar particle ϕ_1 and ϕ_2 and a mass μ. Here, $\mu^2 < 0$ shall be admitted as an infinite number of quanta in order to satisfy the equation $\phi_1^2 + \phi_2^2 = -\mu^2/\lambda = v^2$. In terms of the shifted fields η and ξ, the quantum field is clarified as $\phi_0 = \frac{1}{\sqrt{2}}\,[(v+\eta) + i\xi]$, and the variable derivative of the Lagrangian then confirm the

$$\begin{aligned} \text{Kinetic term :} \quad \mathcal{L}_{\text{kin}}(\eta,\xi) &= (D^\mu\phi)^\dagger(D^\mu\phi) \\ &= (\partial^\mu + ieA^\mu)\phi^*\,\left(\partial_\mu - ieA_\mu\right)\phi \end{aligned} \tag{13.28}$$

Final term to the second order: $V(\eta,\xi) = \lambda\,v^2\eta^2$. Thus the full Lagrangian can be written as

$$\begin{aligned} \mathcal{L}_{\text{kin}}(\eta,\xi) = {} & \frac{1}{2}\,\left(\partial_\mu\eta\right)^2 - \lambda v^2\eta^2 + \frac{1}{2}\,\left(\partial_\mu\xi\right)^2 - \frac{1}{4}\,F_{\mu\nu}F^{\mu\nu} + \frac{1}{2}\,e^2v^2A_\mu^2 \\ & - evA_\mu(\partial^\mu\xi) + \text{int.terms}. \end{aligned} \tag{13.29}$$

Here, η is massive, ξ is massless (as before), μ is the mass for the quanta, and A_μ is defined up to a term $\partial_\mu\alpha$, as is the evidence for Eq. (13.27). Naturally, A_μ and ϕ can be changed spontaneously, so Eq. (13.28) could be rewritten to confirm the deliberation of the thermal photon particles spectrum within the quantum field of earth surface:

$$\mathcal{L}_{\text{scalar}} = (D^{\mu}\phi)^{\dagger}(D^{\mu}\phi) - V(\phi^{\dagger}\phi)$$

$$= (\partial^{\mu} + ieA^{\mu})\frac{1}{\sqrt{2}}(v+h)\left(\partial_{\mu} - ieA_{\mu}\right)\frac{1}{\sqrt{2}}(v+h) - V\left(\phi^{\dagger}\phi\right) \tag{13.30}$$

$$= \frac{1}{2}\left(\partial_{\mu}h\right)^{2} + \frac{1}{2}e^{2}A_{\mu}^{2}(v+h)^{2} - \lambda v^{2}h^{2} - \lambda v h^{3} - \frac{1}{4}\lambda h^{4} + \frac{1}{4}\lambda h^{4}. \tag{13.31}$$

The redeveloped term of the Lagrangian of the scalar field thus revealed which is the Higgs Boson quantum field that could be initiated to produce thermal photons on to the earth surface plane.

To determine this heating photon conversion into the earth surface plane, I further calculated the isotropic distributions of kinetic on the differential cones with respect to the angle θ from the vertical axis. The differential between θ and $\theta + d\theta$ is $\frac{1}{2}\sin\theta d\theta$. The differential photon density at energy $\in$ and angle θ is then given by

$$dn = \frac{1}{2}n(\in)\sin\theta d \in d\theta \tag{13.32}$$

Naturally, the speed of function of the high-energy photons is being estimated as $c(1 - \cos\theta)$ with respect to the absorption of photon per unit area considering the following equations

$$\frac{d\tau_{\text{abs}}}{dx} = \int\int \frac{1}{2}\sigma n\,(\in)(1 - \cos\theta)\sin\theta d \in d\theta. \tag{13.33}$$

Rewriting these variables as integral over s instead of θ, by (13.31) and (13.33), I have determined

$$\frac{d\tau_{\text{abs}}}{dx} = \pi r_0^2\left(\frac{m^2c^4}{E}\right)^2 \int_{\frac{m^2c^4}{E}}^{\infty} \in^{-2} n(\in)\,\overline{\varphi}[s_0(\in)]de, \tag{13.34}$$

where

$$\overline{\varphi}[s_0(\in)] = \int_{1}^{s_0(\in)} s\overline{\sigma}\,(s)ds, \quad \overline{\sigma}(s) = \frac{2\sigma(s)}{\pi r_0^2}. \tag{13.35}$$

This result defines the dimensional variable $\overline{\varphi}$ and dimensionless cross section $\overline{\sigma}$. The variable $\overline{\varphi}\,[s_0]$ is calculated based on a detailed graphical frame for $1 < s_0 < 10$. I calculated $\overline{\varphi}$ by a functional asymptotic calculation

$$\overline{\varphi}[s_0] = \frac{1+\beta_0^2}{1-\beta_0^2}\ln\omega_0 - \beta_0^2\ln\omega_0 - \ln^2\omega_0 - \frac{4\beta_0}{1-\beta_0^2} + 2\beta_0 + 4\ \ln\omega_0\ln\ (\omega_0+1)$$
$$- L(\omega_0),$$

where $s_O - 1 \ll 1$ *or* $s_O \gg 1$,

$$\beta_0^2 = \frac{1-1}{s_0}, \quad \omega_0 = \frac{(1+\beta_0)}{(1-\beta_0)}, \quad \text{and } L(\omega_0) = \int_1^{\omega_0} \omega^{-1}\ln(\omega+1)d\omega. \tag{13.36}$$

The last integral can be written as

$$(\omega+1) = \omega\left(\frac{1+1}{\omega}\right), \quad L(\omega_0) = \frac{1}{2}\ln^2\omega_0 + L'(\omega_0),$$

where

$$\begin{aligned} L'(\omega_0) &= \int_1^{\omega_0} \omega^{-1}\ln\left(1+\frac{1}{\omega}\right)d\omega, \\ &= \frac{\pi^2}{12} - \sum_{n=1}^{\infty}(-1)n^{-1}n^{-2}\omega_0^{-n} \end{aligned} \tag{13.37}$$

This confirms presentation of the heating photons deliberation readily shall allow the calculation of $\overline{\varphi}[s_0]$ to determine the accurate value of heating photon production s_0. Therefore, the correct functional asymptotic formula is expressed as follows to confirm the heating photon production:

$$\begin{aligned} \overline{\varphi}[s_0] &= 2s_0(\ln 4s_0 - 2) + \ln 4s_0(\ln 4s_0 - 2) - \frac{(\pi^2-9)}{3} \\ &+ s_0^{-1}\left(\ln 4s_0 + \frac{9}{8}\right) + \cdots \quad (s_0 \gg 1), \end{aligned} \tag{13.38}$$

$$\begin{aligned} \overline{\varphi}[s_0] &= \left(\frac{2}{3}\right)(s_0-1)^{\frac{3}{2}} + \left(\frac{5}{3}\right)(s_0-1)^{\frac{5}{2}} - \left(\frac{1507}{420}\right)(s_0-1)^{\frac{7}{2}} \\ &+ \cdots \quad (s_0 - 1 \ll 1). \end{aligned} \tag{13.39}$$

The function $\frac{\overline{\varphi}[s_0]}{(s_0-1)}$ is revealed in Fig. 13.5 for $1 < s_0 < 10$; at larger s_0, it defines a normal logarithm function of s_0. The energy-law spectra of the thermal photon is thus written in the form $n(\in) \infty \in_m$ for two systems in pristine states and for a system with photonic band heating in an earth surface plane.

Therefore, the solar irradiance absorption spectrum shall confirm a peak-energy cutoff with $m > 0$, which is the derived form of the thermal photon spectra at a peak-energy cutoff.

Thus, the formation of the spectrum is clarified as follows:

$$n(\in) = D\in^{\beta}, \ \in < \in_m, \quad \beta \leq 0 \tag{13.40}$$

$$= \ 0, \ \in > \in_m \tag{13.41}$$

Thus, I have determined

$$\frac{d\tau_{\text{abs}}}{dx} = \pi r_0^2 D\left(\frac{m^2c^4}{E}\right)^{1+\beta} \times \begin{cases} 0, & E < E_m, \\ F_\beta(\sigma_m), & E > E_m, \end{cases} \tag{13.42}$$

where

$$\sigma_m = \frac{E}{E_m} = \frac{\in_m E}{m^2c^4}, \tag{13.43}$$

$$F_\beta(\sigma_m) = \int_1^{\sigma_m} s_0^{\beta-2}\overline{\varphi}[s_0]ds_0. \tag{13.44}$$

Again, by Eqs. (13.40) and (13.41), I have obtained the asymptotic forms

$$\begin{aligned} \beta = 0: &\quad F_\beta(\sigma_m) \to A_\beta + \ln^2\sigma_m - 4\ln\sigma_m + \cdots, \\ \beta \neq 0: &\quad F_\beta(\sigma_m) \to A_\beta + 2\beta^{-1}\sigma_m^\beta\left(\ln 4\sigma_m - \beta^{-1} - 2\right) + \cdots, \quad \sigma_m > 10 \end{aligned} \tag{13.45}$$

$$\text{All } \beta: \quad F_\beta(\sigma_m) \to \left(\frac{4}{15}\right)(\sigma_m - 1)^{\frac{5}{2}} + \left[\frac{2(2\beta+1)}{21}\right](\sigma_m - 1)^{\frac{7}{2}} + \cdots, \quad \sigma_m - 1 \ll 1. \tag{13.46}$$

Figure 13.8 plots $\sigma_m^{-\beta}F_\beta(\sigma_m)$ for $\beta = 0$–3.0 A_β in $0.5 - A_\beta$ intervals that determine the integral of the region of plane [2, 43]. The value is clarified as $A_\beta = 8.111$ ($\beta = 0$), 13.53 ($\beta = 0.5$), 9.489 ($\beta = 1.0$), 15.675 ($\beta = 1.5$), 34.54 ($\beta = 2.0$), 85.29 ($\beta = 2.5$), and 222.9 ($\beta = 3.0$). Consequently, I have calculated the thermal photons in terms of photonic spectra for both negative and positive indexes:

$$\begin{aligned} n(\in) &= 0, \quad \in < \in_0 \\ &= C_\in^{-\alpha} \ \text{ or } \ D_\in^\beta, \ \in_0 < \in < \in_m \\ &= 0, \quad \in > \in_m. \end{aligned} \tag{13.47}$$

I then obtain

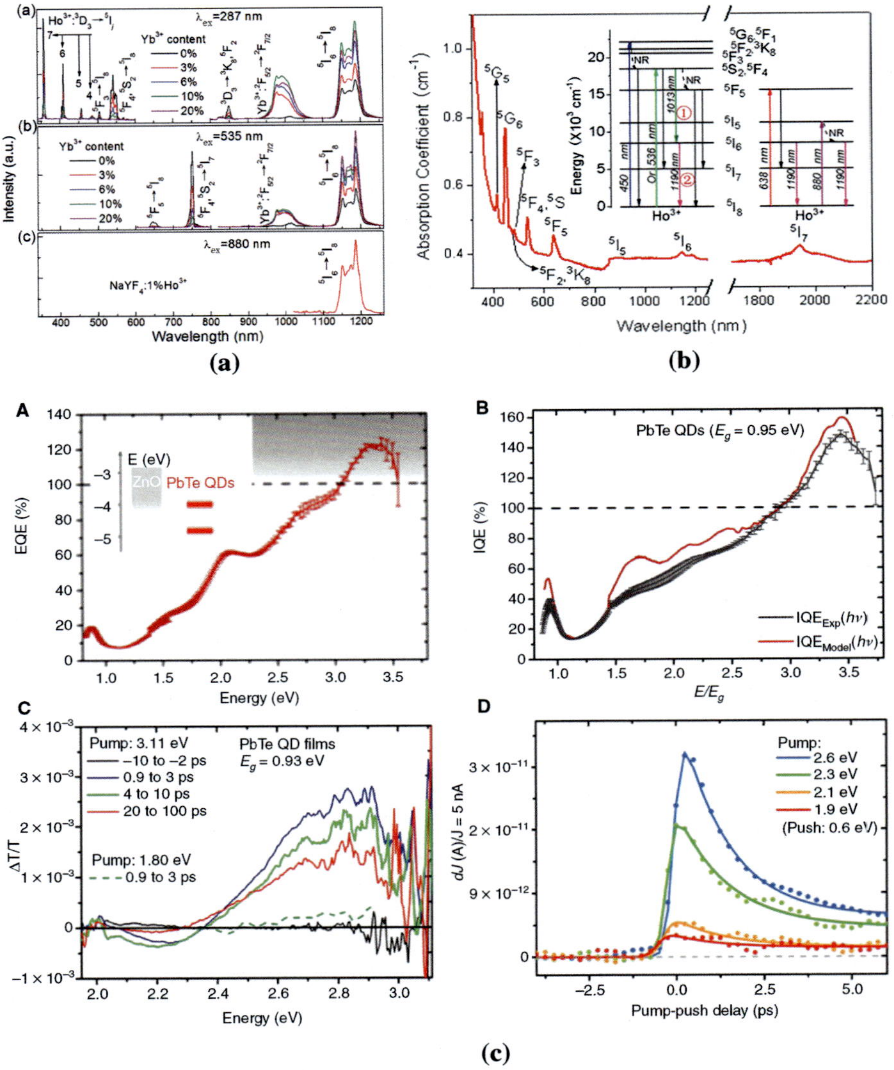

Fig. 13.8 Shows the photon receptor where electron is being heated in the surface of earth considering (1) transforming the curve step, (2) reflection symmetry step for the determination of magnetic-field-induced heating photon production. (3) Shows the functional coincidence rate of heating photon production considering the detuning parameters into the band structure of earth plane [16, 37, 41]

$$\left(\frac{d\tau_{\text{abs}}}{dx}\right)_{\alpha} = \pi r_0^2 C\left(\frac{m^2c^4}{E}\right)^{1-\alpha} \times \begin{cases} 0, & E < E_m, \\ [F_{\alpha}(1) - F_{\alpha}(\sigma_m)], & E_m < E < E_0, \\ [F_{\alpha}(\sigma_0) - F_{\alpha}(\sigma_m)], & E > E_0, \end{cases} \tag{13.48}$$

$$\left(\frac{d\tau_{\text{abs}}}{dx}\right)_{\beta} = \pi r_0^2 D\left(\frac{m^2c^4}{E}\right)^{1+\beta} \times \begin{cases} 0, & E < E_m, \\ [F_{\beta}(\sigma_m)], & E_m < E < E_0, \\ [F_{\beta}(\sigma_m) - F_{\beta}(\sigma_0)], & E > E_0, \end{cases} \tag{13.49}$$

In these functional variables, the heating photon spectra on the earth surface plane can be properly defined by asymptotic formula. The term $\Gamma_{\gamma}^{\text{LPM}}$ defines the photonic obligation to the irradiated light per unit area [4, 11, 21]:

$$\Gamma_{\gamma} \equiv \frac{dn_{\gamma}}{dVdt}. \tag{13.50}$$

Adding all the contributions $\Gamma_{\gamma}^{\text{LPM}}$, the rate of thermal photon production is being confirmed as $O(\alpha_{\text{EM}}\alpha_{\text{s}})$. Here, it has been calculated by the following equation for the polarized emitted rate $\Gamma_{\gamma}^{\text{LPM}}$ at the thermodynamically controlled equilibrium of the plasma surface at temperature T and photophysical reaction μ:

$$\frac{d\Gamma_{\gamma}^{\text{LPM}}}{d^3k} = \frac{d_{\text{F}}q_{\text{s}}^2\alpha_{EM}}{4\pi^2k}\int_{-\infty}^{\infty}\frac{dp_{\parallel}}{2\pi}\int\frac{d^2p_{\perp}}{(2\pi)^2}A\left(p_{\parallel},k\right)\ \text{Re}\left\{2p_{\perp}\cdot f\left(p_{\perp};p_{\parallel},k\right)\right\}, \tag{13.51}$$

where d_{F} presents the variable strategy of the photon particles [N_{c} in SU(N_{c})], q_{s} presents the Abelian charge of the photonic quark, $k \equiv |k|$, and the kinetic functional mode A($p_{\parallel}$, k) are being irradiated particles which are expressed by

$$A\left(p_{\parallel},k\right) \equiv \begin{cases} \dfrac{n_{\text{b}}\left(k+p_{\parallel}\right)\left[1+n_{\text{b}}\left(p_{\parallel}\right)\right]}{2p_{\parallel}\left(p_{\parallel}+k\right)}, & \text{scalars} \\ \dfrac{n_{\text{f}}\left(k+p_{\parallel}\right)\left[1-n_{\text{f}}\left(p_{\parallel}\right)\right]}{2\left[p_{\parallel}\left(p_{\parallel}+k\right)\right]^2}\left[{p_{\parallel}}^2+\left(p_{\parallel}+k\right)^2\right], & \text{fermions} \end{cases} \tag{13.52}$$

with

$$n_{\mathrm{b}}(p) \equiv \frac{1}{\exp\left[\beta(p-\mu)\right]-1}, \quad n_{\mathrm{f}}(p) \equiv \frac{1}{\exp\left[\beta(p-\mu)\right]+1}. \tag{13.53}$$

The function $f(p_{\perp}; p_{\|}, k)$ in Eq. (13.51) is integrated to resolve the below equation that suggested that the heating photon production by earth surface is feasible [21–23]:

$$\begin{aligned} 2p_{\perp} &= i\delta E f\left(p_{\perp};p_{\|},k\right) + \frac{\pi}{2} C_{\mathrm{F}} g_{\mathrm{s}}^{2} m_{\mathrm{D}}^{2} \int \frac{d^{2}q_{\perp}}{(2\pi)^{2}} \frac{dq_{\|}}{2\pi} \frac{dq^{0}}{2\pi} 2\pi\delta\left(q^{0}-q_{\|}\right) \\ &\times \frac{T}{|q|}\left[\frac{2}{\left|q^{2}-\Pi_{\mathrm{L}}(Q)\right|^{2}} + \frac{\left[1-\left(q^{0}/\left|q_{\|}\right|\right)^{2}\right]^{2}}{\left|(q^{0})^{2}-q^{2}-\Pi_{\mathrm{T}}(Q)\right|^{2}}\right] \\ &\times \left[f\left(p_{\perp};p_{\|},k\right) - f\left(q+p_{\perp};p_{\|},k\right)\right] \end{aligned} \tag{13.54}$$

In Eq. (13.54), C_{F} represents the quark [$C_{\mathrm{F}} = (N_{\mathrm{c}}^{2} - 1)/2N_{\mathrm{c}} = 4/3$ in QCD], m_{D} represents the lead-order Debye mass, and δE is considered as the energy variable among quasi-particles, which determines the photonic emissions and the state of the thermodynamically temperature equilibrium:

$$\delta E \equiv k^{0} + E_{p}\operatorname{sign}\left(p_{\|}\right) - E_{p+k}\operatorname{sign}\left(p_{\|}+k\right). \tag{13.55}$$

For an SU(N) gauge model with N_{s} complex scalars and N_{f} Dirac fermions, the Debye mass in the fundamental here is thus represented by the following equation [32]

$$m_{\mathrm{D}}^{2} = \frac{1}{6}(2N + N_{\mathrm{s}} + N_{\mathrm{f}})g^{2}T^{2} + \frac{N_{\mathrm{f}}}{2\pi^{2}} g^{2}\mu^{2}. \tag{13.56}$$

In order to conduct accurate calculation further, the heating photon energy emission rate is $p_{\|} > 0$; I therefore matriculated the distributions of $n(k + p_{\|})[1 \pm n(p_{\|})]$ in the integral which contains $A(p_{\|}, k)$ in Eq. (13.51), which confirms the distribution of heating photon production and it is expressed by the following equation:

$$n_{\mathrm{b}}(-p) = -[1 + \bar{n}_{\mathrm{b}}(p)], \qquad n_{\mathrm{f}}(-p) = [1 - \bar{n}_{\mathrm{f}}(p)], \tag{13.57}$$

where $n(p) \equiv 1/[e\beta(p + \mu) \mp 1]$ is the definite anti-particle functions; thus, the variable $A(p'', k)$ in this interval can be expressed by

$$A\left(p_{\|}, k\right) \equiv \begin{cases} \dfrac{n_{\mathrm{b}}\left(k-\left|p_{\|}\right|\right) \bar{n}_{\mathrm{b}}\left(\left|p_{\|}\right|\right)}{2\left|p_{\|}\right|\left(k-\left|p_{\|}\right|\right)}, & \text{scalars,} \\ \dfrac{n_{\mathrm{f}}\left(k-\left|p_{\|}\right|\right) \bar{n}_{\mathrm{f}}\left(\left|p_{\|}\right|\right)}{2\left[\left|p_{\|}\right|\left(k-\left|p_{\|}\right|\right)\right]^{2}}\left[p_{\|}{}^{2}+\left(k-\left|p_{\|}\right|\right)^{2}\right], & \text{fermions.} \end{cases} \tag{13.58}$$

Thus, the energy E_p of a hard quark with momentum $|p|$ is explicitly given by

$$E_p = \sqrt{p^2 + m_\infty^2} \simeq |p| + \frac{m_\infty^2}{2|p|} \simeq \left|p_{\|}\right| + \frac{p_\perp^2 + m_\infty^2}{2\left|p_{\|}\right|}, \tag{13.59}$$

Here the asymptotic thermal "mass" is

$$m_\infty^2 = \frac{C_{\mathrm{f}} g^2 T^2}{4}. \tag{13.60}$$

Substituting the explicit form of E_p into definition (13.60), it is thus expressed

$$\delta E = \left[\frac{p_\perp^2 + m_\infty^2}{2}\right]\left[\frac{k}{p_{\|}\left(k + p_{\|}\right)}\right]. \tag{13.61}$$

In the above, it has been derived from the explicit forms of Eqs. (13.52) and (13.55). Figure 13.8 plots shows the leading-order heating photoemission rates that confirms the maximize the heating photon generation into the given the electron time-of-flight of the heating-state photons into the earth surface plane to heat the earth naturally at a comfort level (Figs. 13.8 and 13.9).

Conclusions

The tremendous fluctuation of global temperature on Earth surface throughout out the year is indeed problematic as they contribute to cause adverse thermal dynamics on Earth surface. Consequently, it creates uncomfortable condition for all living being to live on it comfortably. Interestingly, this research revealed a mechanism to reform solar photons into cool state photons (*HcP*$^-$) by integrating the Bose–Einstein (*B–E*) photonic distribution theory to cool the earth surface naturally during the hot season. This cooling-state photon could also be reconverted into the thermal state photons (*HtP*$^-$) by the application of Higgs Boson [BR ($H \rightarrow \gamma\gamma^-$)] quantum fields of

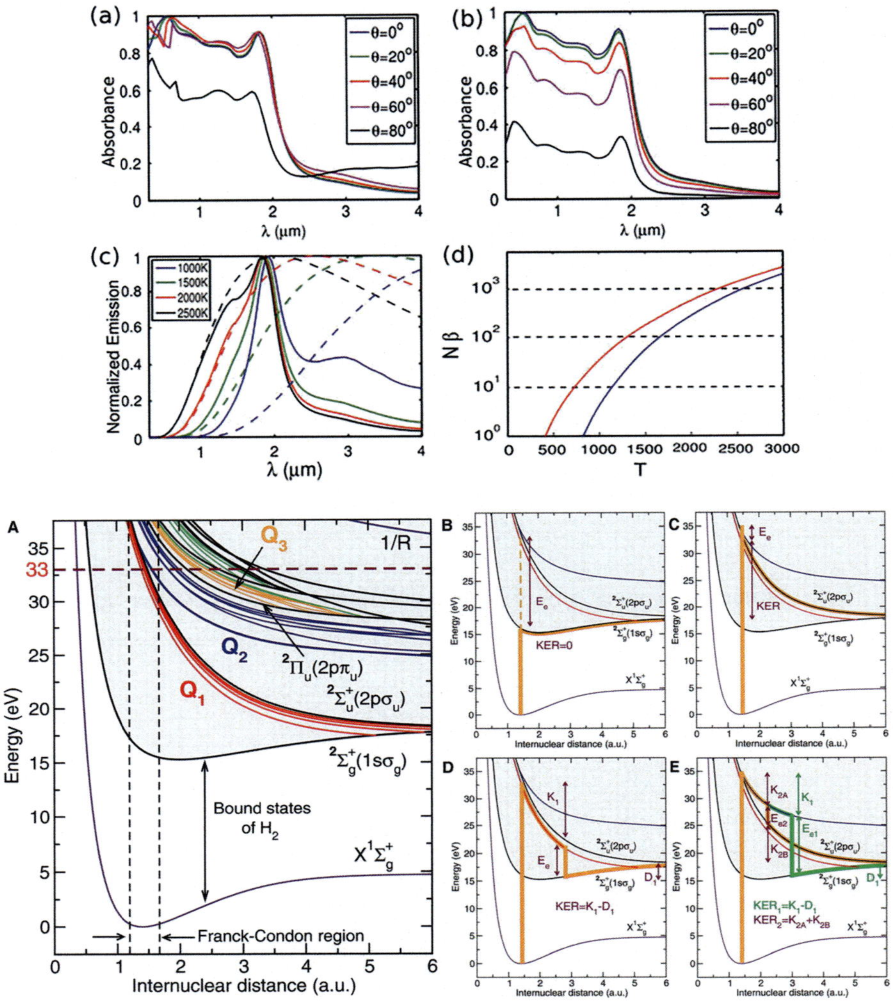

Fig. 13.9 (1) The heating photon formation vs wavelengths at various spectrum of photonic amplitude. (b) The combination time for functional of heating roll-off frequency. (c) The amplitude for the functional frequencies for various temperatures. (d) Heating thermal measurements into the Higgs boson quantum field. (2) Calculative demonstration of the wide-angle absorbance thermal spectra into the earth surface for (A) internuclear distance of heating photons (B–E) polarized incident radiation at different incidence angles considering heating photon emission spectrum (per unit frequency) to heat the earth plane. (*Source*: F. Martin et al. 2007. Science)

earth surface to heat earth surface naturally during the cold season. This natural cooling and heating mechanism of Earth surface would indeed be the cutting-edge science to control global temperature to mitigate eventually the global climate and environmental crisis dramatically.

References

1. A.K. Agger, A.H. Sørensen, Atomic and molecular structure and dynamics. Phys. Rev. A **55**, 402–413 (1997)
2. G. Baur, K. Hencken, D. Trautmann, S. Sadovsky, Y. Kharlov, Dense laser-driven electron sheets as relativistic mirrors for coherent production of brilliant X-ray and γ-ray beams. Phys. Rep. **364**, 359–450 (2002)
3. G. Baur, K. Hencken, D. Trautmann, Revisiting unitarity corrections for electromagnetic processes in collisions of relativistic nuclei. Phys. Rep. **453**, 1–27 (2007)
4. U. Becker, N. Grün, W. Scheid, K-shell ionisation in relativistic heavy-ion collisions. J. Phys. B: At. Mol. Phys. **20**, 2075 (1987)
5. A. Belkacem, H. Gould, B. Feinberg, R. Bossingham, W.E. Meyerhof, Semiclassical dynamics and relaxation. Phys. Rev. Lett. **71**, 1514–1517 (1993)
6. N.D. Benavides, P.L. Chapman, Modeling the effect of voltage ripple on the power output of photovoltaic modules. IEEE Trans. Ind. Electron. **55**, 2638–2643 (2008)
7. B. Boukhezzar, H. Siguerdidjane, Nonlinear control with wind estimation of a DFIG variable speed wind turbine for power capture optimization. Energy Convers. Manag. **50**, 885–892 (2009)
8. V. Cardoso, Quasinormal modes of Schwarzschild black holes in four and higher dimensions. Phys. Rev. D **69**, 044004 (2004)
9. A.N. Celik, N. Acikgoz, Modelling and experimental verification of the operating current of mono-crystalline photovoltaic modules using four- and five-parameter models. Appl. Energy **84**, 1–15 (2007)
10. W. Chihhui, Metamaterial-based integrated plasmonic absorber/emitter for solar thermophotovoltaic systems. J. Opt. **14**, 024005 (2012)
11. W. De Soto, S.A. Klein, W.A. Beckman, Improvement and validation of a model for photovoltaic array performance. Sol. Energy **80**, 78–88 (2006)
12. J.S. Douglas, H. Habibian, C.-L. Hung, A.V. Gorshkov, H.J. Kimble, D.E. Chang, Quantum many-body models with cold atoms coupled to photonic crystals. Nat. Photonics **9**, 326–331 (2015)
13. J. Eichler, T. Stöhlker, Radiative electron capture in relativistic ion-atom collisions and the photoelectric effect in hydrogen-like high-Z systems. Phys. Rep. **439**, 1–99 (2007)
14. H. Faida, J. Saadi, Modelling, control strategy of DFIG in a wind energy system and feasibility study of a wind farm in Morocco. IREMOS **3**, 1350–1362 (2010)
15. T. Ghennam, E.M. Berkouk, B. Francois, A vector hysteresis current control applied on three-level inverter. Application to the active and reactive power control of doubly fed induction generator based wind turbine. IREE **2**, 250–259 (2007)
16. C. Gopal, M. Mohanraj, P. Chandramohan, P. Chandrasekar, Renewable energy source water pumping systems—a literature review. Renew. Sustain. Energy Rev. **25**, 351–370 (2013). https://doi.org/10.1016/j.rser.2013.04.012
17. M.C. Güçlü, J. Li, A.S. Umar, D.J. Ernst, M.R. Strayer, Electromagnetic lepton pair production in relativistic heavy-ion collisions. Ann. Phys. **272**, 7–48 (1999)
18. N. Gupta, S.P. Singh, S.P. Dubey, D.K. Palwalia, Fuzzy logic controlled three-phase three-wired shunt active power filter for power quality improvement. IREE **6**, 1118–1129 (2011)
19. M.F. Hossain, Solar energy integration into advanced building design for meeting energy demand. Int. J. Energy Res. **40**, 1293–1300 (2016a)
20. M.F. Hossain, Production of clean energy from cyanobacterial biochemical products. Strat. Plan. Energy Environ. **3**, 6–23 (2016b)
21. M.F. Hossain, Theory of global cooling. Energy Sustain. Soc. **6**(1–5), 24 (2016c)
22. M.F. Hossain, Green science: Advanced building design technology to mitigate energy and environment. Renew. Sustai. Energy Rev. ***81***, *3051–3060* (2017a)
23. M.F. Hossain, Design and construction of ultra-relativistic collision PV panel and its application into building sector to mitigate total energy demand. J. Build. Eng. ***9***, *147–154* (2017b)

24. M.F. Hossain, Invisible transportation infrastructure technology to mitigate energy and environment. Energy Sustain. Soc. **7**, 27 (2017c)
25. M.F. Hossain, Green science: Independent building technology to mitigate energy, environment, and climate change. Renew. Sustain. Energy Rev. *73*, *695–705* (2017d)
26. M.F. Hossain, Application of advanced technology to build a vibrant environment on planet mars. Int. J. Environ. Sci. Technol. **14**(12), 2709–2720 (2017e)
27. M.F. Hossain, *Sustainable Design and Build* (Elsevier, Amsterdam, 2018a), p. 468. ISBN: 9780128167229
28. M.F. Hossain, Green science: Advanced building design technology to mitigate energy and environment. Renew. Sustain. Energy Rev. **81**(2), 3051–3060 (2018b)
29. M.F. Hossain, Photonic thermal energy control to naturally cool and heat the building. Appl. Therm. Eng. **131**, 576–586 (2018c)
30. M.F. Hossain, Green science: Decoding dark photon structure to produce clean energy. Energy Rep. **4**, 41–48 (2018d)
31. M.F. Hossain, Photon application in the design of sustainable buildings to console global energy and environment. Appl. Therm. Eng. **141**, 579–588 (2018e)
32. M.F. Hossain, Transformation of dark photon into sustainable energy. Int. J. Energy Environ. Eng. **9**, 99–110 (2018f)
33. M.F. Hossain, Bose-Einstein (*B-E*) photonic energy structure reformation for cooling and heating the premises naturally. Adv. Therm. Eng. **142**, 100–109 (2018g)
34. M.F. Hossain, Global environmental vulnerability and the survival period of all living beings on earth. Int. J. Environ. Sci. Technol. (2018h). https://doi.org/10.1007/s13762-018-1722-y
35. M.F. Hossain, Photon energy amplification for the design of a micro PV panel. Int. J. Energy Res. (2018i). https://doi.org/10.1002/er.4118
36. M.F. Hossain, N. Fara, Integration of wind into running vehicles to meet its total energy demand. Energy Ecol. Environ. **2**(1), 35–48 (2016)
37. S.C. Huot, Photon and dilepton production in supersymmetric Yang-Mills plasma. J. High Energy Phys. (2006). https://doi.org/10.1088/1126-6708/2006/12/015
38. E. Kamal, M. Koutb, A.A. Sobaih, B. Abozalam, An intelligent maximum power extraction algorithm for hybrid wind-diesel-storage system. Int. J. Electr. Power Energy Syst. **32**, 170–177 (2010)
39. M. Laine, Thermal 2-loop master spectral function at finite momentum. J. High Energy Phys. (2013). https://doi.org/10.1007/JHEP05%282013%29083
40. L. Langer, S.V. Poltavtsev, I.A. Yugova, M. Salewski, D.R. Yakovlev, G. Karczewski, T. Wojtowicz, I.A. Akimov, M. Bayer, Access to long-term optical memories using photon echoes retrieved from semiconductor spins. Nat. Photonics **8**, 851–857 (2014)
41. Q. Li, D.Z. Xu, C.Y. Cai, C.P. Sun, Recoil effects of a motional scatterer on single-photon scattering in one dimension. Sci. Rep. **3**, 3144 (2013)
42. F. Martin, Single photon-induced symmetry breaking of H2 dissociation. Science **315**, 629 (2007)
43. B. Najjari, A.B. Voitkiv, A. Artemyev, A. Surzhykov, Simultaneous electron capture and bound-free pair production in relativistic collisions of heavy nuclei with atoms. Phys. Rev. A **80**, 012701 (2009)
44. J. Park, H. Kim, Y. Cho, C. Shin, Simple modeling and simulation of photovoltaic panels using Matlab/simulink. Adv. Sci. Technol. Lett. **73**, 147–155 (2014)
45. T. Pregnolato, E.H. Lee, J.D. Song, S. Stobbe, P. Lodahl, Single-photon non-linear optics with a quantum dot in a waveguide. Nat. Commun. **6**, 8655 (2015)
46. A. Reinhard, T. Volz, M. Winger, A. Badolato, K.J. Hennessy, E.L. Hu, A. Imamoğlu, Strongly correlated photons on a chip. Nat. Photonics **6**, 93–96 (2012)
47. B. Robyns, B. Francois, P. Degobert, J.P. Hautier, *Vector Control of Induction Machines* (Springer-Verlag, London, 2012)
48. K.G. Sharma, A. Bhargava, K. Gajrani, Stability analysis of DFIG based wind turbines connected to electric grid. IREMOS **6**, 879–887 (2013)

49. G. Sivasankar, V.S. Kumar, Improving low voltage ride through of wind generators using STATCOM under symmetric and asymmetric fault conditions. IREMOS **6**, 1212–1218 (2013)
50. A. Soedibyo, F.A. Pamuji, M. Ashari, Grid quality hybrid power system control of microhydro, wind turbine and fuel cell using fuzzy logic. IREMOS **6**, 1271–1278 (2013)
51. J.J. Soon, K.S. Low, Optimizing photovoltaic model parameters for simulation, in *IEEE International Symposium on Industrial Electronics* (2012), pp. 1813–1818.
52. R. Szafron, A. Czarnecki, High-energy electrons from the muon decay in orbit: Radiative corrections. Phys. Lett. B **753**, 61–64 (2016)
53. M.S. Tame, K.R. McEnery, Ş.K. Özdemir, J. Lee, S.A. Maier, M.S. Kim, Quantum plasmonics. Nat. Phys. **9**, 329–340 (2013)
54. Y.T. Tan, D.S. Kirschen, N. Jenkins, A model of PV generation suitable for stability analysis. IEEE Trans. Energy Convers. **19**, 748–755 (2004)
55. M.W.Y. Tu, W.M. Zhang, Non-Markovian decoherence theory for a double-dot charge qubit. Phys. Rev. B **78**, 235311 (2008)
56. G. Wang, K. Zhao, J. Shi, W. Chen, H. Zhang, X. Yang, Y. Zhao, An iterative approach for modeling photovoltaic modules without implicit equations. Appl. Energy **202**, 189–198 (2017)
57. Y.F. Xiao, M. Li, Y.C. Liu, Y. Li, X. Sun, Q. Gong, Asymmetric Fano resonance analysis in indirectly coupled microresonators. Phys. Rev. A **82**, 065804 (2010)
58. W.B. Yan, H. Fan, Single-photon quantum router with multiple output ports. Sci. Rep. **4**, 4820 (2014)
59. L. Yang, S. Wang, Q. Zeng, Z. Zhang, T. Pei, Y. Li, L.M. Peng, Efficient photovoltage multiplication in carbon nanotubes. Nat. Photonics **5**, 672–676 (2011)
60. W.M. Zhang, P.Y. Lo, H.N. Xiong, M.W.Y. Tu, F. Nori, General non-Markovian dynamics, in *A Novel Modeling Method for Photovoltaic Cells, 35th Annual IEEE Power Electronics Specialists Conference, Aachen, Germany*, ed. by W. Xiao, W. G. Dunford, A. Capal, (IEEE, Piscataway, 2004), pp. 1950–1956
61. Y. Zhu, X. Hu, H. Yang, Q. Gong, On-chip plasmon-induced transparency based on plasmonic coupled nanocavities. Sci. Rep. **4**, 3752 (2014)

Index

A

Abelian local symmetries, 233
Activated photons, 102
Alternating current (AC), 66
Anaerobic co-digestions of domestic biowaste, 65
Astrodynamical laws, 28
Atmospheric CO_2 concentration increasing rate, 222, 223, 225

B

Bioenergy, 66
Biogas (CH_4), 66
Bioreactor, 58–60, 65, 66, 70
Bio-waste into electricity energy
 circuit cell's temperature, 68
 circuit device, 61
 complex circuit modeling and open circuit voltage, 63
 conversion mechanism, 62
 electricity energy and temperature, 63
 electricity generation model, 64
 I–V relationship the circuit cell, 68
 MATLAB simulation, 67
 methanogenesis, 61
 methanogenesis pathway, 65
 STC of temperature, 63
 temperature of circuit panel, 62
 wastewater treatment mechanism, 60
Blackbody solar irradiation, 169
Boltzmann's constant, 22, 23, 34, 69, 80, 230
Bose–Einstein (B-E) photon distribution mechanism, 78, 96
 cooling mechanisms, 102
 cooling photon emission panels, 102
Bose–Einstein photon distribution theory, 42
Bose–Einstein photon dynamics, 228
Bose–Einstein photonic structure
 electricity conversion, PV panel, 52–54
 extreme relativistic conditions, 42, 54
 HNEPs, 42
 photon dynamic transformation, 42–44
 photon production proliferation, 47
 PV modeling, 45, 46
Bremsstrahlung radiation (BR), 102
Building, water transportation sectors (BWT), 5

C

Carbon nanotubes (CNT), 59, 60
Cartesian coordinate system, 127
Chlorofluorocarbons (CFCs), 228
Clean energy, 41
Clean energy technology, 38, 101
Climate control to global environmental equilibrium
 cooling mechanism of earth surface (*see* Cooling mechanism of earth surface)
 heating mechanism of earth surface (*see* Heating mechanism of earth surface)
CO_2 emissions, 220–222
CO_2 sequestration, 221, 222
CO_2 sink, 221–223
Co-digestions of domestic biowaste, 65
Computational fluid dynamic (CFD), 176

M. F. Hossain, *Sustainable Design for Global Equilibrium*,
https://doi.org/10.1007/978-3-030-94818-4

Conventional cooling technologies, 101
Conventional energy consumption, 16, 41
Conventional heating and cooling systems, 78, 101
Conventional heating mechanism, 228
Cooling mechanism
  DOS, 108, 109
  Markovian master equation, 108
  nano-point breaks, 102
  NS cells, 105
  PB, 110
  PBG, 102, 111
  photon probability density, 103
  photonic band structure/energy conversion modes, 109
  solar irradiance receptor, 103, 104
Cooling mechanism of earth surface
  cooling photons core–shell multidimensional energy state levels, 240
  DOS photons, 237, 238
  EDOS photons, 237
  Fock state cooling photons, 241
  I–V characteristics of cool photons, 230, 232
  nano-point defect and PBG waveguide, 228
  photon band structures, 239
  photon energy level diagram, 231
  photon transformation mode, 239
  photonic probable density, 229
  thermal conductivity, 233
COVID-19, 203, 204, 208, 217
Cyclobutane pyrimidine dimers (CPD), 215

D

Debye mass, 95
Density of states (DOS), 110
Density of states (DOS) photon, 237, 238
Deoxyribonucleic acid (DNA), 209, 210, 213, 215–217
Desulfovibrio, 61
Dewar isomers, 215
DGVM simulation, 9
Direct cooling ionization, 231
Direct cooling photons, 231
Disinfection (DIS) system, 59
Dissociative photons, 231
Domestic biowaste, 58, 65
Domestic energy demand, 66
Doubly fed induction generator (DFIG), 151, 152, 155, 159–162, 164
d-q synchronous voltage equation, 169
Dynamic global environmental modeling (DGVM), 5, 220, 224

E

Einstein's photon energy fluctuation dissipation, 32
Einstein's photon fluctuation dissipation, 214
Electricity energy generation, 23, 25, 152, 169
Electromagnetic radiation (EM radiation), 207
Electron energy level of hydrogen, 19
Energy conversion, 151, 152, 158, 159, 164, 168–170
Energy, building, water, and transportation (EBWT), 3, 4
Environmental vulnerability, 220
Expected density of state (EDOS) photons, 237

F

First order perturbation theory, 60
Flying transportation technology
  aerodynamic characteristics, 184
  aerodynamic traits, 178
  angles, 184
  angular momentum, 184
  battery modeling, 182, 183, 197, 198
  climate change, 175
  construction of roads, 175
  conversion of wind energy, 180
  electrical subsystem, 191, 193, 194
  flying vehicle's performance, 176
  generator, 181, 182
  generator modeling, 194–196
  infrastructure, 175
  k-omega, 177
  physical model, 176
  principles, 184
  propulsive and levitative force model, 176
  simulation phase, 184
  transport velocity model, 176
  wind energy conversion, 190, 191
  wind energy modeling, 186–189
  wind energy modeling sequence, 178–180
Fock state cooling photons, 89, 241
Fock state photon number, 32
Fossil energy, 228
Fossil fuel reserve, 16
Fuzzy controller, 164
Fuzzy Lucidity Checker (FLC), 190

G
Global CO_2 emissions, 4, 10, 219, 222, 223, 225
Global energy demand, 38
Global fossil fuel energy consumption, 16
Global heating and cooling mechanism, 228
Goldstone scalar particles, 233, 235
Green science and technology
  CO_2 emissions, 5, 6
  CO_2 sink, 6, 7
  EBWT, 3
  environment, 4, 10
  G_{ATM}, 7
  global CO_2 emission, 4
  results
    CO_2 emissions, 7, 10
    CO_2 sink, 8
    G_{ATM}, 8
  sustainable world, 4
Greenhouse gases (GHGs), 139
Grid power control, 151
Growth Rate of the Atmospheric CO_2 Concentration (GATM), 7

H
Heating mechanism
  asymptotic forms, 116
  electromagnetic field, 112, 113
  functional asymptotic calculation, 114
  kinetic function, 117
  Lagrangian function, 106, 107, 113
  magnetic-field-induced photon production, 112
  photon energy emission rate, 118, 119
  photon spectrum, 115, 117
  thermal spectra, 120
  transformation, 108
Heating mechanism of earth surface
  differential photon density, 243
  heating energy-level photons, 234
  heating photon formation vs wavelengths spectrum of photonic amplitude, 250
  heating photon production, 244, 248
  heating photon spectra, 247
  Higgs quantum field, 233
  Lagrangian equation, 242
  photon receptor, 246
  radical fields, 235
  scalar field, 242
  scalar field spaces, 235
Heaviside step function, 165
Helium, 78, 80, 84, 96, 102
Higgs Boson BR(H $\rightarrow \gamma\gamma^-$) quantum field, 78, 96
Higgs Boson electromagnetic field, 105, 233
Higgs Bosons ($H \rightarrow \gamma\gamma^-$) quantum field, 89, 91, 228, 243, 249
Higgs quantum field, 233
Hossain cooling photon (HcP^-), 79, 97, 230, 249
Hossain equation, 229
Hossain Static Electric Force (HSEF), 144
Hossain thermal photon (HtP^-), 249

L
Lagrangian equation, 233
Lagrangian function, 106
Load resistant factor design (LRFD) bioreactor, 58

M
Markov factor, 29
MATLAB 9.0 Classical Multidimensional Scaling, 19
MATLAB algorithm, 220, 221, 224
MATLAB Simulink, 152, 160
MATLAB software, 17, 229
MATLAB/Simulink software package, 42
Maximum power point tracking (MPPT), 21, 153, 179
Maxwell's rule, 165
Methane, 61, 66
Methane-generation efficiency, 66
Methanogenesis, 58, 61, 65
Modern Solar Photovoltaics Energy panel, 54

N
Natural cooling and heating technology, 102
Net electricity energy generation from solar irradiance, 23, 25, 34, 35, 37
Net photon energy, 29, 32, 33
Net photophysical current generation into earth surface, 21
Net solar energy on earth calculation
  cartesian coordinate system, 18
  conceptual circuit diagram of whole earth surface, 21
  emission of solar energy, 19
  MPPT, 21
  net current flow into earth surface, 22
  net photon energy, 32

Net solar energy on earth calculation (*cont.*)
- optimum working point for energy generation per unit area, 28, 31
- PBE, 28, 33
- PBG, 29, 33
- P–N junction superconductor, 21
- pyranometer, 19
- solar constant, 17
- solar irradiance, 19
- solar zenith angle, 26, 30
- structural composition of photon and rate of energy, 33

Non-ionizing radiation, 208
Non-series (NS) cells, 80, 105, 135

O

Oceanic CO_2 sequestration, 221
Ozone layer, 77, 78, 96, 97, 101

P

Permanent synchronous generator (PMSG), 191
Perpetual magnet synchronous dynamos (PMSGs), 194
Peukert's coefficient, 183
Peukert's Law, 197
Phe photon receptor, 246
PHOTOMETER, 208
Photon band edges (PBEs), 33, 42, 78, 102, 228
Photon density, 243
Photon emission panel, 78, 80, 82
Photon energy
- bandgap energy, 131
- Boltzmann's constant, 134
- conventional energy deposition, 125
- DC, 128
- electricity conversion, 136–138
- energy cost, 137
- I–V curve, 135
- photocurrent, 132, 133
- photophysical reactions, 134
- PV-generation efficiency, 130
- PV panel, 126, 127
- PV panel functional temperature, 136
- solar energy, 126
- solar thermal energy, 129
- sustainable energy technology, 139

Photon energy level diagram, 230, 231
Photon energy mode, 205
Photon energy proliferation, 206
Photon production proliferation
- Dos dimensional modes, 47, 48
- Einstein's photon fluctuation, 51, 53
- heating photon energy formation, 49
- HnP^- production, 47
- photonic irradiance, 50
- Pv cells, 50

Photonic band structure, 213
Photonic bandgap (PBG), 33, 41, 102
Photonic bandgap energy, 230
Photonic probable density, 229
Photonic radiations (PR), 228
Photonic thermal control to cool buildings naturally
- contour maps of transmissivity of single photon, 78, 79
- diode current and/or saturation current, 80
- DOS and PDOS, 84, 88
- I–V relationship, 80, 81
- nano-breakpoints, 78
- photon nano structure, 78
- photon-generated current, 82
- photonic band structure and mode for energy conversion, 87
- proliferation of dynamic photons in PV cells, 89, 90
- reservoir-induced PB photon self-energy correction, 87
- single-diode mode, 79
- two-diode model of solar irradiance receptor, 79, 80

Photonic thermal control to heat buildings naturally
- corrective functional asymptotic formulas, 92
- covariant derivative, 83
- Debye mass, 95
- differential photon density, 91
- *Lagrangian* for complex scalar field, 89
- magnetic-field-induced photon, 96, 97
- power spectral density of resonator amplitude, 98
- quantum field, 90
- scalar field, 83
- special transformation rule for scalar field, 89
- transformation from energy level into two-diode feed semiconductors, 84, 85

Photons' bandgap (PBG) waveguides, 78, 87, 88, 228, 240, 241
Photophysics radiation, 59
Photophysics radiation application, 147
Photovoltaic (PV), 41

Photovoltaic (PV) modeling, 45
Photovoltaic semiconductor, 42
Physio-biological processes of global oceans, 221
Power backup model, 183
Projected density of states (PDOS), 84, 88, 110
Pyranometer, 19
Pyrheliometer, 19

Q
Quantum electrodynamics (QED), 42, 78, 102, 228

R
Ribonucleic acid (RNA), 209, 213, 215–217

S
Scalar field of Earth surface, 37
Simulink Power Systems, 193
Sludge, 58–61, 65
SODIS system, 147
Solar constant, 17
Solar energy in transportation
  electrical subsystem, 159, 160, 169–171
  modeling of solar energy
    cooling photonic dynamics of proliferation energy, 167
    DOS, 165, 166
    energy conversion modes and photonic band structure, 167
    maximum energy intake in solar panels, 157
    PV system model, 157
  solar energy conversion, 158, 159, 168
Solar energy scalar field on Earth surface, 37
Solar irradiance, 204, 205, 207, 210–212, 215
Solar panel transportation vehicle, 165
Solar radiation, 17–19, 21, 22, 34, 35, 126, 127, 134
Solar zenith angle, 26, 30
Squirrel cage induction generator (SCIG), 162
Squirrel confine initiation generator (SCIG), 188
Stefan–Boltzman laws, 127

T
Tip speed ratio (TSR), 153
Total CO_2 sequestration, 222
Total cooling photonic energy systems, 231
Toxic level of CO_2, 225
Traditional heating technology, 77
Transpiration mechanism
  clean energy, 143
  electrostatic force analysis, 147, 148
  ground water, 143
  HSEF, 144–146
  Plants, 143
  in site water treatment, 146, 147
  static electricity, 146
  UV application, 149
Transportation technology, 4, 5
Transportation vehicle, 152, 156, 157, 164, 165, 168, 172

U
Ultraviolet germicidal irradiation (UVGI)
  application, 204
  electromagnetic radiation, 207, 208, 213, 215
  killing pathogens, 208–210, 215, 216
  production by exterior glazing wall skin
    formation of, 211
    photon energy dynamic rate, 207
    photon energy mode, 205
    proliferation state, 214
    solar energy penetration, 205
    spectrum dynamics, 212
    temperature, 214

W
Wastewater treatment mechanism, 60
Weisskopf-Winger approximation mechanism, 26
Weisskopf–Winger theory, 237
Wind and solar power, 151, 152, 172
Wind energy conversion system (WECS), 180
Wind energy in transportation
  conversion of wind energy, 164
    pitch angle and wind speed, 156
    rotor speed and power coefficient, 156
    stages and main components, 156
  electrical subsystem, 159, 160, 169
  wind turbine modeling
    DFIG, 152
    drive train model, 162
    electronic torque, 162
    gap flux linkage, 163
    mass moment of inertia, 155
    MPPT control, 161, 162
    optimum rotor velocity, 153

Wind energy in transportation (*cont.*)
power coefficient, 163
rotor side, 163
schematic representation, 154
SCIG, 162
stator side, 163
TSR, 153
turbine power, 153
velocity of wind, 153
voltage dips and wind speed signals, 161